Conservation Technology

Conservation Technology

EDITED BY

Serge A. Wich

School of Biological and Environmental Sciences,
Liverpool John Moores University, UK

Alex K. Piel

Department of Anthropology,
University College London, UK

OXFORD

UNIVERSITY PRESS

Great Clarendon Street, Oxford, OX2 6DP,
United Kingdom

Oxford University Press is a department of the University of Oxford.
It furthers the University's objective of excellence in research, scholarship,
and education by publishing worldwide. Oxford is a registered trade mark of
Oxford University Press in the UK and in certain other countries

Published in the United States of America by Oxford University Press
198 Madison Avenue, New York, NY 10016, United States of America

British Library Cataloguing in Publication Data
Data available

Library of Congress Control Number: 2021936415

ISBN 978–0–19–885024–3 (hbk.)
ISBN 978–0–19–885025–0 (pbk.)

DOI: 10.1093/oso/9780198850243.001.0001

Printed and bound by
CPI Group (UK) Ltd, Croydon, CR0 4YY

Preface

The idea for this book developed during an MSc field course that we teach in Tanzania, where Alex directs a long-term primate field site in the Issa valley. For nearly a decade at Issa, we have used various technologies to establish and increase chimpanzee (and other wildlife) detections and monitoring. Alex started there many years ago for PhD work by deploying a custom-designed and built passive acoustic monitoring (PAM) system, integrated with radios to conduct real-time monitoring of chimpanzee pant hoots, in an effort to localize callers to improve habituation efforts. Later on, we initiated a collaboration on the usage of drones in western Tanzania to detect chimpanzee nests. We first trained together in the Netherlands, before Alex and Fiona Stewart flew their first missions using fixed-wings on a region-wide chimpanzee survey. While initial flights were literally bumpy, now drones are a regular part of various conservation questions being asked not just in Tanzania, but also across great ape distribution.

Issa now integrates and relies on PAM, camera traps, drones, digital data collection on tablets, DNA collection through primate faeces, and uses machine learning to analyse images from camera traps and drones. The field site is not unique in its use of technology, and hence we felt that students, colleagues, and collaborators at other sites around the world would find a book that describes these and other conservation technologies useful. We realized that even though there are already books that describe specific technologies such as camera traps, machine learning, or drones, no one volume covers a wide array of field-friendly technologies. We hope that the book is not only applicable for students, but also for managers of conservation areas who more often than not have to monitor a large number of animal species and vegetation with limited resources and where results are directly fed into policies and practices. Determining what options there are available to do so and what each of these options entails in terms of data collected, analyses, costs, durability, and specific scientific questions that can be answered with the data is not easy. We hope that this volume helps those managers and conservation practitioners to make such choices. Similarly, we feel that this book will be of use for colleagues who are developing or initiating new research projects and seek an overview of technologies that are commonly being used in conservation.

Acknowledgements

We would like to thank Liverpool John Moores University for the support of the drone lab that has been instrumental for our drone work, training the next generation of conservationists, and for its support in using and developing technology for conservation in general. We would also like to thank the UCSD/Salk Center for Academic Research and Training in Anthropogeny (CARTA), which has supported primate research and conservation efforts in the Issa valley.

We would like to thank Lian Pin Koh, Peter Wrege, Ammie Kalan, Jorge Ahumada, Oliver Wearn, Lochran Traill, Erin Vogel, Rich Bergl, Danijela Puric-Mladenovic, Stefano Mariani, Chris Gordon, Jan van Gemert, Josh Veitch-Michaelis, Liana Chua, and Koen Arts for kindly reviewing the chapters.

We want to thank the Oxford University Press team for offering us the chance to work with them on this book, especially Ian Sherman and Charles Bath who provided excellent guidance throughout the process.

Serge is thankful for the support from Tine, Amara, and Lenn during the writing of this book and accepting my absence during the many field trips over the past years to work with some of the technologies featured in this book. I hope it will contribute to more and better conservation of wildlife and their habitats.

Alex is grateful for the patience and support offered by Fiona Stewart and his favourite primates Finlay and Caelan. May these technologies allow for the continued conservation of wildlife for your children to observe and appreciate as well.

Contents

4 Acoustic sensors   **53**

Anne-Sophie Crunchant, Chanakya Dev Nakka, Jason T. Isaacs, and Alex K. Piel

5 Camera trapping for conservation   **79**

Francesco Rovero and Roland Kays

12 Digital surveillance technologies in conservation and their social implications

Trishant Simlai and Chris Sandbrook

13 The future of technology in conservation

Margarita Mulero-Pázmány

List of Contributors

Briana Abrahms Center for Ecosystem Sentinels, Department of Biology, University of Washington, Seattle. WA, USA

Herizo Andrianandrasana Durrell Wildlife Conservation Trust Madagascar Programme and Ministry of the Environment and Sustainable Development, Antananarivo, Madagascar

Rob. G. Appleby The Centre for Planetary Health and Food Security, Griffith University and Wild Spy Pty Ltd, Brisbane, Australia

Richard A. Bergl Conservation, Education, and Science Department, North Carolina Zoo, Asheboro, NC, USA

Kate Cornelsen Centre for Ecosystem Science, School of Biological, Earth and Environmental Sciences, University of New South Wales, Sydney, Australia

Drew T. Cronin Conservation, Education, and Science Department, North Carolina Zoo, Asheboro, NC, USA

Anne-Sophie Crunchant School of Biological and Environmental Sciences, Liverpool John Moores University, UK and Greater Mahale Ecosystem Research and Conservation Project (GMERC), Tanzania

Anthony Dancer Zoological Society of London, London, UK

Chanakya Dev Nakka IBM iX, Bangalore, India

K. Whitney Hansen Environmental Studies Department, University of California, Santa Cruz, CA, USA

Mike Hudson Durrell Wildlife Conservation Trust, Jersey, Channel Islands and Institute of Zoology, Zoological Society of London, London, UK

Jason T. Isaacs California State University, Channel Islands, CA, USA

Samual M. Jantz Department of Geographical Sciences, University of Maryland, MD, USA

Lucas Joppa Microsoft AI for Earth, WA, USA

Neil R. Jordan Taronga Conservation Society Australia, Sydney, Centre for Ecosystem Science, School of Biological, Earth and Environmental Sciences, University of New South Wales, Australia and Botswana Predator Conservation, Maun, Botswana

Roland Kays Department of Forestry and Environmental Biology, North Carolina State University, Raleigh, USA and North Carolina Museum of Natural Sciences, NC, USA

Erin E. Kane Department of Anthropology, Boston University, Boston, MA, USA

Andrew J. King Department of Biosciences, Swansea University, Swansea, UK

Cheryl D. Knott Department of Anthropology and Department of Biology, Boston University, USA

Barney Long Re:Wild, Austin, TX, USA

Steven N. Longmore Astrophysics Research Institute, Liverpool John Moores University, Liverpool, UK

Antony J. Lynam Center for Global Conservation, Wildlife Conservation Society, New York, NY, USA

Edward McLester School of Biological and Environmental Sciences, Liverpool John Moores University, Liverpool, UK

Dan Morris Microsoft AI for Earth, WA, USA

Margarita Mulero-Pázmány School of Biological and Environmental Sciences, Liverpool John Moores University, Liverpool, UK

Jeff Muntifering Save the Rhino Trust, Namibia

Caitlin A. O'Connell Department of Biological Sciences, Human and Evolutionary Biology Section, University of Southern California, Los Angeles, CA, USA

Jonathan Palmer Center for Global Conservation, Wildlife Conservation Society, New York, NY, USA

Antoinette J. Piaggio United States Department of Agriculture, Animal Plant Health Inspection Service, Wildlife Services, National Wildlife Research Center, Fort Collins, CO, USA

Alex K. Piel Department of Anthropology, University College London, London, UK and Greater Mahale Ecosystem Research and Conservation Project (GMERC), Tanzania

Lilian Pintea Jane Goodall Institute, USA

Benjamin J. Pitcher Taronga Conservation Society Australia and Department of Biological Sciences, Macquarie University, Sydney, Australia

Kasim Rafiq Environmental Studies Department, University of California, Santa Cruz, USA and Center for Ecosystem Sentinels, Department of Biology, University of Washington, Seattle. WA, USA and Botswana Predator Conservation, Maun, Botswana

Francesco Rovero Department of Biology, University of Florence, Florence, Italy and MUSE—Museo delle Scienze, Trento, Italy

Chris Sandbrook Department of Geography, University of Cambridge, Cambridge, UK

Amy M. Scott Department of Anthropology, Boston University, Boston, MA, USA

Trishant Simlai Department of Geography, University of Cambridge, Cambridge, UK

Tri Wahyu Susanto Department of Biology, National University, Indonesia

Serge A. Wich School of Biological and Environmental Sciences, Liverpool John Moores University, UK and Institute for Biodiversity and Ecosystem Dynamics, University of Amsterdam, Amsterdam, the Netherlands

CHAPTER 1

Conservation and technology: an introduction

Alex K. Piel and Serge A. Wich

Those of us who study wildlife and simultaneously the threats to it find ourselves at the intersection of two unprecedented ages: the Information or Digital Age—where electronic devices and digital data govern the flow of information—and the Anthropocene—where human activities alter the dynamics to life, including climate, ocean, and forests (Steffen et al., 2007; Joppa, 2015). That the natural environment is being altered by anthropogenism is undisputed; that technological innovations can assist conservation biologists to support solutions to these global problems, however, is not. The subject of this book is what technologies are being used in conservation practice, to address what types of questions and what are the associated challenges to their deployment.

Applicable technology must address fundamental problems in the field. As such, the process begins with a simple set of key questions: What do conservation biologists need to know to support conservation management and policy? Which data, resolution, or parameters are not being captured that need to be? What existing technology can help obtain the type and quality of data needed? How do we enhance data capture? From a management perspective, can we identify not just hotspots of biological interest, but particular key resources (specific trees, water sources) rather than broader areas or general vegetation types (Allan et al., 2018)? Finally, if technology offers progress towards these answers, there is a subsequent set of important questions to

consider, namely the efficacy and efficiency of any new, innovative method and, relatedly, the ethical and practical implications of using new devices (Ellwood et al., 2007).

Conservationists require a suite of data to address key questions on animal presence, distribution, habitat integrity, and the threats to species and entire ecosystems. Broadly, conservationists use these data to assess biodiversity, monitor ecosystems, investigate population dynamics, and study behaviour, all in response to an increasingly large and global anthropogenic footprint. We often want to do this longitudinally, especially to examine trends and change over time, and also geographically, across increasingly vast landscapes. Moreover, long-term projects are faced with how to digitize and standardize data collection protocols while maintaining interobserver reliability, all the while confronted with rotating staff and often the training of volunteers/interns/students. Alas, conservation science involves not only knowing what data are needed, but what tools can help acquire them and further, how to store, process, and analyse them once procured. The final, and perhaps most crucial step is providing results in a way that facilitates actions by decision-makers (Chapter 2).

Specifically, conservation scientists need data on species diversity, counts of individual animals, migration patterns, habitat integrity, resource availability and distribution, and information on animal health, for example. These data inform on

Alex K. Piel and Serge A. Wich, *Conservation and technology: an introduction*. In: *Conservation Technology*. Edited by Serge A. Wich and Alex K. Piel, Oxford University Press.
© Oxford University Press 2021. DOI: 10.1093/oso/9780198850243.003.0001

landscape-wide questions about animal abundance and distribution, as well as disease risks. Higher-resolution data on group- and even individual-level behaviour are important. Disease influences demography (birth and death rate), group composition, and individual behaviour (Prentice et al., 2014). With wildlife disease and zoonoses especially a widespread concern for conservationists (Deem et al., 2008), identifying pathogens and conducting diagnostics for evaluating animal health are critical. Historically, these processes (1) required collection and subsequent export of animal tissue or by-products (e.g. faeces), (2) were costly in terms of money and time, and (3) often resulted in damaged shipments and contaminated or ruined samples. There is thus great demand for mobile labs to process samples in the field.

Together, data on wildlife presence, habitat, behaviour, health, and threats provide snapshots of species or habitat conservation status, which are useful for establishing a baseline understanding and the subsequent need for management. Ultimately, though, effective biodiversity conservation requires fast and effective means of assessing how species diversity or numbers shift in response to anthropogenic changes like settlement expansion, conversion of forest to agriculture, and poaching among many others. Traditionally, data from camera traps and drones were collected in isolation; contemporary integration of technology allows the simultaneous processing of multiple types of data, with networks of various sensors collecting data across platforms (Turner, 2014). Scientists can then combine data collected from more recent technologies like biologgers with more traditional ecological data, (e.g. temperature, rainfall) to analyse relationships—often in near real-time—to better understand the behavioural consequences of a changing world (Wall, 2014; Kays et al., 2015). Knowing the type and severity of threats has implications for the urgency needed to address them as well. The variability of animal responses further influences management strategies. What animals eat, where they go, and how they behave more broadly (e.g. activity budgets, grouping patterns) in response to reduced habitat availability, human presence, or introduced species has bearing on what steps can be taken to abate imminent threats.

Given the importance of ecological monitoring for nearly any conservation project, conservation scientists have long been interested in animal presence, distribution, and density. Traditionally, census and monitoring measures were conducted via capture-recapture approaches or conventional ground (e.g. reconnaissance) surveys, both of which are almost entirely dependent on staff or student availability and capacity, with handwritten documentation traditionally stored in data books (Verma et al., 2016). In almost all cases, data collectors were required to be in the areas where wildlife lives, raising questions about disturbance and generally the role of the observer in influencing animal behaviour and movement (Kucera & Barrett, 2011; Nowak et al., 2014). Moreover, the larger the data collection team, the greater the concern for interobserver reliability, whether because of variability in researcher knowledge, experience, bias, or training. Meanwhile, especially for long-term projects with permanent researcher presence on site, data books piled up on bookshelves with armies of undergraduate students or interns recruited to digitize them for subsequent analyses.

One of the most commonly used methods for data collection, both traditionally and through the present for censusing various taxa, is via line transects (Yapp, 1956). In this method, researchers walk a straight line, often of random bearing, and record the perpendicular distance of all direct or indirect wildlife observed from the line. Thorough descriptions of line transect use can be found in various reviews (Chapter 5; Krebs, 1999; Buckland et al., 2001; Thomas et al., 2010). This method of sampling can be practical, effective, and inexpensive. By calculating the distance walked and the width of the area visible to observers, researchers can estimate population densities of species of interest. Line transects have been globally applied to various taxa, predominantly used with terrestrial (Varman & Sukumar, 1995; Plumptre, 2000) and marine (de Boer, 2010) mammals, and primates specifically (Brugiere & Fleury, 2000; Buckland et al., 2010). Initially used for ground surveys, this method is now common for estimating the population sizes of large mammals (e.g. African herbivores—Jachmann, 2002), where vast areas need to be covered to sufficiently survey distribution (Trenkel et al., 1997). Aerial line transects have also been used for birds (Ridgway, 2010) and marine species as well (Miller et al., 1998).

When estimating population densities directly is not possible, scientists have historically used methods that reveal relative measures. This is particularly useful with birds, for example, where the number of calls per unit area provides a relative measure of population density. Also known as point counts, the same approach is sometimes used for roadside tallies of wildlife as well (reviewed in Schwarz & Seber, 1999). These types of surveys are based on the proportion of observed evidence to search effort and the assumption that change over time is attributable to population increase or decrease. Despite their pervasive use, point counts are not suitable for assessing relative abundance of rare or cryptic species (Ralph et al., 1995) given the typically low sample size.

Animal presence and distribution are generally dependent on habitat quality, which is another important metric for conservation scientists. Not only is there a growing empirical evidence that habitat quality influences occupancy of various taxa, from plants (Honnay et al., 1999) to primates (Arroyo-Rodríguez & Mandujano, 2009; Marshall, 2009; Foerster et al., 2016), but also quality can influence species spatial dynamics, especially in fragmented landscapes (Thomas et al., 2001; reviewed in Mortelliti et al., 2010). Evaluating habitat quality is especially useful when identifying the ecological constraints that influence fitness is impossible (Johnson, 2005).

The earliest ways to measure habitat were via ground surveys, with an attempt to identify those features that facilitate or prevent animal presence (Forbey et al., 2017). Methods included counting or trapping animals, collecting faeces, and generally mapping key landmarks. These can include the counting of important resources, such as food and water sources, nest or sleeping sites, natural barriers (e.g. rivers), and anthropogenic footprints like agriculture and settlements. They can also include measurements of habitat integrity that further affect animal movement, such as fragmentation, canopy cover, and species diversity (Johnson, 2005). Historically these data have been collected from ground-truthing, including specimen collection for diversity indices, manual tree measurements for gauging canopy height, and mapping landmarks

from recces. Like most of these traditional methods, new technology has enhanced the scope and quality of data that can be collected and thus improved dramatically how we understand wildlife habitat quality (Forbey et al., 2017).

Analysis of land cover from manned aircraft was initially a common method to assess landscape-wide patterns of habitat quality and habitat change. From the 1970s, satellite imagery became the preferred means of generating these data (Roy et al., 1992; Serneels et al., 2001) and remote sensing was identified as a reliable method for predicting fine-scale habitat quality and use by individual species (e.g. sage grouse—Homer et al., 1993). Remote sensing has been used to inform on vegetation types and proportions (Curran, 1980), especially useful for specialist species or else in cases where humans target specific tree species for extraction (Kuemmerle et al., 2009; Brandt et al., 2012).

Eventually, how, when, and where animals navigate these habitats when observers are not around emerged as central questions in conservation science. Examples include how extensive crop raiding is by cryptic species (e.g. sloth bears—Joshi et al., 1995) and evaluations of translocated (or reintroduced) species (e.g. dormouse—Bright & Morris, 1994). Locating and tracking individual animals using global navigation satellite systems (GNSS) relied on cumbersome, expensive, and heavy radio collars (Rasiulis et al., 2014). Early versions of GNSS collars stored data in-house and relied on collar removal to retrieve data. Moreover, especially for larger animals, pioneering studies had negative results for conservation. Alibhai and colleagues (Alibhai et al., 2001; Alibhai & Jewell, 2001) described a direct relationship between immobilization schedule (the number of darting episodes) and reduced inter-calf interval in black rhinos, raising scientific and ethical issues in the early tests of this technique. Subsequent deployments were plagued with system inaccuracies, human and transmission error (sometimes up to 1000 m—Habib et al., 2014), and disturbance to the collared individual. Besides radio collars, 'geolocators', which were attached to animals, used the time of day to calculate an animal's position, and so were prone to large error margins as much as tens of kilometres from the animal's actual location (Weimerskirch & Wilson, 2000). Other problems abounded.

Traditional systems have faced challenges on numerous dimensions, from limited spatial and temporal coverage to issues concerning data recording, storage, transmission, and interobserver reliability (Verma et al., 2016). Technology is addressing those limitations, sometimes multiple at a time. A look at the number of peer-reviewed works on the topic of conservation technology shows a clear trend over the last few decades that uses the term, 'Conservation Technology' (Figure 1.1).

Recent developments in conservation technology improve the speed and type of data capture, including the quality, quantity, and reliability of this process, as well as the ways and speed that data are analysed. For the study of large, especially terrestrial species, ground teams have long been limited by the time required to cover vast areas as well as the danger and logistical challenges of surveying in remote areas. Ground surveys continue to be used and offer the advantage of identifying small-scale threats (e.g. individual snares) that other methods will miss, but increasingly, ground teams are now complemented, if not replaced entirely, with remote sensing methods. The current volume details these methods, as well as the implications for their use. Today, capturing data about species presence/absence and habitat change across large spatial scales is currently conducted using either medium/high-resolution satellite imagery (Chapter 2) or drones (Chapter 3). As the applications and uses of these methods are

driven, and limited, by the sensors their vehicles carry, we discuss them both here. More, regardless of operator, the ultimate objective of imaging is similar across platforms, for example, to provide spatially specific analysis and spatially distributed data on targets that exhibit recognizable signatures.

The most common type of remote sensing from either satellites or drones is multispectral, whereby sensors assess the radiation (or brightness) emitted from surface areas on earth. The cost of high resolution (sub-metre) imagery and sensors continue to decline as the diversity of applications expands in conservation. For example, LiDAR (light detection and ranging) on low flying aircraft, but also from space-borne satellites and drones, estimates above-ground carbon stocks and overall biomass, ecosystem structure, and broadly provides critical data for environmental management (Urbazaev et al., 2018; Vaglio Laurin et al., 2020). Advances in hyperspectral remote sensing—multispectral sensors with hundreds of bands across a near-continuous range, including visible, infrared, ad electromagnetic spectrum—have further potential, being able to identify fine-scale features like habitat variation at the level of subtle vegetative or soil differences (Turner et al., 2003; Shive et al., 2010). Satellites with hyperspectral sensors offer the ability to remotely map specific plant species and thus ecosystem resilience to anthropogenism (Underwood et al., 2003; Li et al., 2014). Finally, increasingly, and especially with drones, users are

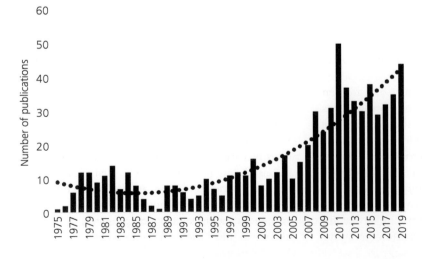

Figure 1.1 The number of scientific publications on conservation and technology published between 1975 and 2019. Results reflect a search for title words 'Conservation' and 'Technology' in Scopus (http://www.scopus.com). The data search was conducted on 27 April 2020.

also applying thermal-sensitive sensors—detecting temperature differentials—which identify anything from fires to people and animals across large landscapes (Chapter 3). For conservationists, thermal imaging sensors are providing data to answer questions about species distribution, counts, and the location of fires (Chapter 3).

Satellites and drones offer views from above and thus suffer from an inability to access ground-based biodiversity if those are covered by vegetation or are too small. Two of the most common methods to capture ground-truthing data are camera traps (Chapter 4) and acoustic sensors (Chapter 5). Both have been used for nearly a century in conservation studies. Even more so than satellite or drone imagery, cameras and acoustic sensors capture especially species composition and diversity across a sampled area. Historically, data captured from these devices were used predominantly to build species lists. However, analytical and technological developments in both have transformed their applicability to conservation-relevant questions. For example, with both methods researchers can now calculate animal density from either DISTANCE sampling (acoustics: Marques et al., 2013, camera trapping: Cappelle et al., 2019) or from spatially explicit capture–recapture analyses (acoustics: Efford & Fewster, 2013; camera trapping: Després-Einspenner et al., 2017). Additionally, the hardware of both cameras and acoustic sensors has also improved. Commercial cameras with 3/4G-enabled networking capability allow devices to transmit image captures to base stations where image-recognition tools help distinguish poachers from primates (CITE). Acoustic sensors interfaced with local mobile phone networks also offer near real-time transmission of detected sounds (Aide et al., 2013).

Besides merely detecting animal movement, many movement-related questions concern directionality, speed, and fitness costs. For example, biologgers now reveal animal movement in three dimensions, integrating accelerometers, barometers, and gyroscopes (Allan et al., 2018). Once limited to terrestrial systems, biologgers are now employed to investigate the physiological mechanisms that influence life history and population changes, even for globally distributed populations.

They can sense and document nearly all aspects of an animal's movements, from when a fish opens and closes its mouth (Viviant et al., 2014) to the nutritional geometry that influences marine mammal diving behaviour (Machovsky-Capuska et al., 2016).

For the most part, sensors are small, lightweight, satellite-based, and often do not require invasive methods in deployment (Bograd et al., 2010). Large species from the tropics (Yang et al., 2014) to smaller species at the poles (Kuenzer, 2014) can be monitored from space using satellite technology, expanding the geographic scope through which we investigate related questions. As Wilmers and colleagues (2015) summarized in a review of biologging devices, 'Nearly all biological activity involves change of one kind or another. Increasingly these changes can be sensed remotely'. The last decades have transformed our ability from merely identifying animal locations, to now monitoring their movement, social interactions, and reconstructing behavioural, energetic, and physiological states from afar. Chapter 6 describes some of these innovative biologging tools that are improving the resolution and type of data that we gather from animal-borne devices.

In addition to information extracted from loggers, data also come indirectly from evidence left behind by animals; for example, hair (Mowat & Strobeck, 2000; Macbeth et al., 2010), faeces (Muehlenbein et al., 2012), and tools (Stewart et al., 2018). The emerging field of conservation physiology (Wikelski & Cooke, 2006), which centres around neuro-endocrine stress indicators and their influences on behaviour, can reveal wildlife responses to environmental change, from forest loss to water temperature to tourism (Ellenberg et al., 2007; reviewed in Acevedo-Whitehouse & Duffus, 2009). Faeces-extracted DNA has also long been an important tool for conservation scientists. DNA analyses can reveal abundance and survivorship (Sitka deer—Brinkman et al., 2011; pronghorn—Woodruff et al., 2016), population dynamics, and density variability across sites (coyotes—Morin et al., 2016), among other aspects. Historically, DNA extraction and hormonal assays were conducted in laboratories far removed from field stations and required researchers to preserve, store, and export samples. These restrictions have increased in recent years, combined with pressure to build local capacity of range country scientists in molecular analyses (O'Connell et al., 2019). Chapter 7 explores field labs and how much of this analytical work can be conducted with portable field kits.

As this volume makes clear, conservationists have traditionally employed morphological and behavioural data collection techniques using direct observations, as well as sensors placed on satellites, cameras, acoustic units, and even animals themselves. Moreover, for decades an additional focus of censusing populations, assessing health, and tracking movement using indirect evidence has focused on the DNA extracted from hair or faeces from individual organisms. However, by inventorying far larger amounts of DNA in the environment, that is, eDNA, many more questions can be addressed, namely about past and present biodiversity (Beng & Corlett, 2020). Environmental DNA is animal genetic material originating from the hair, skin, faeces, or urine of animals but that has degraded and can be extracted from water, soil, or sediment (Thomsen & Willerslev, 2015; Beng & Corlett, 2020). Like most of the aforementioned techniques, using eDNA is non-invasive and sample collection requires little specialist technical or taxonomic knowledge (like the others as well, analyses are highly technical and complex). Moreover, sampling entire systems increases the likelihood of capturing DNA of cryptic or elusive species, as well as for those in which morpho-types are similar and thus potentially difficult to decipher from observation only. Finally, unlike using, for example, drones or acoustic sensors, when rain or wind, respectively, impairs sampling, eDNA collection is not constrained by weather conditions (Thomsen & Willerslev, 2015).

Beng and Corlett (Beng & Corlett, 2020) summarize the conservation applications of eDNA, from being a fast, efficient means of monitoring population dynamics on community or species levels as well as a means to map their distribution across vast spatial scales, to identifying biological invasions and assessing the status of eradication measures. eDNA is also useful in revealing past and present biodiversity and trophic patterns within seawater, freshwater, and even permafrost material dating back tens of thousands of years (Willerslev

et al., 2014). In that sense, eDNA offers a temporal range that other methods cannot and with prices declining is a non-invasive way to vastly increase the spatial scale of biodiversity assessment, while simultaneously maintaining—if not increasing—species-specific resolution.

The expansion of technological tools for conservation practice is not limited to *what* data are collected but also *how* they are collected. Traditionally, paper and pen sufficed for behavioural observations of wildlife and also for recording survey data. The proliferation of smartphones/tablets and applications associated with data collection and integration have immeasurably improved the now seemingly antiquated process of recording data by hand for eventual transcription into digital format. The appearance/vanish rate of apps in app/play stores make a comprehensive chapter impossible to write. Nonetheless, Chapter 9 describes some of the more common applications relevant for conservation scientists, compares usability, price, and applications. Nearly all offer the advantage of being custom designable, streamlined for specific data collection, and cloud accessible for remote access to ground-truthed data. Perhaps the most well-known app, and most pervasively used in species protection/conservation management is SMART, to which we devote an entire chapter (Chapter 10), including numerous case studies (Hötte et al., 2016) that demonstrate the role the software has played in improving the efficiency and efficacy of patrols and overall species protection (Wilfred et al., 2019).

With the growth in data collection techniques and storage capacities have come better, faster, and more automated ways of sieving through data to identify patterns and relationships between variables of interest. Historically, large data sets were plagued with delays between data acquisition, analysis, and subsequent interpretation. Now, we are in the midst of another wave of transformational improvement, with automated processing of big data using computer vision and/or deep learning to streamline data management and pattern extraction (Miao et al., 2019). Computer vision is an interdisciplinary field of artificial intelligence (AI), whereby computers use pattern recognition to decipher and interpret the visual world, also known as machine learning and object classification. Machine learning techniques have transformed the extent to which we can train computers to identify objects, patterns, species, and faces, among others (Norouzzadeh et al., 2018). These data can help identify species and in some cases, count individuals (Seymour et al., 2017) at near-identical accuracy levels to humans, but far faster. Moreover, resulting data can inform—if not guide—ground teams that may need to act urgently. For example, radar and optical satellites can provide daily scanning of tropical forests at 5–20 m resolution and relay results that reveal illegal logging in near-real-time (Lynch et al., 2013) to ground teams; pictures and coordinates can be sent to smartphones for immediate action.

Resolution in behavioural, physiological, and ecological data is prompting a parallel surge in partnerships to bridge a data-rich scientific world with industries that can help support necessary analyses. Microsoft's 'AI for Earth' platform uses computer vision techniques to classify wildlife species and deep learning to automate survey data. A similar collaboration between Hewlett-Packard and Conservation International resulted in 'Earth Insights'—software developed to monitor endangered species (Joppa, 2015). Near real-time results of this type of information across a wide scale, whether on animal (Wall, 2014) or poacher (Tan et al., 2016) movements, offer managers the potential to act immediately to mediate threats and protect wildlife. That may result in increased deployment of specific deterrents (e.g. fences for elephants), targeted patrols, or additional surveillance to vulnerable areas. In summary, whereas the ethical considerations for these technological tools are far behind in development (see next), in some ways, the analytical tools are way ahead, fuelled by partnerships and interdisciplinary collaborations (Wilber et al., 2013; Sheehan et al., 2020). The importance, diversity, and contributions of computer vision are discussed in Chapter 11.

As with nearly all technological revolutions, the social tools that are required to accompany such innovations lag behind the physical tools themselves. Drone imagery, acoustic sensors, and camera traps are often designed to reveal data on elusive or cryptic animals. Either incidentally or maliciously, metadata that may compromise people, security, and privacy are thus vulnerable to being exposed. The progress needed to advance a parallel advancement in social ethics requires people not just familiar with the technology and its potential, but also philosophy and sociology. Chapter 12 explores questions about who has access to these data, how should they be shared, and how we weigh scientific, environmental, and empirical benefits against social costs.

We hope to have demonstrated the vast reach of technology in contemporary conservation challenges, with continuing improvements in sensor quality and capability, as well as power and data storage, among others—all while prices generally decline. Improvements in data resolution are key to expanding the types of questions that we can ask. For example, sub-one metre resolution of satellite imagery now allows for questions into botanical diversity, absolute abundance, and indices of productivity for individual plants. Application of such imagery is useful when identifying specific threats (e.g. expansion of oil palm trees—Srestasathiern & Rakwatin, 2014) or to quantify mosaic habitats (Gibbes et al., 2010), too. The final chapter (13) describes the future of conservation technology, highlighting how incremental enhancements of current methods and also innovative methods will guide how we protect biodiversity going forward.

In closing, a comprehensive review of the techniques and associated scientific questions at the core of conservation technology is beyond the scope of any single volume. Instead, here we have selected what we consider to be some of the most commonly used, important, and applicable tools in conservation technology. In each chapter, the authors are experts in the field and review what is known and what is current. We use case studies to exemplify how the tools are applied and discuss the limitations as well as what lies ahead. We focus on some recent, pioneering methods of data collection and also developments to more established ways of data collection. We hope that this volume captures the innovative ways that technology contributes to, improves on, and ultimately drives contemporary conservation practice.

References

Acevedo-Whitehouse, K., & Duffus, A. L. J. (2009). Effects of environmental change on wildlife health.

Philosophical Transactions of the Royal Society B: Biological Sciences, 364, 3429–3438.

Aide, T. M., Corrada-Bravo, C., Campos-Cerqueira, M., Milan, C., Vega, G., & Alvarez, R. (2013). Real-time bioacoustics monitoring and automated species identification. *PeerJ*, 1, e103.

Alibhai, S. K. and Jewell, Z. C. (2001). Hot under the collar: the failure of radio-collars on black rhinoceros *Diceros bicornis*. *Oryx*, 35(4), 284–288.

Alibhai, S. K., Jewell, Z. C., & Towindo, S. S. (2001). Effects of immobilization on fertility in female black rhino (*Diceros bicornis*). *Journal of Zoology*, 253(3), 333–345.

Allan, B. M., Nimmo, D. G., Ierodiaconou, D., Vanderwal, J., Pin Koh, L., & Ritchie, E. G. (2018). Futurecasting ecological research: the rise of techno-ecology. *Ecosphere*, 9(5), e02163.

Arroyo-Rodríguez, V., & Mandujano, S. (2009). Conceptualization and measurement of habitat fragmentation from the primates' perspective. *International Journal of Primatology*, 30(3), 497–514.

Beng, K. C., & Corlett, R. T. (2020). Applications of environmental DNA (eDNA) in ecology and conservation: opportunities, challenges and prospects, biodiversity and conservation. *Springer Netherlands*, 29, 2089–2121.

Bograd, S., Block, B. A., Costa, D. P., & Godley, B. J. (2010). Biologging technologies: new tools for conservation. Introduction. *Endangered Species Research Species Research*, 10, 1–7.

Brandt, J. S., Kuemmerle, T., Li, H., Ren, G., Zhu, J., & Radeloff, V. C. (2012). Using Landsat imagery to map forest change in southwest China in response to the national logging ban and ecotourism development. *Remote Sensing of Environment*, 121, 358–369.

Bright, P. W., & Morris, P. A. (1994). Animal translocation for conservation: performance of dormice in relation to release methods. *Journal of Applied Ecology*, 31(4), 699–708.

Brinkman, T. J., Person, D. K., Chapin, S. F., Smith, W., & Hundertmark, K. J. (2011). Estimating abundance of Sitka black-tailed deer using DNA from fecal pellets. *Journal of Wildlife Management*, 75(1), 232–242.

Brugiere, D., & Fleury, M. C. (2000). Estimating primate densities using home range and line transect methods: a comparative test with the black colobus monkey *Colobus satanas*. *Primates*, 41(4), 373–382.

Buckland, S. T., Anderson, D. R., Burnham, K. P., Laake, J. L., Borchers, D. L., & Thomas, L. (2001) *Introduction to Distance Sampling: Estimating Abundance of Biological Populations*. Oxford University Press, Oxford, UK.

Buckland, S. T., Plumptre, A. J., Thomas, L., & Rexstad, E. A. (2010). Design and analysis of line transect surveys for primates. *International Journal of Primatology*, 31(5), 833–847.

Cappelle, N., Després-Einspenner, M.-L., Howe, E. J., Boesch, C., & Kühl, H. S. (2019). Validating camera trap distance sampling for chimpanzees. *American Journal of Primatology*, 81(3), 1–9.

Curran, P. (1980). Mulispectral remote sensing of vegetation amount. *Progress in Physical Geography*, 4(3), 315–341.

De Boer, M. (2010). Spring distribution and density of minke whale Balaenoptera acutorostrata along an offshore bank in the central North Sea. *Marine Ecology Progress Series*, 408, 265–274.

Deem, S. L., Karesh, W. B., & Weisman, W. (2008). Putting Theory into Practice: wildlife Health in Conservation. *Conservation Biology*, 15(5), 1224–1233.

Després-Einspenner, M. L., Howe, E. J., Drapeau, P., & Kühl, H. S. (2017). An empirical evaluation of camera trapping and spatially explicit capture-recapture models for estimating chimpanzee density. *American Journal of Primatology*, 79(7), 1–12.

Efford, M. G., & Fewster, R. M. (2013). Estimating population size by spatially explicit capture-recapture. *Oikos*, 122(6), 918–928.

Ellenberg, U., Setiawan, A. N., Cree, A., Houston, D. M., & Seddon, P. J. (2007). Elevated hormonal stress response and reduced reproductive output in Yellow-eyed penguins exposed to unregulated tourism. *General and Comparative Endocrinology*, 152, 54–63.

Ellwood, S. A., Wilson, R. P., & Addison, A. C. (2007). Technology in conservation: a boon but with small print. In Macdonald, D., & Service, K. (eds). *Key Topics in Conservation Biology*. Blackwell Publishing, Oxford, UK (pp. 156–172).

Foerster, S., Zhong, Y., Pintea, L., Murray, C. M., Wilson, M. L., Mjungu, D. C., & Pusey, A. E. (2016). Feeding habitat quality and behavioral trade-offs in chimpanzees: a case for species distribution models. *Behavioral Ecology*, 27, 1004–1016.

Forbey, A., Patricelli, G. L., Delparte, D. M., et al. (2017). Emerging technology to measure habitat quality and behavior of grouse: examples from studies of greater sage-grouse. *Wildlife Biology*, SP1. https://doi.org/10.2981/wlb.00238.

Gibbes, C., Sanchayeeta, A., Rostant, L. V., Southworth, J., & Youliang, Q. (2010). Application of object based classification and high resolution. *Remote Sensing*, 2(12), 2748–2772.

Habib, B., Shrotriya, S., Sivakumar, K., et al. (2014). Three decades of wildlife radio telemetry in India: a review. *Animal Biotelemetry*, 2(1), 1–10.

Homer, C. G., Edwards, Jr, T. C., Ramsey, D., & Price, K. P. (1993). Use of remote sensing methods in modelling sage grouse winter habitat. *Journal of Wildlife Management*, 57(1), 78–84.

Honnay, O., Hermy, M., & Coppin, P. (1999). Impact of habitat quality on forest plant species colonization. *Forest Ecology and Management*, 115, 157–170.

Hötte, M. H. H., Kolodin, I. A., Bereznuk, S. L., et al. (2016). Indicators of success for smart law enforcement in protected areas: a case study for Russian Amur tiger (*Panthera tigris altaica*) reserves. *Integrative Zoology*, 11(1), 2–15.

Jachmann, H. (2002). Comparison of aerial counts with ground counts for large African herbivores. *Journal of Applied Ecology*, 39(5), 841–852.

Johnson, M. D. (2005). Habitat quality: a brief review for wildlife biologists. *Transactions of the Western Section of the Wildlife Society*, 41(July), 31–41.

Joppa, L. N. (2015). Technology for nature conservation: an industry perspective. *Ambio*, 44(4), 522–526.

Joshi, A. R., Garshelis, D. L., & Smith, J. L. D. (1995). Home ranges of sloth bears in Nepal: implications for conservation. *Journal of Wildlife Management*, 59(2), 204–214.

Kays, R., Crofoot, M. C., Jetz, W., & Wikelski, M. (2015). Terrestrial animal tracking as an eye on life and planet. *Science*, 348(6240), aaa2478.

Krebs, J. R. (1999). *Ecological Methodology*. Addison-Wesley Educational Publishers, Inc, Menlo Park, CA.

Kucera, T. E., & Barrett, R. H. (2011). A history of camera trapping. In O'Connell, A. F., Nichols, J., & Karanth, K. U. (eds). *Animal Ecology: Methods and Analyses*. Berlin Heidelberg, Germany, Springer (pp. 9–26).

Kuemmerle, T., Chaskovskyy, O., Knorn, J., et al. (2009). Forest cover change and illegal logging in the Ukrainian Carpathians in the transition period from 1988 to 2007. *Remote Sensing of Environment*, 113(6), 1194–1207.

Kuenzer, C. (2014). Earth observation satellite sensors for biodiversity monitoring: potentials and bottlenecks. *International Journal of Remote Sensing*, 35, 6599–6647.

Li, Z., Xu, D., & Guo, X. (2014). Remote sensing of ecosystem health: opportunities, challenges, and future perspectives. *Sensors*, (1), 21117–21139.

Lynch, J., Maslin, M., Balzter, H., & Sweeting, M. (2013). Choose satellites to monitor deforestation. *Nature*, 496, 293–294.

Macbeth, B. J., Cattet, M., Stenhouse, G. B., Gibeau, M. L., & Janz, D. M. (2010). Hair cortisol concentration as a non-invasive measure of long-term stress in free-ranging grizzly bears (*Ursus arctos*): considerations with implications for other wildlife. *Canadian Journal of Zoology*, 88(10), 935–949.

Machovsky-Capuska, G. E., Priddel, D., Leong, P. H. W, et al. (2016). Coupling bio-logging with nutritional geometry to reveal novel insights into the foraging behaviour of a plunge-diving marine predator. *New Zealand Journal of Marine and Freshwater Research*, 50(3), 418–432.

Marques, T. A., Thomas, L., Martin, S. W., et al. (2013). Estimating animal population density using passive acoustics. *Biological Reviews*, 88(2), 287–309.

Marshall, A. J. (2009). Effect of habitat quality on primate populations in kalimantan: gibbons and leaf monkeys as case studies. In Gursky, S. & Supriatna, J. (eds). *Indonesian Primates*. Berlin Heidelberg, Germany, Springer (pp. 157–177).

Miao, Z., Gaynor, K. M., Wang, J., et al. (2019). Insights and approaches using deep learning to classify wildlife. *Scientific Reports*, 9(1), 1–9.

Miller, K. E., Ackerman, B. B., Lefebvre, L. W., & Clifton, K. (1998). An evaluation of strip-transect aerial survey methods for monitoring manatee populations in Florida. *Wildlife Society Bulletin*, 26(3), 561–570.

Morin, D. J., Kelly, M. J., & Waits, L. P. (2016). Monitoring coyote population dynamics with fecal DNA and spatial capture – recapture. *Journal of Wildlife Management*, 80(5), 824–836.

Mortelliti, A., Amori, G., & Boitani, L. (2010). The role of habitat quality in fragmented landscapes: a conceptual overview and prospectus for future research. *Oecologia*, 163(2), 535–547.

Mowat, G., & Strobeck, C. (2000). Estimating population size of grizzly bears using hair capture, DNA profiling, and mark-recapture analysis. *Journal of Wildlife Management*, 64(1), 183–193.

Muehlenbein, M. P., Ancrenaz, M., & Sakong, R. (2012). Ape conservation physiology: fecal glucocorticoid responses in wild pongo pygmaeus morio following human visitation. *PLoS One*, 7(3), 1–10.

Norouzzadeh, M. S., Nguyen, A., Kosmala, M., et al. (2018). Automatically identifying, counting, and describing wild animals in camera-trap images with deep learning. *Proceedings of the National Academy of Sciences of the United States of America*, 115(25), E5716–E5725.

Nowak, K., le Roux, A., Richards, S. A., Scheijen, C. P. J., & Hill, R. A. (2014). Human observers impact habituated Samango monkeys' perceived landscape of fear. *Behavioral Ecology*, 25(5), 1199–1204.

O'Connell, M. J., Nasirwa, O., Carter, M., et al. (2019). Capacity building for conservation: problems and potential solutions for sub-Saharan Africa. *Oryx*, 53(2), 273–283.

Plumptre, A. J. (2000). Monitoring mammal populations with line transect techniques in African forests. *Journal of Applied Ecology*, 37(2), 356–368.

Prentice, J. C., Marion, G., White, P. C. L., Davidson, R. S., & Hutchings, M. R. (2014). Demographic processes drive increases in wildlife disease following population reduction. *PLoS One*, 9(5), e86563.

Ralph, J. C., Sauer, J. R., & Droege, S. (1995) *Monitoring Bird Populations by Point Counts*. Pacific Southwest Research Station, Albany, CA.

Rasiulis, A. L., Festa-Bianchet, M., Couturier, S., & Côté, S. D. (2014). The effect of radio-collar weight on survival of migratory caribou. *Journal of Wildlife Management*, 78(5), 953–956.

Ridgway, M. S. (2010). Line transect distance sampling in aerial surveys for double-crested cormorants in coastal regions of Lake Huron. *Journal of Great Lakes Research*, 36(3), 403–410.

Roy, P. S., Moharana, S. C., Prasad, S. N., & Singh, I. J. (1992). Vegetation analysis and study of its dynamics in Chandaka Wildlife Sanctuary (Orissa) using aerospace remote sensing. *Photonirvachak*, 20(4), 223–235.

Schwarz, C. J., & Seber, G. A. F. (1999). Estimating animal abundance: review III. *Statistical Science*, 14(4), 427–456.

Serneels, S., Said, M. Y., & Lambin, E. F. (2001). Land cover changes around a major east African wildlife reserve: the Mara Ecosystem (Kenya). *International Journal of Remote Sensing*, 22(17), 3397–3420.

Seymour, A. C., Dale, J., Hammill, M., Halpin, P. N., & Johnston, D. W. (2017). Automated detection and enumeration of marine wildlife using unmanned aircraft systems (UAS) and thermal imagery. *Scientific Reports*, 7(2), 1–10.

Sheehan, E. V., Bridger, D., Nancollas, S. J., & Pittman, S. J. (2020). PelagiCam: a novel underwater imaging system with computer vision for semi-automated monitoring of mobile marine fauna at offshore structures. *Environmental Monitoring and Assessment*, 192(1), 11.

Shive, J. P., Pilliod, D. S., & Peterson, C. R. (2010). Hyperspectral analysis of Columbia spotted frog habitat. *Journal of Wildlife Management*, 74(6), 1387–1394.

Srestasathiern, P., & Rakwatin, P. (2014). Oil palm tree detection with high resolution multi-spectral. *Remote Sensing*, 6(10), 9749–9774.

Steffen, W., Crutzen, P. J., & McNeill, J. R. (2007). The Anthropocene: are humans now overwhelming the great forces of nature. *AMBIO: A Journal of the Human Environment*, 36(8), 614–621.

Stewart, F. A., Piel, A. K., Luncz, L., et al. (2018). DNA recovery from wild chimpanzee tools. *PLoS One*, 13(1), 1–13.

Tan, T. F. Teoh, S. S., & Yen, K. (2016). Embedded human detection system based on thermal and infrared sensors for anti-poaching application. In IEEE Conference on Systems, Process and Control (ICSPC) (pp. 37–42).

Thomas, J. A., Bourn, N. A., Clarke, R. T., et al. (2001). The quality and isolation of habitat patches both determine where butterflies persist in fragmented landscapes. *Proceedings of the Royal Society B: Biological Sciences*, 268, 1791–1796.

Thomas, L., Buckland, S. T, Rexstad, E. A, et al. (2010). Distance software: design and analysis of distance sampling surveys for estimating population size. *Journal of Applied Ecology*, 47(1), 5–14.

Thomsen, P. F., & Willerslev, E. (2015). Environmental DNA—an emerging tool in conservation for monitoring past and present biodiversity. *Biological Conservation*, 183, 4–18.

Trenkel, V. M., Buckland, S. T., McLean, C., & Elston, D. A. (1997). Evaluation of aerial line transect methodology for estimating red deer (Cervus elaphus) abundance in Scotland. *Journal of Environmental Management*, 50(1), 39–50.

Turner, W., Spector, S., Gardiner, N., Fladeland, M., Sterling, E., & Steininger, M. (2003). Remote sensing for biodiversity science and conservation. *Trends in Ecology and Evolution*, 18(6), 306–314.

Turner, W. (2014). Sensing biodiversity. *Science*, 346, 301–303.

Underwood, E., Ustin, S., & Dipietro, D. (2003). Mapping nonnative plants using hyperspectral imagery. *Remote Sensing of Environment*, 86, 150–161.

Urbazaev, M., Thiel, C., Cremer, F., et al. (2018). Estimation of forest aboveground biomass and uncertainties by integration of field measurements, airborne LiDAR, and SAR and optical satellite data in Mexico. *Carbon Balance and Management*, 13(1), 5.

Vaglio Laurin, G., Puletti, N., Grotti, M., et al. (2020). Species dominance and above ground biomass in the Białowieża Forest, Poland, described by airborne hyperspectral and lidar data. *International Journal of Applied Earth Observation and Geoinformation*, 92, 102178.

Varman, K. S., & Sukumar, R. (1995). The line transect method for estimating densities of large mammals in a tropical deciduous forest: an evaluation of models and field experiments. *Journal of Biosciences*, 20, 273–287.

Verma, A., Van Der Wal, R., & Fischer, A. (2016). Imagining wildlife: new technologies and animal censuses, maps and museums. *Geoforum*, 75, 75–86.

Viviant, M., Monestiez, P., & Guinet, C. (2014). Can we predict foraging success in a marine predator from dive patterns only? Validation with prey capture attempt data. *PLoS One*, 9(3), e88503.

Wall, J. (2014). Novel opportunities for wildlife conservation and research with real-time monitoring. *Ecological Applications*, 24, 593–601.

Weimerskirch, H., & Wilson, R. P. (2000). Oceanic respite for wandering albatrosses. *Nature*, 406, 954–956.

Wikelski, M., & Cooke, S. J. (2006). Conservation physiology. *Trends in Ecology & Evolution*, 21(2), 38–46.

Wilber, M. J., Sceirer, W., Boult, T., et al. (2013). Animal recognition in the Mojave desert: vision tools for field biologists. In Proceedings of IEEE Workshop on Applications of Computer Vision (pp. 206–213).

Wilfred, P., Kayeye, H., Magige, F. J., Kisingo, A., & Nahonyo, C. L. (2019). Challenges facing the introduction of SMART patrols in a game reserve, western Tanzania. *African Journal of Ecology*, 57(4), 523–530.

Willerslev, E., Davison, J., Moora, M., et al. (2014). Fifty thousand years of Arctic vegetation and megafaunal diet. *Nature*, 506(7486), 47–51.

Wilmers, C. C., Nickel, B., Bryce, C. M., et al. (2015). The golden age of bio-logging: how animal-borne sensors are advancing the frontiers of ecology. *Ecology*, 96(7), 1741–1753.

Woodruff, S. P. Lukacs, P. M., Christianson, D., & Waits, L. P. (2016). Estimating Sonoran pronghorn abundance and survival with fecal DNA and capture–recapture methods. *Conservation Biology*, 30(5), 1102–1111.

Yang, Z., Wang, T., Skidmore, A. K., de Leeuw, J., Said, M. Y., & Freer, J. (2014). Spotting East African mammals in open savannah from space. *PLoS One*, 9(12), e115989.

Yapp, W. B. (1956). The theory of line transects. *Bird Study*, 3(2), 93–104.

CHAPTER 2

From the cloud to the ground: converting satellite data into conservation decisions

Lilian Pintea, Samuel M. Jantz, and Serge A. Wich

2.1 Introduction

Some of the major threats to biodiversity and ecosystem services are conversion, modification, and fragmentation of wildlife habitats by human activities and changes in land cover and land use (Laurance, 1999; Brooks et al., 2002; Sanderson et al., 2002; Groom et al., 2006; Hansen et al., 2013). Developments in remote sensing, geographic information systems (GIS), mobile and cloud computing provide the opportunity to systematically and cost-effectively monitor land cover, land-use changes, and threats (Rose et al., 2015). Using observations from Earth Observing (EO) satellites, it is now possible to obtain detailed spatial and temporal data on land cover and characteristics (such as tree height) of ecosystems that are both locally relevant and consistent in spatial and radiometric resolutions across multiple scales—from a village to a global level (Hansen et al., 2013; Jantz et al., 2018).

However, there are still challenges in incorporating EO products into conservation management practice (Whitten et al., 2001; Pintea et al., 2002; Green, 2011; Salafsky, 2011; Matzek et al., 2014; Buchanan et al., 2015; Toomey et al., 2017). A critical step in conservation is the use of information by decision-makers to change policies, behaviours, and business practices that drive species loss (Whitten et al., 2001; Toomey et al., 2017). The main barriers to the use of information provided by remote sensing are often not technological.

The barriers can include a lack of clearly defined management questions and a lack of budget for training and transfer of technology, as well as a lack of providing results to decision-makers and organizations that manage the environment, wildlife, and other natural resources (Sayn-Wittgenstein, 1992). There is an urgent need to not only monitor land-cover changes at multiple geographic scales, but also to understand how the latest EO satellite data, technologies, and tools could be converted into more relevant, cost-effective, and actionable information products that could lead to the reduction of threats to biodiversity (Pintea et al., 2002; Green, 2011; Buchanan et al., 2015). Information products are scientific results designed with specific audiences and management contexts in mind and within specific windows of opportunity (Rose, 2018) to support policy and conservation decisions.

These challenges are often described in the conservation literature as research-implementation gaps by which scientific information accumulates but is not incorporated into management decisions and actions (Matzek et al., 2014). Solutions have focused on improving communication and designing better information products. However, research from social psychology to organizational management has convincingly shown that empirical evidence is only a minor factor in influencing human decision-making and behavioural change (Pielke, 2007; Owens, 2012; Newell et al., 2014). While developing and sharing better information products is important, larger amounts of, higher resolution, or timely data do not inevitably lead to increased use by decision-makers and ultimately to conservation outcomes. Recent understandings of the science–policy interface from communication, science, and technology studies suggest that information and knowledge should not be seen as a product to be transferred from researchers to users but rather a 'process of relating that involves negotiation of meaning among partners' (Roux et al., 2006; p. 11; Pielke, 2007; Owens, 2012). As a result, Toomey et al. (2017) urged conservation practitioners to adjust from trying to solve research-implementation gaps to creating more research-implementation spaces where researchers, decision-makers, and other partners engage, collaborate, identify, and understand how data, information, and knowledge are produced, by whom and for whom.

Finally, with 65% of the world's land (Rights and Resources Initiative, 2015) and more than 80% of Earth's biodiversity under indigenous or local community customary ownership care and use (UN, 2014; Raygorodetsky, 2018), it is important to consider how EO applications, knowledge, and tools are perceived and filtered through existing local beliefs, traditions, values, experiences, and concerns along with the capacity and resources available to use that information (Toomey et al., 2017). This chapter provides a short introduction to EO technologies that are being used in conservation and a case study using the Jane Goodall Institute (JGI) chimpanzee conservation efforts in Tanzania as an example of creating research-implementation spaces where EO data and results have been developed and used with the local communities, researchers, and district and regional governments.

2.2 New technology

2.2.1 Remote sensing and EO satellites: hardware and software

Conservation is inherently spatial and complex (Game et al., 2013). It requires using geographic data to understand the current and changing distributions of species, the extent of species' habitats and its change, and ecosystem processes along with human activities to design and implement strategic actions that will minimize or eliminate the most important threats to biodiversity (Groves et al., 2003). Geographic data, or spatially referenced data, are any data that have a location associated with them. Geospatial technologies are a combination of three major groups of technologies involved in the collection, manipulation, storage, management, analysis, visualization, and communication of these geographic data and information. They include (1) global navigation satellite systems (GNSS), (2) GIS, and (3) remote sensing (RS) (Longley et al., 2011).

RS is defined as 'the science and art of obtaining information about an object, area, or phenomenon through the analysis of data acquired by a device that is not in contact with the object, area or phenomenon under investigation' (Lillesand et al., 2004). RS data are acquired using a variety of sensors mounted on ground structures (e.g. handheld platform, towers, vehicles), airborne (e.g. drones, aircraft, balloons) or spaceborne platforms (such as satellites orbiting the Earth). This chapter will focus mostly on the use of spaceborne EO sensors; other chapters focus on drones (Chapter 3), acoustic sensors (Chapter 4), and camera traps (Chapter 5).

There are two major types of RS sensors: passive and active. A digital camera mounted on a drone is an example of a passive sensor that works by detecting reflected sunlight from the object, while active sensors such as LiDAR (light detection and ranging) and SAR (synthetic aperture radar) provide their own energy to illuminate the object or scene they observe (Jensen, 2014).

Active and passive sensors can be divided into non-imaging (e.g. radiometers) and imaging sensors (e.g. imaging scanners) that are analogue or digital and acquired on a range of spectral and spatial resolutions (Table 2.1). Understanding the trade-offs, benefits, and limitations of various categories of platform/sensor combinations and matching these with the conservation problem are critical to the development of cost-effective RS applications. For example, ground-based platforms (such as handheld devices, tripods, towers, and cranes) are useful for collecting detailed data from a single point and are often used as ground-truthing; however, they cannot provide the aerial perspective over a larger region that sensors

Table 2.1 Example of the taxonomy of remote sensors

Category	Subcategory
Platforms	Ground—Airborne—Spaceborne
Sensors	Active—Passive
Output	Non-Imaging—Imaging
Image output	Analogue—Digital
Spectrum range	Visible—Infrared—Thermal—Microwave
Spatial resolution	High <5 m—Medium >5 m <50 m—Low > 50 m

placed on aerial and spaceborne platforms provide (Turner et al., 2003). Similarly, drones provide very high-resolution imagery on a flexible schedule, but could be cost-prohibitive to cover larger sites compared with satellite images. For example, Maxar is the first company to deliver 30 cm native and 15 cm high definition synthetic (using image resampling algorithms) satellite imagery[1] that could be collected along a swath width of 13.1 km at nadir (i.e. looking straight down), covering a total of 171.61 km^2 just in one image. A typical commercial level fixed-wing drone such as an eBee can deliver imagery accuracy below 3 cm but can only fly for about 1 hour on one battery charge and cover around 220 ha (2.2 km^2) while flying at 120 m asl, which is the approximate legal altitude ceiling in many countries[2] (Wich & Koh, 2018). This means that it will take approximately 78 flights to cover an area captured by just one Maxar satellite image, but at a five times higher resolution. Of course, flying higher will increase coverage area but will result in decreased resolution. This will require substantial planning, time, equipment, and resources while, in contrast, high-resolution satellite imagery is becoming increasingly available at lower or no cost for conservation NGOs. Ultimately, the choice of sensor should be driven by the management question as a very high resolution might be required for some questions. In contrast, lower resolution might be sufficient for others (Wich & Koh, 2018).

GNSS is a satellite-based radio navigation system that allows accurate determination of geographical locations anywhere on Earth. It is a growing system that currently includes the Global Positioning System (GPS) by the United States, and other fully operational systems such as Global Navigation Satellite System (GLONASS) by the Russian Federation, BeiDou Navigation Satellite System (BDS) by the People's Republic of China, and Galileo, a global GNSS owned and operated by the European Union[3] When combined with data collection apps such as Open Data Kit (ODK) (discussed in Chapter 9) or Esri's ArcGIS Survey 123, GNSS-enabled handheld devices such as smartphones and tablets enable users to collect, temporarily store, and share georeferenced field data, including photos and sounds.

Finally, GIS could be defined as a computer system for capturing, storing, checking, integrating, manipulating, analysing, and displaying geospatial data (Chorley, 1988). It is more than computer software and includes hardware, software, data, methods, people, and institutional processes (Fu & Sun, 2010). Therefore, in the context of conservation, GIS integrates GNSS-enabled mobile and RS technologies to collect, capture, manage, manipulate, integrate, analyse, model, display, and communicate geographic data about species, habitats, and ecosystems, human activities, or threats driving biodiversity loss and conservation efforts. These components can be joined in one computer desktop or a network of computers via a local area network (LAN) or increasingly today via a Web GIS that uses web servers and cloud technologies (Fu & Sun, 2010).

2.2.2 Earth-observing satellites: analysis

There are almost 700 EO satellites in orbit that currently enable scientists and conservation practitioners to study the Earth and its processes on a range of spatial, radiometric, and temporal scales.[4] The sensor or combination of sensors that can be leveraged depends on the conservation target parameters. Coarse-resolution sensors tend to have high revisit frequencies and can provide information on global to regional land change and ecosystem functioning.

---

[1] https://blog.maxar.com/earth-intelligence/2020/introducing-15-cm-hd-the-highest-clarity-from-commercial-satellite-imagery
[2] https://www.sensefly.com/drone/ebee-geo/

[3] https://www.gps.gov/systems/gnss/
[4] https://www.thegeospatial.in/earth-observation-satellites-in-space

The Advanced Very High Resolution Radiometer (AVHRR), a meteorological satellite operated by the National Oceanic and Atmospheric Administration (NOAA), has been providing coarse resolution (1–8 km) imagery in 4–5 bands in the visible, near-infrared, and thermal wavelengths for over 30 years with consistent global coverage (Townshend, 1994). Due to its coarse spatial resolution, AVHRR data have primarily been used to study the Earth System and its processes. For example, AVHRR data have been used to provide clear evidence that the effects of anthropogenic climate change are accelerated in northern latitudes (Myneni et al., 1997), quantify the rates of tropical deforestation and associated carbon emissions (DeFries et al., 2002), and show that global tree cover has increased over that past 34 years but is still declining in the tropics (Song et al., 2018). AVHRR data have also been indispensable for coral reef conservation management. Since 2000, a global user base has relied on NOAA's Coral Reef Watch to provide near real-time alerts of coral reef bleaching events as well as forecasts of future bleaching risk (Liu et al., 2006).

Since 1999, the Moderate Resolution Imaging Spectrometer (MODIS) has provided daily global coverage at 250 m–1 km spatial resolution with 36 bands that encompass the visible, infrared, and thermal wavelengths (Townshend & Justice, 2002). Lessons learned from using AVHRR data enabled the development of several science products that have conservation relevance, such as measures of vegetation productivity, fire detection, land-cover, and land-cover change (Justice et al., 2002). Defries et al. (2005) used MODIS data to show that while protected areas across the tropics have experienced some forest loss within their boundaries, forest is disappearing at a much faster rate just outside their boundaries, leaving them increasingly isolated. In Belize, fires detected from MODIS alerted the Friends for Conservation and Development (FCD) of forest clearing on Chiquibul National Park's western portion. FCD followed up with aerial and ground reconnaissance and confirmed that over 200 acres of forest had been illegally cleared by Guatemalan farmers (Davies et al., 2015).

A current paradigm in conservation biology is the preservation of remnant habitat patches (Pimm & Brooks 2013), as well as preserving or establishing linkages between patches to enable the flow of individuals and genes between subpopulations (Bennett, 2003). An accurate characterization of patches and corridors is pivotal and at 250 m resolution, MODIS is unable to fulfil this role. The Landsat series of satellites has been collecting imagery since 1972, mostly at 30 m resolution for most bands but earlier satellites were coarser in resolution with fewer bands. However, prior to 1999 there was no systematic global acquisition strategy. Before 2008, individual Landsat scenes needed to be purchased (Woodcock et al., 2008), which hampered the widespread use of the imagery. The current Landsat archive contains millions of images requiring approximately 1.3 million gigabytes of storage space. In order to use this imagery for regional to global scale analysis, all the images need to have atmospheric effects removed and be assessed for quality systematically. To accomplish this, RS scientists developed methods that leverage the recent increase in computing power to mine the entire Landsat archive for good quality images and systematically remove the effects of the atmosphere, enabling them to create products relevant for conservation monitoring. For example, the Global Land Analysis and Discovery (GLAD) lab at the University of Maryland, College Park has released and is maintaining a forest loss data set derived from 30 m Landsat imagery that has global accuracy but also has relevance at the local scale (Hansen et al., 2013). Tracewski et al. (2016) used the GLAD forest loss product to quantify forest loss within the ranges of over 11,000 forest-dependent species. Their analysis resulted in hundreds of species on the IUCN Red List of Threatened Species having their extinction risks increased because of inferred rapid population declines from either habitat loss or because of a restricted area of occupancy owing to scant forest cover remaining within their range. Joshi et al. (2016) used the GLAD product to assess forest loss within 76 landscapes prioritized for tiger (*Panthera tigris*) conservation over 13 Asian countries. They found much less forest loss occurred than anticipated over the 2001–2014 period, but expansion of oil-palm plantations did result in extensive habitat loss. Results from this study highlight the value of satellite-based monitoring for prioritizing the protection of key habitats and corridors.

2.2.2.1 Detecting animals from space

Species distribution and abundance are important parameters to estimate when attempting to conserve a particular species. Landsat imagery has been used to indirectly detect animals for many years. For example, Löffler and Margules (1980) were able to estimate the distribution of the hairy-nosed wombat (*Lasiorhinus latifrons*) in South Australia due to its mound-building activity, which created areas of bare ground visible from space. The stark contrast of faecal stains on ice enabled Fretwell and Trathan (2009) to map Emperor Penguin (*Aptenodytes forsteri*) colonies across Antarctica. More recently, very high resolution (VHR, <1 m) commercial satellites have enabled the detection of individual animals from space. In a pilot study, Yang et al. (2014) created an algorithm to automatically count individual wildebeests (*Connochaetes taurinus*) and zebras (*Equus quagga*) in the Maasai Mara National Reserve, Kenya, using 50 cm pan-sharpened imagery from the GeoEye-1 satellite. On Rowley Island in the Canadian Arctic, Stapleton et al. (2014) used 50 cm WorldView-2 and 65 cm QuickBird imagery to count individual polar bears (*Ursus maritimus*). They used the count data to estimate bear abundance and found it produced an estimate similar to abundance estimated from a line transect aerial survey conducted a few days prior to the satellite imagery acquisition, 94 (95% CI: 92–105) versus 102 (95% CI: 69–152) individuals, respectively. In a follow-up to their previous study, Fretwell et al. (2012) used QuickBird imagery to estimate the number of Emperor Penguin breeding pairs within the colonies detected from the analysis of Landsat imagery. They used linear regression to combine data from 11 ground-truthing stations with individuals detected from the imagery to produce the first estimate for the total number of breeding pairs in Antarctica.

2.2.2.2 Applications for near real-time monitoring and alerts

Platforms hosted in the cloud can benefit conservation efforts. The most prominent among them is Global Forest Watch,[5] which hosts several data sources relevant for conservation. Users can delineate an area of interest and receive a notice via e-mail if forest loss or a fire is detected. Moreover, the Forest Watcher mobile application enables the offline use of Global Forest Watch's (GFW) spatial data.[6] Users can set up an area to monitor and download deforestation and fire alerts, then take this information to monitor and manage forests while in the field directly from a smartphone or tablet and help investigate and report what they find, enabling immediate action on the ground regardless of internet connectivity (Petersen & Pintea, 2017). The frequency of notification depends on which alert is of interest to a user. The Fire Information for Resource Management System (FIRMS) uses MODIS data to create fire alerts that are available within hours of satellite overpass, which could result in multiple alerts per day, albeit at the 1 km scale (Davies et al., 2015). If a user wishes to monitor, for example, illegal logging in a protected area, an alert sensitive to smaller-scale disturbances is needed. The GLAD team developed an alert for humid tropical forests at 30 m resolution that is available as new Landsat imagery is acquired, which means users can be updated weekly, but the prevalence of clouds in the region of interest could result in a delayed alert (Hansen et al., 2016a; Petersen & Pintea, 2017; Palminteri, 2017; Pintea et al., 2019).

2.3 Case study: converting EO data into chimpanzee conservation decisions in Tanzania

2.3.1 Creating research-implementation spaces using conservation standards

Chimpanzees *(Pan troglodytes)* have been listed as Endangered on the IUCN Red List since 1996. The major threats to chimpanzee survival are: (1) habitat loss, degradation, and fragmentation from the incompatible conversion of forests and woodlands to agriculture, logging for timber and charcoal, mining, and human settlements; (2) illegal commercial

---

[5] https://www.globalforestwatch.org, GFW

[6] https://play.google.com/store/apps/details?id=com.forestwatcher

bushmeat hunting and trade; (3) disease; and (4) the illegal pet trade (Humle et al., 2016).

The most eastern distribution of the chimpanzee range is in Tanzania (Kano, 1971). Tanzania is home to two of the longest ongoing long-term chimpanzee research sites in the world (Gombe Stream and Mahale Mountains National Parks) that provided pioneering accounts of chimpanzee behaviour (Wilson et al., 2020; Nakamura et al., 2015). However, 91% of the chimpanzee range in Tanzania is outside these two national parks and on lands managed by local communities, village, district, and regional governments (Pintea unpublished data). Owing to this, JGI started to engage and work with village and district governments as key stakeholders in chimpanzee conservation through its community-led development approach called Tacare in 1994 (Pintea, 2016).

Tacare is a rural development approach where the focus is on improving people livelihoods through nature-based solutions with conservation as one of the outcomes. It addresses the threats to chimpanzees, other wildlife, and habitats by assuring that local people and institutions are not only engaged, but also own and drive the development and conservation efforts in their landscapes. It recognizes that local communities are the most connected to and depend directly on ecosystem services (Chancellor et al., 2020). Tacare includes listening to communities about their needs and priorities and actively connecting nature-based solutions (Maes & Jacobs, 2017) with the conservation objectives and planning at the community scale, such as facilitating village land-use plans that assure land tenure and enable community development, while also contributing to conservation.

As part of its Tacare approach, JGI uses 'Open Standards for the Practice of Conservation' or 'Conservation Standards' (CMP, 2013) to identify chimpanzee and other conservation needs, target and prioritize conservation actions, and measure success. Conservation Standards is a science-based and collaborative planning approach that uses adaptive management to help focus conservation decisions and actions on clearly defined objectives and prioritized threats and measures success in a manner that enables adaptation and learning over time (CMP, 2013). JGI uses Conservation Standards internally

to guide decisions on its institutional strategies and allocation of resources and with partners to develop a common understanding of the conservation needs, threats, and cooperate on joint implementation of conservation actions from village to regional scales. Since 2005 JGI and partners have relied on Conservation Standards to make strategic decisions and guide chimpanzee conservation efforts in Tanzania, Uganda, DRC, Republic of Congo, Burundi, Liberia, Guinea, and Senegal.

Key questions facing stakeholders engaged in the Conservation Standards process are: 'How species and habitats are doing and what are the major threats to their survival?', 'What actions are needed to minimize or eliminate those priority threats?', and 'Are our actions effective in minimizing or eliminating the most important threats?' (CMP, 2007). In addressing these questions, the Conservation Standards approach is oriented around a five-step project management cycle adopted for conservation: assess, plan, implement, analyse and adapt, and share (CMP, 2020). In the next sections, we will use the Conservation Standards project cycle to illustrate how JGI and partners have used EO data and results to guide decisions and support the development and implementation of chimpanzee conservation strategies and actions in Tanzania.

2.3.2 Assess

A Conservation Standards cycle starts with defining the project's purpose, teams, and articulating geographic scope, a vision of what the project hopes to achieve, and the conservation targets on which the plan will focus. It also includes making sense of the project's context, including identifying threats, opportunities, and key stakeholders (CMP, 2020).

2.3.2.1 Defining geographic scope and developing basemaps

The use of EO data and geospatial technologies starts with identifying, downloading, cleaning, and compiling all relevant geospatial data and RS imagery into a geodatabase to produce a series of basemaps. This geodatabase serves as the foundation for other geospatial applications and provides geographical context to the conservation planning process. It includes multiple baseline features or raster layers defining administrative boundaries, protected areas, settlements, land tenure, land cover/land use, topography along with elevation, slope, and shaded relief, hydrological features like streams and watersheds, and transportation networks.

Geospatial layers are usually compiled from multiple sources, including individual researchers, national and international mapping agencies, and global databases such as the World Database on Protected Areas (WDPA)[7] or OpenStreetMap.[8] Global and regional data sets could be of various quality that could impact their use as part of a local conservation planning process. These include major roads that do not exist when validated on the ground, protected areas in wrong locations, administrative boundaries that do not overlap or correspond to actual locations on the ground, missing settlements, and other errors. National data sets have similar issues. Many countries in Africa are struggling to develop operational national spatial data infrastructures (SDI), policies, or standards to improve the quality and availability of geospatial data (Lance et al., 2013).

Tanzania is still at the beginning of developing an SDI for the country (Mansourian et al., 2015). Therefore, all conservation planning efforts in Tanzania used participatory mapping with key project stakeholders, including local communities, government officers, scientists, and conservation practitioners, to ground-truth, validate, and update spatial data to meet specific conservation project requirements. Participatory mapping can be done in focal groups using sketching on printed basemaps or by projecting the map and sketching on the wall while digitizing onscreen during the conservation planning workshops.

Since 2002 JGI has developed a Participatory Remote Sensing (PRS) model (Pintea, 2006) as part of its Tacare approach. It combines Participatory Rural Appraisal (PRA) with very high-resolution satellite imagery below 1-metre resolution, enabling local people to record their knowledge, values, and perspectives of their communities by mapping water sources, wildlife migrations, land tenure, land

uses, vegetation, traditional belief sites, and others. JGI's PRS approach uses satellite imagery from Maxar or Planet to print paper maps at the village scales between 1:1000 to 1:5000. The institute is currently exploring the potential uses of drones as a complementary source of imagery to support smaller-scale participatory community mapping efforts.

Local validation by local communities and decision-makers of the existing EO and geospatial data and basemaps had three important benefits in Tanzania. First, it improved the quality of data to support specific conservation planning needs by recording and combining local and expert knowledge with the latest GIS/RS data. Second, it enabled key decision-makers to connect and start collaborating early with researchers. Finally, it supported a research-implementation space that increased stakeholder's awareness of the potential and limitations of spatial data, developed trust and facilitated the actual use of information products derived from geospatial technologies later on in the conservation process.

2.3.2.2 Mapping chimpanzee distribution

The Conservation Standards process involves the selection of a limited number of conservation targets or specific species, such as chimpanzee communities, populations, or habitats, to represent biodiversity and the ultimate aims of the project (CMP, 2020). At this stage, EO data and associated geospatial technologies can be used to map the current, historic and, in some cases, anticipated future extent of the conservation targets under different scenarios.

The purpose of this step is to help project teams identify and agree on clear common goals and how to spend their limited time and resources to achieve those goals in the face of uncertainty and lack of complete information (Salafsky et al., 2002). EO data and geospatial technologies enable teams to model and map conservation targets' distribution and status while selecting the best data and models available with the management and planning questions.

For example, in the case of the Greater Gombe Ecosystem Conservation Action Plan (GGE-CAP) (Figure 2.1), stakeholders had access to decades

---

[7] https://www.protectedplanet.net/
[8] https://www.openstreetmap.org/

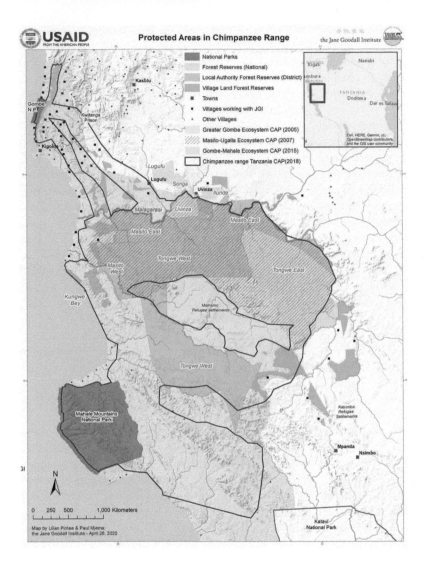

Figure 2.1 Geographic scope of Greater Gombe, Masito-Ugalla, and Gombe-Mahale Ecosystem Conservation Action Planning areas, overlaid with protected areas and chimpanzee range in western Tanzania.

of long-term Gombe research data that enabled detailed definition and mapping of the chimpanzee conservation targets. Using data location points and expert knowledge, researchers used GIS to estimate and map community home ranges or areas where chimpanzees move in search of food and other resources. These range maps were then overlaid with habitat change maps of both inside and outside Gombe derived from 1972 Landsat MSS and 1999 Landsat ETM+ satellite images at 85-metre resolution and ground-truthed with digitized and orthorectified 1974 panchromatic aerial photos and 1-metre resolution Ikonos satellite images acquired

in 2001 (Pintea et al., 2002; Pintea, 2007). These maps illustrated to stakeholders that deforestation and an increase in settlements outside the national park have had an unequal effect on the Gombe chimpanzee communities.

The Kasekela community range is the largest and individuals from this community only use areas inside the national park. Kasekela has also been the least affected by the changes outside the park (Pusey et al., 2007). In contrast, both the Mitumba and a third community, Kalande chimpanzees, were trapped between the Kasekela community and the park boundaries. Historically, both the Mitumba and Kalande chimpanzees travelled outside of the park (Pusey et al., 2007). However, due to deforestation, largely driven by the conversion of forests and woodlands to oil palm, farmland, and settlements, these two communities lost that habitat (Pintea, 2007; Pusey et al., 2007).

The combination of long-term data and spatial insights acquired using GIS and RS products allowed stakeholders to discuss and ultimately agree that each of the three chimpanzee communities should represent a separate conservation target, as each was impacted by different threats depending on their spatial relationship to the park boundary, and thus required different conservation strategies and actions. This illustrates the challenges of identifying a conservation target and the importance of data to inform those decisions. In some cases, focusing on threats such as habitat loss (that could be easier to gather data on using EO technologies) could provide a preliminary insight that can then be verified with chimpanzee habitat use and ranging studies.

However, most of the conservation planning efforts outside long-term research sites do not have access to such detailed long-term data. In these cases, EO data and associated geospatial technologies could be useful in helping teams to match the best available data and models with specific project and planning needs. For example, stakeholders working on developing a national Chimpanzee Conservation Action Plan for Tanzania (TAWIRI, 2018) (Figure 2.1) agreed to focus on two conservation targets: chimpanzee populations and chimpanzee habitats. Decision-makers identified two major types of locations describing chimpanzee targets: (1) known chimpanzee areas with confirmed chimpanzee presence, and (2) areas predicted to be potentially suitable for chimpanzees.

JGI has collaborated with a network of partners to systematically survey and map chimpanzee distribution outside national parks since 2003 (Ogawa et al., 2006, Moyer et al., 2006). Evidence of chimpanzee presence was compiled from surveys, including reconnaissance missions (Piel et al., 2015) and crowdsourced citizen-science village forest monitoring efforts (Pintea, 2016). All these locations were integrated into the same project geodatabase using Esri's ArcGIS and buffered by a 5-km radius as agreed by stakeholders to represent known chimpanzee areas.

Areas predicted to be potentially suitable for chimpanzees in Tanzania were mapped using 12 environmental predictors largely derived from EO satellite images to annually model habitat suitability at 28.5-metre resolution. Tanzania distribution map was extracted from a model that covered the entire chimpanzee range in Africa (Jantz et al., 2016). The suitability model's primary RS derived data include annual composites of per cent canopy cover, canopy height, and top of canopy reflectance for Landsat ETM+ bands 3, 4, 5, and 7. For all variables except canopy height, methods for deriving these layers globally are reported in Hansen et al. (2013). Canopy height was obtained using Landsat ETM+ and height estimates from waveforms returned from the Geoscience Laser Altimeter System (GLAS) (Hansen et al., 2016b).

To produce the final conservation target maps, chimpanzee presence and habitat suitability data were overlaid with the IUCN chimpanzee range[9] and then edited and classified into more spatially detailed polygons of core ranges and corridors using expert knowledge and participatory mapping. Other conservation action planning efforts facilitated by JGI using Conservation Standards went through a similar process. For example, in the Eastern DRC Conservation Action Plan for Grauer's gorilla (*Gorilla beringei graueri*) and chimpanzees (Maldonado et al., 2012), stakeholders agreed to use presence within a 5×5 km grid as evidence of the target distribution, combined with expert participatory mapping of populations sketched as hand-drawn polygons on the basemaps. EO data and geospatial technologies were essential in these processes not only to collect, clean, organize, analyse, and visualize target population, model habitat suitability, and identify existing data gaps, but also to support the creation of research-implementation spaces and dialogue among stakeholders that helped reach a decision how to interpret existing incomplete scientific results and focus on using existing data to improve conservation efforts and actions.

---

[9] https://www.iucnredlist.org/resources/spatial-data-download

2.3.2.3 Mapping chimpanzee population and habitat viability

Once planning teams have identified conservation targets, a key step is to define their current and desired status or viability because it provides a baseline against which change can be measured (CMP, 2020). Viability assessments of conservation targets that include species, habitats, or ecosystem process help project teams build a set of hypotheses to guide conservation and research. It begins by identifying key ecological attributes (KEAs) for each of the targets. A KEA is *an aspect of a target's biology or ecology that if present, defines a healthy target and if missing or altered, would lead to the outright loss or extreme degradation of that target over time* (TNC, 2007).

The conservation standards group KEAs go into three indicator classes:

- **Size** is a measure of the area or abundance of the conservation target's occurrence (e.g. acres of habitat).
- **Condition** is a measure of the biological composition, structure and biotic interactions that characterize the occurrence (e.g. presence of key species).
- **Landscape context** is an assessment of the target's environment, including ecological processes and regimes that maintain the target occurrence such as flooding, fire regimes, and many other kinds of natural disturbance, and connectivity such as species targets having access to habitats and resources or the ability to respond to environmental change through dispersal or migration (e.g. average distance in km between habitat patches) (TNC, 2007).

EO data and associated geospatial technologies can be used to define, measure, and continually monitor and update some of the KEAs. Guided by research or management questions, chimpanzee habitats can be mapped at different resolutions using various remote sensors and methods.

For example, in Gombe, calibrated and normalized Landsat MSS and ETM+ satellite imagery at 85-metre resolution allowed researchers to map long-term changes in the vegetation's greenness using NDVI and reference these against changes in chimpanzee feeding behaviour between 1972 and 1999 (Pintea, 2007; Pintea et al., 2012). However, 30-metre Landsat ETM+ data were not sufficient to support research questions on chimpanzee hunting behaviour. Higher-resolution vegetation maps classified from 4-metre Ikonos multispectral satellite imagery were required to demonstrate that chimpanzee hunts were both more likely to occur and succeed in woodland and semideciduous forest than in evergreen forest, emphasizing the importance of visibility and prey mobility (Gilby et al., 2006).

In the GGE-CAP, vegetation maps derived from 2-metre QuickBird and WorldView satellite imagery allowed mapping of individual scattered trees. This resolution informed detailed sweep surveys of a corridor mosaic north of Gombe, enabling JGI to map the distribution of scattered trees and patchy habitats important for chimpanzees, differentiating them from oil palm, banana, and other crops in a human-dominated landscape (Wilson et al., 2020). Although these high-resolution EO applications provided detailed habitat data, they only covered the GGE-CAP area of 677 km^2. It also required technical personnel to commit extensive time and resources to pre-process, ground-truth, and develop very high-resolution classification products from EO data.

Such application of high-resolution EO data was prohibitive for the Tanzanian Chimpanzee Conservation Action plan (TAWIRI, 2018), which to cost-effectively monitor annually the entire chimpanzee range and corridors in Tanzania required an area of 17210 sq km. The planning team opted for the lower resolution yet globally freely available tree canopy cover and tree cover loss products derived from the 28.5-metre Landsat imagery annually updated by the GLAD, University of Maryland since 2000 (Hansen et al., 2013).

These medium-resolution EO data were combined with a habitat suitability model and other geospatial data to support decision-makers to agree on three habitat health indicators that could be cost-effectively and operationally monitored using 28.5-metre GLAD products:

- Size—per cent of evergreen forest and woodland loss with tree cover >25% in suitable chimpanzee habitat relative to 2000 baseline (2001 onward);

- Condition—per cent of evergreen forest loss with tree cover >65% in suitable chimpanzee habitat relative to 2000 baseline (2001 onward);
- Landscape context—proximity to people (distance from any human-made features such as roads, houses, and farms).

Despite their medium resolution, these EO-derived products were actionable because decision-makers defined thresholds that clearly marked how to interpret each habitat health indicator using a four-scale rating from the Conservation Standards process (TNC, 2007; CMP, 2007). After comparing different thresholds of forest and woodland loss, government decision-makers, conservation NGO staff, and chimpanzee researchers used their expert opinion and agreed to rank chimpanzee habitat health as (Figure 2.2):

- Very Good—if suitable for chimpanzee forest and woodland loss is <1% compared to 2000

baseline. This is an ecologically desirable status and requires little intervention for maintenance.
- Good—if suitable for chimpanzee forest and woodland loss is between 1% and 2.5% compared to 2000 baseline. This is within an acceptable range of variation; some intervention required for maintenance.
- Fair—if suitable for chimpanzee forest and woodland loss is between 2.5% and 5% compared to 2000 baseline. This is outside an acceptable range of variation and requires conservation or restoration intervention.
- Poor—if suitable for chimpanzee forest and woodland loss is >5% compared to 2000 baseline. Restoration increasingly difficult; may result in the destruction of chimpanzee habitats.

To further enable EO data use, JGI developed a dynamic habitat viability geospatial tool in ArcGIS Pro that effectively engages decision-makers in KEA discussions by enabling collaboration and generate

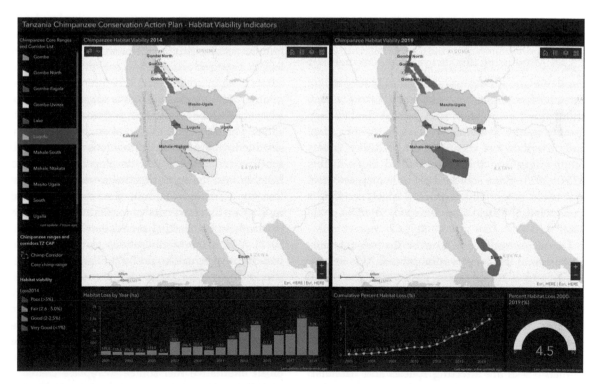

Figure 2.2 Example of an interactive online dynamic dashboard and web map developed using Esri's ArcGIS platform to visualize the status and trends in habitat viability indicators developed to inform the implementation of the Tanzania Chimpanzee Conservation Action Plan (TAWIRI 20, 2018). The map on the left shows habitat viability in chimpanzee core ranges and corridors in 2014 compared to 2019, on the right.

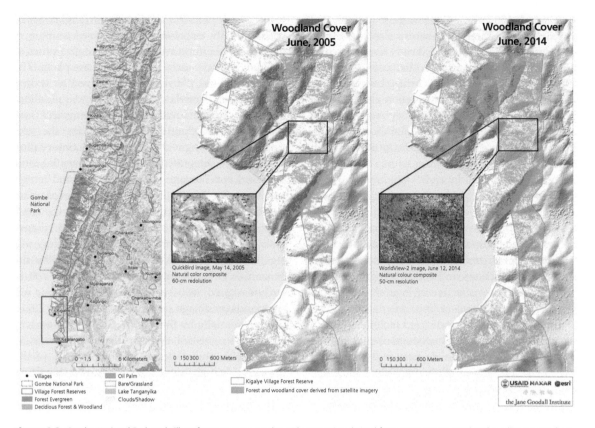

Figure 2.3 Gombe National Park and village forest reserves overlay with vegetation derived from 2014, 60-cm QuickBird satellite images from Maxar (left) with Kigalye village in the orange polygon. Natural regeneration of miombo woodlands in Kigalye Village Forest Reserve between 2005 and 2014 as detected by QuickBird and WorldView-2 satellite images.

Forest Reserve had 370 hectares of woodlands (70% of its total area), while in 2005—when the village reserve was created—its woodlands had decreased to 156 hectares, just 42% of the forested area recorded in 1972. The imagery also shows that by 2014, community efforts had increased these woodlands to 302 hectares, or about 82% of the forested area documented in 1972 (Pintea, 2016).

Successful natural regeneration in Kigalye and other village forest reserves in the GGE was the result of long-term efforts that included developing and implementing a systematic conservation action plan following Conservation Standards (GGE-CAP, 2009), all part of the larger TACARE long-term project operating in the area since 1994 (Pintea, 2011; Wilson et al., 2020). The implementation stage of the action plan combined elements of GeoDesign (Pintea, 2016) and use

of EO data, community mapping, ODK, Survey 123, and other geospatial technologies to inform and facilitate a participatory land-use planning process owned and driven by the local communities.

JGI's community mobile platform has been successfully deployed and used in Tanzania, the Republic of Congo, the Democratic Republic of Congo, and Uganda to support the implementation of a variety of conservation strategies from law enforcement to forest and wildlife monitoring. JGI also introduced and supported ODK use by partners that further adapted the app for other applications such as behaviour and ecological research (Chapter 9). In 2014, the system was expanded with a new app called Forest Watcher, co-developed by JGI in partnership with Google Earth Outreach and World Resources Institute (WRI).

The Forest Watcher mobile app brings the dynamic online forest monitoring and alert systems of GFW offline and into the field. It helps local communities, protected area rangers, and concerned citizens to monitor areas of interest, view deforestation and fire alerts, navigate to a point to investigate, and record their observations, regardless of connectivity (Pintea et al., 2016; Petersen & Pintea, 2017). In western Uganda, private forest owners' associations used the Forest Watcher app to support accountability and transparency among its members and inform the collective management of the private forests in the Budongo-Bugoma corridor. Meanwhile, Uganda Wildlife Authority rangers and managers used the app to identify and stop encroachment in Kibale National Park (Pintea et al., 2019).

2.3.4 Analyse and adapt

Interactions between chimpanzee communities, habitats, and threats occur in dynamic and complex systems. In most cases, viability and threat indicators are developed with incomplete scientific knowledge and should be viewed as hypotheses that need to be continuously tested by new research studies. Conservation Standards encourages practitioners to regularly analyse their project data as they are gathered and convert those data into useful information that could help project teams reflect and adapt their strategies and actions accordingly (CMP, 2020).

In the case of the Tanzania Chimpanzee Conservation Action Plan (TAWIRI, 2018) described earlier, it was decided that habitat viability indicators would be updated annually using EO data. This method has enabled decision-makers to analyse the status and trends in habitat viability and adapt conservation strategies as part of the plan's implementation. The latest analysis of the habitat size viability indicator showed that in 2014, when the work on the national plan had started, most of the large chimpanzee core ranges were in good health (Figure 2.2). Four years later, the trends in indicators illustrate a disturbing acceleration in forest and woodland loss, revealing that some core chimpanzee habitats (e.g. Lugufu) degraded from 'Good' to 'Fair' condition, while others (e.g. Masito and Ugalla) are on the trajectory to follow Lugufu in the next few years.

It is often difficult during planning workshops or meetings with decision-makers to reserve enough time to communicate complex scientific results to a non-scientific audience (Lilian Pintea personal communication). Therefore, simple habitat health indicators developed using Conservation Standards methods enabled managers to interpret EO data quickly and easily and focus on the main problems that need immediate attention and action, such as areas that lost viability or experience an increase in threats, and do something about it.

EO habitat viability indicators combined with very high-resolution satellite imagery showed that habitat loss is especially high in riverine and evergreen forests, which are critical for the viability and survival of chimpanzees in the region. Further analysis of chimpanzee habitat loss data combined with field and aerial surveys identified livestock and large-scale pastoralist migrations as a new source and driver of threats to chimpanzee habitats in Tanzania. These insights from habitat viability maps led to new investments and approaches to better protect evergreen and riverine forests, like behaviour change campaigns to improve soil fertility, along with an increased focus on livestock migrations and data-driven law enforcement using a decision support and alert system (DSAS).[12]

2.3.5 Share

This final step in the Conservation Standards project cycle involves sharing lessons, giving and receiving feedback, and promoting a learning culture with project teams and the broader community of partners and stakeholders (CMP, 2020). To enable adaptive management and learning throughout the entire conservation decision-making process, conservation organizations can take advantage of the WebGIS and cloud platforms to integrate their diverse geospatial data, tools, and users.

JGI leverages and uses Esri's ArcGIS cloud as its geospatial science platform with specific tools, data workflows, and models organized under different Decision Support Systems (DSS). It allows JGI to cost-effectively connect and scale-up field data collections across multiple projects and partners with

---

[12] https://www.usaid.gov/documents/1860/landscape-conservation-western-tanzania

data analysis, modelling, visualization, and storytelling tools targeted towards specific windows of opportunity for informing decision-makers and improving conservation decisions.

For example, the habitat health DSS developed with the University of Maryland, Esri, and support from the National Aeronautics and Space Administration (NASA) uses Survey 123 by Esri along with other integrated Esri mobile data collection apps to collect, store, manage, and share chimpanzee, other great ape, and human threat data reported by local communities, protected area rangers, researchers, and other partners working with JGI across great ape ranges in Africa. Once uploaded to the habitat DSS, these field data are visualized using dynamic dashboards and available for further modelling and analysis within the system.

The DSS's strength is that it focuses on using actionable management information that is meaningful for decision-makers because it was defined with their participation as part of the conservation action planning processes described earlier. The continuous integration of large remotely sensed data sets to update models as new satellite data are acquired enables the timely provision of information about chimpanzee habitats relevant across a range of scales. The DSS uses the Conservation Standards process as the management framework to define data needs and information products, including statistics, maps, charts, and dashboards, which aggregates raster data into information designed to inform specific decisions.

Figure 2.2 shows an example of habitat viability indicators visualized in a dashboard developed to annually inform decision-makers on the status and trends in habitat indicators that they agreed to monitor as part of the Tanzania Chimpanzee Conservation Action Plan (TAWIRI, 2018). This system developed in collaboration with the University of Maryland, Esri, Conservation Measures Partnership (CMP) and others and funded by NASA is in the process of being transferred for operational use to the Tanzania Wildlife Research Institute (TAWIRI) with support from the United States Agency for International Development (USAID). TAWIRI is the institutional home of the national chimpanzee plan and has the mandate to implement and update the project on behalf of the Government of Tanzania. A similar system using Esri's Protected Area Management (PAM) solutions is being developed to support four Districts in Kigoma and Katavi Regions and help patrol and enforce implementation of a large network of Local Authority Forest Reserves (LAFR). To move forward with the system, JGI first provided internet access to all four District offices and covered hardware, software, training, and maintenance costs.

Finally, DSS includes ArcGIS storytelling tools such as the StoryMaps app that allows users to easily communicate and share geospatial data and maps in a web format to a wider public audience.[13]

2.4 Limitations and constraints

The application of EO satellite data to conservation has several limitations. First of all, cloud coverage strongly limits the availability and overall quality of optical satellite imagery, especially in the tropics (Hilker et al., 2012). Very high-resolution satellite data are also expensive and conservation organizations might find it difficult to access because of the high cost (Boyle et al., 2014). The cost of hardware and software to store, manage, process, and classify the EO data could also be a major constraint for conservation practitioners and local governments (Steering Committee on Space Applications and Commercialization, 2003). Users also need to have significant expertise in data processing and software development to fully utilize RS data (He et al., 2015). Finally, end-users increasingly rely on internet-based workflows to identify, download, and analyse EO data (Sudmanns et al., 2020). These factors mean that many parts of the world that do not have access to high-speed internet are left behind in taking advantage of the disruptive changes that are transforming the RS and GIS fields. In Tanzania, the JGI had to cover the costs of setting up and maintaining internet access in the District government offices to enable District officials and managers to access and use geospatial products, webmaps, and dashboards provided by applications in ArcGIS Online.

---

[13] See TACARE community-driven conservation story (https://arcg.is/DnCqr0) and Gombe 60 (https://arcg.is/H04Db) as examples.

2.5 Social impacts and privacy

Policy development often lags behind advances in technology, which is also true of the new generation of highly detailed RS technologies applied in the absence of legal and policy constraints (Slonecker et al., 1998). This includes the development of new and higher resolution sensors, a shift of the responsibility for data collection and use from the public to the private sector, and expansion of actors into the international arena that further complicates issues because of the need for global policy and regulations (Slonecker et al., 1998).

One of the concerns is regarding individual privacy rights. EO satellite data could be used to provide detailed insights into people's lives, such as predicting the socioeconomic status of individual households (Watmough et al., 2013; Watmough et al., 2019). Court cases in the USA have concluded that the individual's privacy is not protected when the subject or the property is in plain or public view. Therefore observations made from RS platforms do not violate privacy rights (Robila, 2006). However, in the United States, the collection of satellite imagery finer than 0.31 m (native) is restricted. Similar regulations exist in the European Union. On the other hand, these requirements do not apply to Chinese or Indian companies that are rapidly making advancements in high-resolution imaging technology (Coffer, 2020).

Another concern is the misuse of data or derived information. EO data could be used to directly map or predict with relatively high certainty the potential location of an endangered species (He et al., 2015; Jantz et al., 2016). Special consideration should be given that such data does not end up in the hands of poachers or illegal loggers. Therefore, from individuals to governments and from academia to the private sector, everyone should accept full responsibility for their RS activities and how they choose to share, communicate, and use EO data (Wasowski, 1991).

2.6 Future directions

The potential of EO RS data for conservation will continue to increase by the development of new space missions and sensors that will allow mapping and monitoring of global ecosystems at an unprecedented level of detail (He et al., 2015). One of the emerging research topics and conservation opportunities at the interface of biodiversity and RS is spectranomics (Asner & Martin, 2009; Asner & Martin, 2016). New hyperspectral sensors, planned on spaceborne platforms by Planet and NASA, will enable researchers to monitor in detail forest canopy function and composition and connect species functional traits with biodiversity indicators directly. Another clear trend is the emergence of big EO data in the geospatial cloud platforms such as Esri ArcGIS, Microsoft AI for Earth, Maxar GBDX, and Google Earth Engine. These new computing and data storage platforms enable researchers to store, access, process, analyse, collaborate, and share large amounts of EO data at an unprecedented level. Integration of EO RS data with social sensing, mobile, and other sensors and application combined with deep learning algorithms will also improve the accuracy of land cover and land-use change classes (Ma et al., 2019). New satellite constellations provide the means to image the entire planet on a daily basis at high spatial resolution. For example, Planet Labs'[14] constellation of Dove satellites already provides daily imagery globally in the visible and near-infrared domain at 3 m resolution. In order to work with EO data in the cloud, a fast internet connection is required, which usually is not the case in remote project regions. New satellite constellations that provide high-speed internet across the globe could help address this problem. For instance, SpaceX has touted that its StarLink (https://www.starlink.com) constellation will be capable of bringing broadband internet speed to anywhere on Earth. However, the affordability of such an internet connection is still unknown.

2.7 Conclusions

Understanding the trade-offs, benefits, and limitations of various EO technologies, platforms, and sensor combinations and matching them to the problems and realities on the ground

---

[14] https://www.planet.com

is critical to developing cost-effective applications for conservation. The foundation of this lies in understanding the question and what makes it actionable for decision-makers and the barriers and opportunities for the EO satellite data and information to be used. It is also about finding a balance between focusing on new technologies and resource investment that enables conservation practitioners to build capacity to operationally use, maintain, and update these technologies. Finally, and most importantly, conservation practitioners need to recognize that conservation is a social process that engages science, not a scientific process that engages society (Balmford & Cowling, 2006; Adams & Sandbrook, 2013). In the case of chimpanzee conservation in Tanzania, action plans using Conservation Standards combined with JGI's Tacare approach created a series of research-implementation spaces (Toomey et al., 2017) in which local communities, district and regional government officials and managers were able to collaborate and interact with geospatial data and results in a diversity of ways. This enabled JGI and partners to develop geospatial applications and solutions 'with' and not 'for' stakeholders in Tanzania and led to the use of data and technologies as part of locally driven decision-making processes. Informed by spatial insights, these local conservation efforts resulted in expansion of new protected areas managed by village and District governments and the restoration of habitats in some degraded village forest reserves.

2.8 Acknowledgements

This work was supported by funds from NASA, USAID, and the JGI. We thank the Tanzania National Parks (TANAPA), Gombe National Park (GONAPA), TAWIRI, Gombe Stream Research Centre (GSRC), and Tacare field teams of the JGI, Tanzania, for carrying out the work needed to make chimpanzee research and conservation possible. We also thank Esri, Maxar, and Google Earth Outreach for their long-term partnership and support in applying GIS, RS, mobile, and cloud technologies in Tanzania to benefit people, animals, and the environment.

References

Adams, W. M., & Sandbrook, C. (2013). Conservation, evidence and policy. *ORYX*. https://doi.org/10.1017/S0030605312001470

Asner, G. P., & Martin, R. E. (2009). Airborne spectranomics: mapping canopy chemical and taxonomic diversity in tropical forests. *Frontiers in Ecology and the Environment*, 7(5), 269–276.

Asner, G. P., & Martin, R. E. (2016). Convergent elevation trends in canopy chemical traits of tropical forests. *Global Change Biology*, 22(6), 2216–2227.

Balmford, A., & Cowling, R. M. (2006). Fusion or failure? The future of conservation biology. *Conservation Biology*, 20(3), 692–695.

Bennett, A. F. (2003). *Linkages in the Landscape: The Role of Corridors and Connectivity in Wildlife Conservation*. IUCN, Forest Conservation Programme. Available at: https://doi.org/10.2305/iucn.ch.2004.fr.1.en

Boyle, S. A., Kennedy, C. M., Torres, J., Colman, K., Pérez-Estigarribia, P. E., & De La Sancha, N. U. (2014). High-resolution satellite imagery is an important yet underutilized resource in conservation biology. *PLoS One*, 9(1), e86908.

Brooks, T. M., Mittermeier, R. A., Mittermeier, C. G., et al. (2002). Habitat loss and extinction in the hotspots of biodiversity. *Conservation Biology*, 16(4), 909–923.

Buchanan, G. M., Brink, A. B., Leidner, A. K., Rose, R., & Wegmann, M. (2015). Advancing terrestrial conservation through remote sensing. *Ecological Informatics*, 30, 318–321.

Chancellor, R., Rundus, A., Nyiratuza, M., Nyandwi, S., & Aimable, T. (2020). Community-based conservation and chimpanzee research in Gishwati forest, Rwanda. *American Journal of Primatology*. Available at: https://doi.org/10.1002/ajp.23195

Chorley, R. (1988). Some reflections on the handling of geographical information. *International Journal of Geographic Information Systems*, 2(1), 3–9.

CMP (2007). *Open Standards for the Practice of Conservation, Version 1*. Conservation Measures Partnership. Available at: https://www.cbd.int/doc/pa/tools/Open%20standards%20for%20the%20practice%20of%20conservation.pdf

CMP (2013). *Open Standards for the Practice of Conservation, Version 3*. Conservation Measures Partnership. Available at: http://www.iai.int/admin/site/sites/default/files/uploads/2015/08/CMP_Open_Standards_Version_3.0_April_2013.pdf

CMP (2020). *Open Standards for the Practice of Conservation, Version 4*. Conservation Measures Partnership.

Available at: https://conservationstandards.org/wp-content/uploads/sites/3/2020/10/CMP-Open-Standards-for-the-Practice-of-Conservation-v4.0.pdf

Coffer, M. M. (2020). Balancing privacy rights and the production of high-quality satellite imagery. *Environmental Science and Technology*, 54(11), 6453–6455.

Davies, D. K., Murphy, K. J., Michael, K., et al. (2015). The use of NASA LANCE imagery and data for near real-time applications. In Lippitt, C., Stow, D., & Coulter, L. (eds.). *Time-Sensitive Remote Sensing*. Springer, New York, NY (pp. 165–182).

DeFries, R., Hansen, A., Newton, A. C., & Hansen, M. C. (2005). Increasing isolation of protected areas in tropical forests over the past twenty years. *Ecological Applications*, 15(1), 19–26.

DeFries, R. S., Houghton, R. A., Hansen, M. C., Field, C. B., Skole, D., & Townshend, J. (2002). Carbon emissions from tropical deforestation and regrowth based on satellite observations for the 1980s and 1990s. *Proceedings of the National Academy of Sciences of the United States of America*, 99(22), 14256–14261.

Fretwell, P. T., & Trathan, P. N. (2009). Penguins from space: faecal stains reveal the location of Emperor Penguin colonies. *Global Ecology and Biogeography*, 18, 543–552.

Fretwell, P. T., LaRue, M. A., Morin, P., et al. (2012). An emperor penguin population estimate: the first global, synoptic survey of a species from space. *PLoS One*, 7(4), e33751.

Fu, P., & Sun, J. (2010). *Web GIS: principles and applications*. Esri Press, New York, NY.

Game, E. T., Kareiva, P., & Possingham, H. P. (2013). Six common mistakes in conservation priority setting. *Conservation Biology*, 27(3), 480–485.

GGE-CAP (2009). *The Greater Gombe Ecosystem Conservation Action Plan 2009–2039*. USAID. Available at: https://pdf.usaid.gov/pdf_docs/PA00SZ8F.pdf

Gilby, I. C., Eberly, L. E., Pintea, L., & Pusey, A. E. (2006). Ecological and social influences on the hunting behaviour of wild chimpanzees, Pan troglodytes schweinfurthii. *Animal Behaviour*, 72(1), 169–180.

Green, R. E. (2011). *What Do Conservation Practitioners Want from Remote Sensing?* Cambridge Conservation Initiative Report, Cambridge.

Groom, R. E., Meffe, G. K., & Carol, C. R. (2006). *Principles of Conservation Biology*, 3rd ed. Sinauer Associates, Inc./Oxford University Press, Oxford, UK.

Groves, C. (2003). *Drafting a Conservation Blueprint: A Practitioner's Guide to Planning for Biodiversity*. Island Press, Washington, DC.

Hansen, M. C., Krylov, A., Tyukavina, A., et al. (2016a). Humid tropical forest disturbance alerts using Landsat data. *Environmental Research Letters*, 11(3), 034008.

Hansen, M. C., Potapov, P. V., Goetz, S. J., et al. (2016b). Mapping tree height distributions in sub-Saharan Africa using Landsat 7 and 8 data. *Remote Sensing of Environment*, 185, 221–232.

Hansen, M. C., Potapov, P. V., Moore, R., et al. (2013). High-resolution global maps of 21st-century forest cover change. *Science*, 342(6160), 850–853.

He, K. S., Bradley, B. A., Cord, A. F., et al. (2015). Will remote sensing shape the next generation of species distribution models? *Remote Sensing in Ecology and Conservation*, 1(1), 4–18.

Hilker, T., Hall, F. G., Tucker, C. J., et al. (2012). Data assimilation of photosynthetic light-use efficiency using multi-angular satellite data: II model implementation and validation. *Remote Sensing of Environment*, 121, 287–300.

Humle, T., Maisels, F., Oates, J. F., Plumptre, A., & Williamson, E. A. (2016). Pan troglodytes. The IUCN Red List of Threatened Species. Available at: https://doi.org/10.2305/IUCN.UK.2016-2.RLTS.T15933A17964454.en

Jantz, S. M., Pintea, L., Nackoney, J., & Hansen, M. C. (2016). Landsat ETM+ and SRTM data provide near real-time monitoring of chimpanzee (Pan troglodytes) habitats in Africa. *Remote Sensing*, 8(5), 427.

Jantz, S. M., Pintea, L., Nackoney, J., & Hansen, M. C. (2018). Global forest maps in support of conservation monitoring. In Leidner, A. K. (ed.). *Satellite Remote Sensing for Conservation Action*. National Aeronautics and Space Administration, Washington, DC.

Jensen, J. R. (2014). *Remote Sensing of the Environment: An Earth Resource Perspective*, 2nd ed. Pearson Education Limited, Harlow, England.

Joshi, A. R., Dinerstein, E., Wikramanayake, E., et al. (2016). Tracking changes and preventing loss in critical tiger habitat. *Scientific Advances*, 2(4), e1501675.

Justice, C. O., Townshend, J. R. G., Vermote, E. F., et al. (2002). An overview of MODIS Land data processing and product status. *Remote Sensing of Environment*, 83 (1–2), 3–15.

Kano, T. (1971). The chimpanzee of Filabanga, western Tanzania. *Primates*, 12, 229–246.

Lance, K. T., Georgiadou, Y. P., & Bregt, A. K. (2013). Opening the black box of donor influence on Digital Earth in Africa. *International Journal of Digital Earth*, 6, 1–21.

Laurance, W. F. (1999). Reflections on the tropical deforestation crisis. *Biological Conservation*, 91, 109–118.

Lillesand, T., Kiefer, R., & Chipman, J. (2004). *Remote Sensing and Image Interpretation*, 5th ed. John Wiley, New York, NY.

Liu, G., Strong, A. E., Skirving, W. J., & Arzayus, L. F. (2006). Overview of NOAA Coral Reef Watch Program's near-real-time satellite global coral bleaching monitoring activities. Proceedings of the 10th International Coral Reef Symposium.

Löffler, E., & Margules, C. (1980). Wombats detected from space. *Remote Sensing of Environment*, 9(1), 47–56.

Longley, P. A., Goodchild, M. F., Maguire, D. J., & Rhind, D. W. (2011). *Geographic Information Systems & Science*, 3rd ed. Wiley, New York, NY.

Ma, L., Liu, Y., Zhang, X., Ye, Y., Yin, G., & Johnson, B. A. (2019). Deep learning in remote sensing applications: a meta-analysis and review. *ISPRS Journal of Photogrammetry and Remote Sensing*, 152, 166–177.

Maes, J., & Jacobs, S. (2017). Nature-based solutions for Europe's sustainable development. *Conservation Letters*, 10(1), 121–124.

Maldonado, O., Aveling, C., Cox, D., et al. (2012). *Grauer's Gorillas and Chimpanzees in Eastern Democratic Republic of Congo: (Kahuzi-Biega, Maiko, Tayna and Itombwe Landscape) Conservation Action Plan 2012–2022*. Available at: https://www.iucn.org/es/node/21017

Mansourian, A., Lubida, A., Pilesjö, P., Abdolmajidi, E., & Lassi, M. (2015). SDI planning using the system dynamics technique within a community of practice: lessons learnt from Tanzania. *Geo-Spatial Information Science*, 18(2–3), 97–110.

Matzek, V., Covino, J., Funk, J. L., & Saunders, M. (2014). Closing the knowing-doing gap in invasive plant management: accessibility and interdisciplinarity of scientific research. *Conservation Letters*, 7(3), 208–215.

Moyer, D., Plumptre, A. J., Pintea, L., et al. (2006). Surveys of chimpanzees and other biodiversity in Western Tanzania. The Jane Goodall Institute. Available at: https://pages.ucsd.edu/~jmoore/publications/HernandezEtAl2006WCSTanz.pdf

Myneni, R. B., Keeling, C. D., Tucker, C. J., Asrar, G., & Nemani, R. R. (1997). Increased plant growth in the northern high latitudes from 1981 to 1991. *Nature*, 386, 698–702.

Nakamura, M., Hosaka, K., Itoh, N., & Zamma, K. (2015). *Mahale Chimpanzees: 50 Years Of Research*. Cambridge University Press, Cambridge, UK.

Newell, B. R., McDonald, R. I., Brewer, M., & Hayes, B. K. (2014). The psychology of environmental decisions. *Annual Review Environment Resources*, 41(1), 321–340.

Ogawa, H., Moore, J., & Kamenya, S. (2006). Chimpanzees in the Ntakata and Kakungu areas, Tanzania. *Primate Conservation*, 21, 97–101.

Owens, S. (2012). Experts and the environment—the UK royal commission on environmental pollution 1970–2011. *Journal of Environmental Law*, 24(1), 1–22.

Palminteri, S. (2017). An early warning system for locating forest loss. Mongabay Environmental News. Available at: https://news.mongabay.com/2017/11/an-early-warning-system-for-locating-forest-loss/

Petersen, R., & Pintea, L. (2017). Forest watcher brings data straight to environmental defenders. World Resources Institute. Available at: https://www.wri.org/blog/2017/09/forest-watcher-brings-data-straight-environmental-defenders

Piel, A. K., Cohen, N., Kamenya, S., Ndimuligo, S. A., Pintea, L., & Stewart, F. A. (2015). Population status of chimpanzees in the Masito-Ugalla Ecosystem, Tanzania. *American Journal of Primatology*, 77(10), 1027–1035.

Pielke, R. A. (2007). *The Honest Broker: Making Sense of Science in Policy and Politics*. Cambridge University Press, Cambridge, UK.

Pimm, S. L., & Brooks, T. (2013). Conservation: forest fragments, facts, and fallacies. *Current Biology*, 23(24), R1098–R1101.

Pintea, L. (2006). Land use planning in Tanzania. Imaging Notes 21.

Pintea, L. (2007). Applying remote sensing and GIS for chimpanzee habitat change detection, behaviour and conservation, PhD Thesis, University of Minnesota.

Pintea, L. (2011). From Maps to GeoDesign: conserving Great Ape landscapes in Africa. ArcNews. Available at: https://www.esri.com/news/arcnews/summer11articles/from-maps-to-geodesign.html

Pintea, L. (2016). Geodesign restores chimpanzee habitats in Tanzania. ArcNews. Available at: https://www.esri.com/about/newsroom/arcnews/geodesign-restores-chimpanzee-habitats-in-tanzania/

Pintea, L., Bauer, M. E., Bolstad, P. V., & Pusey, A. (2002). Matching multiscale remote sensing data to inter-disciplinary conservation needs: the case of chimpanzees in Western Tanzania. In Pecora 15/Land Satellite Information IV/ISPRS Commission I/FIEOS 2002 Conference Proceedings.

Pintea, L., Mtiti, E.R., Mavanza, M., et al. (2016). 20 years and counting: adaptive management of chimpanzee habitats in the Greater Gombe Ecosystem, Tanzania. In Conservation Measures Partnership Open Standards for the Practice of Conservation. Available at: https://docs.google.com/document/d/1WXt0tSDK-fEIbraoG4jcAL7hVx0MDEap2Lfdtu50a3FM/edit?usp=sharing

Pintea, L., Petersen, R., Akugizibwe, T., Kumanya, A. H., & Bourgault, L. (2019). Forest Watcher mobile app: what makes satellite data actionable for chimpanzee conservation in Uganda. In American Association of Geographers Annual Meeting.

Pintea, L., Pusey, A. E., Wilson, M. L., Gilby, I. C., Collins, D. A., Kamenya, S., & Goodall, J. (2012). Long-term changes in the ecological factors surrounding the chimpanzees of Gombe National Park: impacts on biodiversity and ecosystems. In Plumptre, A. J. (ed.). *Long-Term Changes in Africa's Rift Valley*. Nova Science Publishers, New York, NY (pp. 194–210).

Pusey, A. E., Pintea, L., Wilson, M. L., Kamenya, S., & Goodall, J. (2007). The contribution of long-term research at Gombe National Park to chimpanzee conservation. *Conservation Biology*, 21(3), 623–634.

Raygorodetsky, G. (2018). Indigenous peoples defend Earth's biodiversity—but they're in danger. National Geographic Magazine. Available at: https://www.nationalgeographic.com/environment/article/can-/indigenous-land-stewardship-protect-biodiversity-

Rights and Resources Initiative (2015). Who owns the world's land? A global baseline of formally recognized indigenous and community land rights. Washington, DC. Available at: https://rightsandresources.org/wp-content/uploads/GlobalBaseline_web.pdf

Robila, S. A. (2006). Use of remote sensing applications and its implications to the society. In International Symposium on Technology and Society Proceedings. Available at: https://doi.org/10.1109/ISTAS.2006.4375896

Rose, R. A., Byler, D., Eastman, J. R., et al. (2015). Ten ways remote sensing can contribute to conservation. *Conservation Biology*, 29(2), 350–359.

Rose, D. C. (2018). Avoiding a post-truth world: embracing post-normal conservation science. *Conservation and Society*, 16(4), 518–524.

Roux, D. J., Rogers, K. H., Biggs, H. C., Ashton, P. J., & Sergeant, A. (2006). Bridging the science-management divide: moving from unidirectional knowledge transfer to knowledge interfacing and sharing. *Ecology and Society*, 11(1), 4.

Salafsky, N., Margoluis, R., Redford, K. H., & Robinson, J. G. (2002). Improving the practice of conservation: a conceptual framework and research agenda for conservation science. *Conservation Biology*, 16(6), 1469–1479.

Salafsky, N., Salzer, D., Stattersfield, A. J., et al. (2008). A standard lexicon for biodiversity conservation: Unified classifications of threats and actions. *Conservation Biology*, 22, 897–911.

Salafsky, N. (2011) Integrating development with conservation: a means to a conservation end, or a mean end to conservation? *Biological Conservation*, 144(3), 973–978.

Sanderson, E. W., Jaiteh, M., Levy, M. A., Redford, K. H., Wannebo, A. V., Woolmer, G. (2002). The human footprint and the last of the wild. *Bioscience*, 52(10), 891–904.

Sayn-Wittgenstein, L. (1992). Barriers to the use of remote sensing in providing environmental information. *Environmental Monitoring Assessment*, 20, 159–166.

Slonecker, E. T., Shaw, D. M., & Lillesand, T. M. (1998). Emerging legal and ethical issues in advanced remote sensing technology. *Photogrammetric Engineering & Remote Sensing*, 64(6), 589–595.

Song, X. P., Hansen, M. C., Stehman, S. V., et al. (2018). Global land change from 1982 to 2016. *Nature*, 560, 639–643.

Stapleton, S., LaRue, M., Lecomte, N., Atkinson, S., Garshelis, D., Porter, C., & Atwood, T. (2014). Polar bears from space: assessing satellite imagery as a tool to track arctic wildlife. *PLoS One*, 9(7), e101513.

Steering Committee on Space Applications and Commercialization, N.R.C. (2003). *Using Remote Sensing in State and Local Government Information for Management and Decision Making*. Available at: https://proceedings.esri.com/library/userconf/proc03/p1076.pdf

Sudmanns, M., Tiede, D., Augustin, H., & Lang, S. (2020). Assessing global Sentinel-2 coverage dynamics and data availability for operational Earth observation (EO) applications using the EO-Compass. *International Journal of Digital Earth*, 13(7), 1–17.

TAWIRI (2018). *Tanzania Chimpanzee Conservation Action Plan 2018–2023*. Available at: http://tawiri.or.tz/wp-content/uploads/2018/07/Tanzania-Chimpanzee-Conservation-Action-Plan-2018.pdf

TNC (2007). *Conservation Action Planning Handbook: Developing Strategies, Taking Action and Measuring Success At Any Scale*. The Nature Conservancy, Arlington, VA.

Toomey, A. H., Knight, A. T., & Barlow, J. (2017). Navigating the space between research and implementation in conservation. *Conservation Letters*, 10(5), 619–625.

Townshend, J. R. (1994). Global data sets for land applications from the advanced very high resolution radiometer: an introduction. *International Journal of Remote Sensing*, 15(17), 3319–3332.

Townshend, J. R. G., & Justice, C. O. (2002). Towards operational monitoring of terrestrial systems by moderate-resolution remote sensing. *Remote Sensing of Environment*, 8(1), 351–359.

Tracewski, Ł., Butchart, S. H. M., Di Marco, M., et al. (2016). Toward quantification of the impact of 21st-century deforestation on the extinction risk of terrestrial vertebrates. *Conservation Biology*, 30(5), 1070–1079.

Turner, W., Spector, S., Gardiner, N., Fladeland, M., Sterling, E., & Steininger, M. (2003). Remote sensing for biodiversity science and conservation. *Trends in Ecology & Evolution*, 18(6), 306–314.

United Nations (UN) (2014). *World Conference on Indigenous People.* Available at: https://www.un.org/en/ga/69/meetings/indigenous/#&panel1-1

Wasowski, R. J. (1991). Some ethical aspects of international satellite remote sensing. *Photogrammetric Engineering & Remote Sensing*, 57, 41–48.

Watmough, G. R., Atkinson, P. M., & Hutton, C. W. (2013). Exploring the links between census and environment using remotely sensed satellite sensor imagery. *Journal of Land Use Science*, 8(3), 284–303.

Watmough, G. R., Marcinko, C. L. J., Sullivan, C., et al. (2019). Socioecologically informed use of remote sensing data to predict rural household poverty. *Proceedings of the National Academy of Sciences of the United States of America*, 116(4), 1213–1218.

Whitten, T., Holmes, D., & MacKinnon, K. (2001). Conservation biology: a displacement behavior for academia? *Conservation Biology*, 15(1), 1–3.

Wich, S. A., & Koh, L. P. (2018). *Conservation Drones: Mapping and Monitoring Biodiversity*. Oxford University Press, Oxford, UK.

Wilson, M. L., Lonsdorf, E. V., Mjungu, D. C., et al. (2020). Research and conservation in the greater Gombe ecosystem: challenges and opportunities. *Biological Conservation*, 252, 108853.

Woodcock, C. E., Allen, R., Anderson, M., et al. (2008). Free access to landsat imagery. *Science*, 320(5879), 1011.

Yang, Z., Wang, T., Skidmore, A. K., De Leeuw, J., Said, M. Y., & Freer, J. (2014). Spotting East African mammals in open savannah from space. *PLoS One*, 9, 1–16.

CHAPTER 3

Drones for conservation

Serge A. Wich, Mike Hudson, Herizo Andrianandrasana, and Steven N. Longmore

3.1 Introduction

Global biodiversity has been decreasing sharply over a wide array of taxa. This trend is predicted to continue in the future if there is no reduction in anthropogenic activities such as habitat loss, hunting, greenhouse gas emissions, and use of pesticides (Maxwell et al., 2016; Newbold et al., 2016; Hallmann et al., 2017; Tilman et al., 2017; Lanz et al., 2018; Plumptre et al., 2019). Evaluating whether conservation activities are effective requires data on animal distribution, density, habitat, and threats, all at frequent intervals so that trend analyses can be conducted (White, 2019). For most species on Earth, such data sets do not exist due to the high costs of gathering such data (Grooten & Almond, 2018). Drones have been suggested as one method of collecting these much-needed data, with the potential to reduce costs while increasing data quality when compared with traditional survey techniques (Wich & Koh, 2018).

There are a variety of methods used to obtain data on animal distribution and density (Buckland et al., 2001; MacKenzie, 2006). Researchers often count animals or their signs (e.g. nest, dung) along transects (Buckland et al., 2001). Depending on the species, such transects can be conducted on foot, by ship, by crewed aircraft, or by car (Buckland et al., 2001). For more elusive animals, researchers have used camera traps (Wearn & Glover-Kapfer, 2019) and autonomous acoustic recorders (Spillmann et al., 2015) (see Chapters 4 and 5) that are often placed in grids within an area to potentially derive densities. These methods have

several challenges. Often the confidence intervals around the parameter estimates generated are large, which hampers the detection of change in trend analyses. In addition, survey costs are often high due to the habitually large and difficult-to-access areas that need to be surveyed. Covering such areas on foot to either conduct transects or place camera traps or acoustic recorders is time-consuming and regularly needs large teams for the initial placement of the sensors and subsequent regular retrieval of data. For species that can be surveyed from the air, the costs of hiring aircraft can be prohibitively high, or aircraft might not be available. Similar challenges face the usage of ships as observation platforms.

Equally important for conservation is mapping and classifying animal habitat to monitor change over time. Most of this work is conducted through the analyses of satellite images (Horning et al., 2010), but before the large-scale availability of satellite data, crewed aircraft were also used for mapping (Tomlins & Lee, 1983; Wolf et al., 2014). Even though satellite imagery provides researchers with increasingly higher resolution and more frequent images globally, there can be challenges with using satellite data (see Chapter 2). First, in the humid tropics, cloud cover can hamper the collection of images, especially in certain seasons. Second, in artic areas, the almost continuous darkness throughout parts of the year hampers data collection. Third, some analyses require higher resolution data than satellites offer at present (Anderson & Gaston, 2013). Fourth, where higher resolution data are available, their costs can be prohibitively expensive. Fifth, satellite image analyses often

Serge A. Wich et al., *Drones for conservation*. In: *Conservation Technology*.
Edited by Serge A. Wich and Alex K. Piel, Oxford University Press.
© Oxford University Press 2021. DOI: 10.1093/oso/9780198850243.003.0003

Figure 3.1 From left to right: Fixed-wing drone © Conservation drones, multirotor drone © Serge Wich, hybrid-drone © Wingtra.

require specialist geographic information system (GIS) and/or remote sensing software.

Once conservationists understand the distribution of a species and its habitat, they must strive to understand its threats so that targeted conservation measures can be implemented. One such threat, for which there is currently a lack of data, is the illegal hunting of animals. Hunting occurs for meat or body parts (e.g. rhino horn, elephant tusks). Hunting for meat is widespread worldwide and is generally decimating wildlife (Benítez-López et al., 2019; Ripple et al., 2019). Hunting for body parts focuses on a subset of species such as rhinos and elephants and has led to large reductions in their numbers.[1] Hunting for both meat and body parts appears difficult to reduce. Current methods to reduce poaching of such highly prized species, and poaching in general, include improving law enforcement, promoting alternative livelihoods, improving anti-poaching efforts, and so forth. Still, their effectiveness is limited (Lindsey et al., 2013).

An important question is whether drones can complement or improve on these existing data collection methods for conservation (Wich & Koh, 2018).

3.2 New technology

3.2.1 Data collection: hardware and software

Hardware: Drones can be split into three categories: fixed-wing, multirotor, and hybrid VTOL (vertical take-off and landing) (Figure 3.1). These all come with advantages and disadvantages (Table 3.1).

Irrespective of the drone system, there are three additional important components: the power source, flight controller, and a ground control station (GCS). Although gasoline-powered systems exist, most of the systems flown by conservationists are powered by rechargeable lithium-ion polymer (LiPo) batteries with a small number using lithium-ion (Li-ion). The flight controller is the piece of hardware through which an operator directly controls the movement of the drone. There are several proprietary and open-source flight controllers that vary in their level of complexity. Many of these are highly advanced and integrate several sensors such as global navigation satellite system (GNSS) and inertial measurement unit (IMU), and allow the drone to follow pre-programmed coordinates during a mission. The final component is the GCS and its software. The GCS has a data link with a drone and can be used for various applications such as sending commands to the drone, obtaining data on the location of the drone, its battery status, and seeing images that a camera on a drone is taking. The GCS can be a laptop, tablet, or mobile phone, depending on the specific drone and software.

There are several options available to conservationists who want to obtain a drone. The first is to purchase an off-the-shelf system. There are many off-the-shelf systems available from various companies such as DJI,[2] Parrot,[3] DELAIR,[4] and SenseFly.[5] Some of these companies (DJI and Parrot) provide drones for the general consumer and professional market, whereas others focus purely

---

[1] https://www.savetherhino.org/rhino-info/poaching-stats/

[2] https://www.dji.com
[3] https://www.parrot.com
[4] https://delair.aero/
[5] https://www.sensefly.com/

Table 3.1 Drone categories

	Multirotor	Fixed wing	Hybrid VTOL
Launch area	Small	large	Small
Flight duration	Short (<1 hr)	Long (>1 hr)	Intermediate (30–60 minutes)
Payload	Heavy (few kilograms)	Light (<1 kg)	Light (<1 kg)
Pilot experience	Minimal (although larger systems require more training)	Substantial training	Intermediate

Note: table based on Table 2.1 in (Wich and Koh, 2018) and based on battery-powered systems.

on the professional market (DELAIR and Sensefly). Both consumer and professional-grade drones have been used for conservation work (Wich & Koh, 2018). Some of the systems are sold with sensors that can be swapped depending on the data requirements, and parts of the electromagnetic spectrum that are of interest. There are also companies offering more bespoke configurations based on consumer platforms.[6] The last option is to build a system from components which can be purchased separately or, in the case of research labs, even build the parts (e.g. wings, fuselages).

When deciding to purchase a drone, it is important to think carefully about the location where the drone will be flown and the data that need to be collected. In terms of data that need to be collected, it is important to assess whether a drone would be the best option or whether other methods might yield similar data in a more cost-effective manner (Wich & Koh, 2018). The location is important with respect to flight regulations, privacy issues related to data collection, weather (particularly wind speed in relation to the wind speed that the drone can operate to), presence of fine particles (e.g. sand), telemetry connectivity issues due to vegetation, take-off-and-landing options, and so forth (Duffy et al., 2017). The data that need to be collected will impact the sensor choice (e.g. standard visual spectrum, thermal infrared, multispectral, hyperspectral), the flight length of which the drone needs to be capable, payload capacity, etc. In addition, there are several other aspects to consider, such as whether batteries can be transported internationally, whether in case of an accident or malfunction there are spare parts that can be purchased locally or whether these need to be brought on the trip, whether there is someone who can carry out repairs, and so on.

3.2.2 Analyses

At present, there are three main uses of drones in conservation (mapping, animal counts, and poacher detection) which require two largely different analytical pipelines. For mapping, it is important to obtain full coverage of the area of interest. Hence, flights are often conducted in a grid pattern during which photos are being collected with a high percentage of overlap. The high percentage of overlap is important so that the software that applies the Structure-from-Motion process (SfM) to yield a map will work well (Westoby et al., 2012; Wolf et al., 2014). There are a number of commercially available packages (e.g. Pix4Dmapper,[7] Agisoft Metashape[8]) that have user-friendly workflows to process individual images from grid flights into an orthomosaic. Alternatively, open-source packages such as VisualSfM/CMVS[9] and Microsoft ICE[10] can be used to process images into one large map of the area of interest. These are less user friendly and have fewer features, but they can provide good results as well. Often the resulting orthomosaics are combined with contextual layers in GIS software packages such as ArcGIS, QGIS, and Global Mapper. Some of these programmes have started to integrate the SfM processing into their GIS so that such contextual information can be easily added to the orthomosaics. The maps produced through the SfM process can be further processed for land-cover classification and land-cover change analyses. Some of those analyses can be conducted within the GIS platforms such as ArcGIS and QGIS. Still, often researchers

---

[6] https://dronexpert.nl/en/

[7] https://www.pix4d.com/
[8] https://www.agisoft.com/
[9] http://ccwu.me/vsfm/
[10] https://www.microsoft.com/en-us/research/product/computational-photography-applications/image-composite-editor/

use more specialized remote sensing packages such as ERDAS Imagine or use cloud options such as Google Earth Engine on which orthomosaics can be uploaded and analysed.

For animal counting and poacher detection, the requirement is that objects of interest (e.g. rhinos, poachers, cars) are detected in the individual images. Generally, detection and classification are conducted manually. This is costly and time-consuming and, at present, represents a major challenge for the development of end-to-end workflows where data are collected and analysed in a cost- and time-efficient way. To alleviate this issue, researchers are working on automating the detection and classification of objects (Lhoest et al., 2015; Gonzalez et al., 2016; Longmore et al., 2017). There are a large number of different approaches to automating the detection of animals on images, such as spectral thresholding (Chabot and Bird, 2012), blob detection (Ward et al., 2016), and using local maxima and isolines (Lhoest et al., 2015). A popular approach is to apply neural network approaches in which computer algorithms are trained with a subset of the data, validated, and then used with test data. This approach has led to some promising results for detecting animals and poachers (Maire et al., 2015; Bondi et al., 2018; Bondi et al., 2019).

There are many options for object detection and classification using machine learning for experienced programmers (Lamba et al., 2019). However, to the best of the authors' knowledge, there are no off-the-shelf open-source programs that allow non-programmers to label images, train a number of machine learning algorithms, validate these, and then use a test data set to determine their accuracy in detecting and classifying the objects of interest. There are commercial options available through Google[11] and Microsoft[12] that are relatively user-friendly but might still deter users due to the steep learning curves and high costs involved when processing large numbers of images. Nevertheless, companies such as Microsoft and Google provide computational and web-based services to facilitate the uptake of machine learning at a large scale and facilitate access for non-programmers to such techniques. There are also user-friendly options available, such as Picterra,[13] which require no coding skills and allow for detection and mapping of objects on images acquired with drones.

3.3 Wider applications: a review of what has been done

As mentioned earlier, there are three general aspects of conservation that drones have been used for: habitat mapping and monitoring, animal counting, and anti-poaching efforts. These aspects will be described in the following three sections.

3.3.1 Land-cover classification and land-cover change detection

Knowledge about land-cover types and changes in land-cover are crucial for conservation managers. These maps can provide a manager with details on which land-cover types occur in an area, their spatial extent, and any changes if a series of maps are produced through time. Because of the large spatial extent of many conservation projects, satellites are most commonly used for land-cover mapping. However, sometimes higher resolution is required for specific areas and drones can be particularly useful for such cases. Such information can then be used to determine how animal species might be distributed over the area or where emergent trees might occur (Alexander et al., 2018). As a result, drones are increasingly used for land-cover mapping in a wide variety of landscapes ranging from tropical rainforests, to temperate forests and arctic areas (Wich & Koh, 2018). Because conservation researchers often need to map relatively large areas, they often use fixed-wing drones instead of multirotor drones for land-cover classification studies. However, the use of drones for land-cover classification and land-cover change detection is still restricted to relatively small areas in the range of several square kilometres, depending on the mapped area's complexity. This is due to the relatively small image footprint that cameras on drones have and the need to fly a grid pattern in which the sidelap—the percentage of overlap between images obtained on the parallel flight lines—is very high to allow for processing an accurate orthomosaic. For complex environments, such as tropical forests, 90% sidelap is recommended by Pix4Dmapper to process an accurate orthomosaic. In less complex areas, 60–70% sidelap might be sufficient to produce an accurate orthomosaic. Due to the shorter flight duration of multirotor drones compared with fixed-wing and hybrid VTOL systems, the former maps much smaller areas.

Depending on a study's aim, different sensors can be fitted to the drone so that images with different parts of the electromagnetic spectrum can be obtained. By far the most common sensor is one that captures data in the visual spectrum. The orthomosaics produced with such images often provide very high-resolution information to the user. They can be used to classify land-cover types or detect specific features such as a particular tree species using the orthomosaic and/or the point cloud associated with it (e.g. Reid et al., 2011; Cunliffe et al., 2016; Wich et al., 2018). However, the relatively narrow bandwidth of the visual spectrum limits the options for land-cover classification compared to multispectral sensors. Such cameras can, for instance, have four bands (green, red, red edge, and near-infrared) as well as a standard visual spectrum sensor. This provides more options for land-cover classification but also to determine indices of vegetation health (e.g. Normalized Difference Vegetation Index (NDVI), Green NDVI (GNDVI), etc. (Michez et al., 2016; Assmann et al., 2019)). Pixel- or object-based supervised or unsupervised classification methods that use the reflectance of the various bands are most commonly used for land-cover classification (Laliberte and Rango, 2009; Fraser et al., 2016; Wich et al., 2018). In areas where vegetation types differ structurally, researchers have also been using the point clouds generated during the SfM process to distinguish vegetation types by vegetation height (Cunliffe et al., 2016). Even more options to distinguish features in the landscape are provided by hyperspectral cameras, which have a much higher number of bands than multispectral cameras. However, the costs of these sensors are high (e.g. €40,000, at the time of writing, for a Rikola camera[14]) and as a result they are infrequently used (Mitchell et al., 2012; Mitchell et al., 2016). Selecting a method to classify land-cover types is not straightforward. Several decisions have to be made, such as whether to use a pixel-based or object-based approach and which algorithm to use for the classification (review in Chapter 7 of Wich & Koh, 2018).

3.3.2 Animal counts

Determining the distribution and density of animals and deriving abundances from those data is a key aspect of conservation science. Drones have therefore been extensively used to test whether animals or their signs can be detected (e.g. nests: Chabot & Bird, 2015; Christie et al., 2016; Wich & Koh, 2018). Indeed, we now know that with drones we can detect a relatively large number of species in several vegetation types. Drones equipped with visual spectrum cameras have been the most commonly used system for animal detection, but a growing number of studies have been using thermal sensors to detect animals (Gonzalez et al., 2016; Burke et al., 2018; Rashman et al., 2018; Burke et al., 2019; Scholten et al., 2019; Spaan et al., 2019; Kays et al., 2019). These have been particularly useful in situations when visual spectrum cameras struggle (e.g. low-light, camouflage, partial cover under vegetation). Although the thermal imaging data are promising, there are still challenges in classifying the various species that exist in an area from thermal data. Additionally, individuals detected from the ground and under vegetation are likely to be missed on thermal images (Kays et al., 2019). Kays et al. (2019) suggest that combining flash photography or IR illumination with thermal sensors might reduce such challenges. It remains an issue, though, that individuals under vegetation will likely be missed from the air and that some sort of correction is needed. Visual spectrum data have so far been useful in terrestrial

---

[14] http://vespadrones.com/product/hyperspectral-camera-rikola/

areas, including for various bird species, particularly raptors and species in coastal and marsh areas (Israel & Reinhard, 2017), large mammals such as elephants in savannas (Vermeulen et al., 2013), and deer (Chrétien et al., 2016). Within terrestrial areas, drones have been used in a variety of environments, from arctic regions to tropics and highland plateaus (Duffy et al., 2017), highlighting their versatility.

There are also a growing number of studies on animals in aquatic settings. Several marine mammals can be detected with drones (Hodgson et al., 2013; Koski et al., 2015), sea turtles in clear waters up to a relatively narrow depth (Rees et al., 2018), similarly for sharks (Kiszka et al., 2016; Rieucau et al., 2018; Colefax et al., 2019) and fish species such as salmon (Groves et al., 2016).

Even though detecting animals or their signs is beneficial for conservation efforts by providing valuable data on animal distribution, it is important to determine whether drones can also be used to determine density, and whether these densities are comparable to those obtained with traditional methods such as line transects. A review of studies that used drones to obtain animal density data and compared those to density data from alternative methods was provided by Wich and Koh (Wich & Koh, 2018). It is promising to see some exciting new studies. Guo et al. (2018) used drones to obtain imagery to calculate the density and abundance of domestic and wild herbivore species on the Tibetan Plateau. Another study on large herbivores used drones to determine the abundance of the Tibetan antelope (*Pantholops hodgsonii*) in the Chang Tang National Nature Reserve in China (Hu et al., 2018). Both of these studies used fixed-wing drones. Fixed-wing drones are also being used in other challenging areas such as the South Shetland Islands (Antarctica) to estimate the abundance of chinstrap penguins (*Pygoscelis antarcticus*) (Pfeifer et al., 2019). In this work 14 colonies were studied, and the drone data were combined with previous ground counts to obtain information on population trends. Researchers working in the marine environment continue to use drones to study species such as blacktip reef sharks (*Carcharhinus melanopterus*) (Rieucau et al., 2018). In addition to detecting sharks on videos recorded with the DJI Phantom

II's camera, they used the drone data to distinguish individual sharks and study shoaling patterns.

To determine whether estimating densities of animals or their signs with drones leads to similar results as seen in other methods, the drone and alternative methods need to be conducted simultaneously. Ideally the alternative counts would be absolute, but unfortunately that is usually not possible, so the comparison will usually be between two imperfect methods. Several studies have examined this issue for animals or their signs (review in Wich & Koh, 2018). A study on orangutan nests using a visual spectrum camera showed a significant correlation between ground nest counts and those on images acquired with a fixed-wing drone (Wich et al., 2016a). Similar results were found in a study of chimpanzee nests in Tanzania, during which aerial and ground counts were conducted (Bonnin et al., 2018). In both studies, only a relatively small percentage of the nests found on the ground were detected on the aerial images. This indicates that a relatively large percentage are missed from the air due to nests being inside the tree canopy. At present, it is not clear whether the missed fraction varies for different vegetation types. As a result of this, more work needs to be conducted to assess whether precise and accurate great ape densities can be obtained from nest counts with drones. Researchers have therefore started to investigate whether great apes can be detected directly. During the day, they are mostly inside the canopy or on the ground and virtually impossible to detect on visual spectrum images from the air. As a result, researchers are using thermal sensors to detect the great apes when they are in their night nest, just after dusk or just before dawn. An initial study with orangutans in Sabah provided two promising results. Firstly, all orangutans that were observed in a nest from the ground could be detected in the thermal images acquired with the drone. Secondly, in survey flights in which it was not known where orangutans were, the aerial detections were confirmed to be orangutans that could be located from the ground (Burke et al., 2019). Because the total number of orangutans in the area was not known, a false negative rate could not be calculated. Future studies need to determine a false-negative rate for great ape surveys using drones. Several other recent studies have examined whether thermal sensors provide comparable results as ground counts and all found that results from both methods were comparable (Gooday et al., 2018; Corcoran et al., 2019; Spaan et al., 2019). A study on spider monkeys in Mexico indicated that the thermal data provided higher counts than ground counts. This finding was attributed to the possibility that some animals were missed from the ground due to the spider monkeys being high up in the canopy and thus occluded from ground view by vegetation (Spaan et al., 2019). Impacts of vegetation cover were also found in a study on New Zealand fur seals (*Arctocephalus forsteri*), in which fewer observations were made from the aerial thermal images in areas with high canopy cover compared with open areas because the seals were under the canopy (Gooday et al., 2018). Another study that used an aerial thermal sensor to detect primates made fewer detections from thermal data than were observed from the ground by observers, likely due to the canopy cover and the primates being covered by the vegetation when they were inside the tree canopy (Kays et al., 2019). The same study found it difficult to differentiate the various species of interest from the thermal data. This can be a particular problem when animals are similarly sized. However, when species have different sizes (such as orangutans and proboscis monkeys), size can be used to differentiate them (Burke et al., 2019). To control the number of animals in studies comparing ground with aerial counts, a recent study has used a large number of fake plastic birds that could be counted from the ground as well as from the air. This study showed that the counts made with visual spectrum images taken from a drone were more accurate than ground counts (Hodgson et al., 2018).

The increased usage of drones by conservationists leads to the generation of a large number of images (stills and video). These images are predominantly examined manually by image analysts, a time-consuming and hence costly process due to the costs of human image analysts. It is important to realize how quickly the time required to examine images can grow. A 1-hour flight with a still taken every 2 seconds will lead to 1800 images. If a trained image analyst needed 1 minute to examine each image to detect an animal (or multiple animals), it

would still take 1800 minutes (30 hours) to process the images. Similar issues have been highlighted for other technologies used to obtain images in conservation, such as camera traps (Weinstein, 2018). The additional costs associated with image analysts' time are reducing the cost-effectiveness that drones could potentially offer. As such, it is important to try to automate the detection and classification of animals in drone images (Wich & Koh, 2018). Detection of animals or their signs using automated approaches in images collected from drones or crewed aircraft has been addressed in a relatively small number of studies that almost exclusively dealt with single species (Weinstein, 2018; Wich and Koh, 2018; Eikelboom et al., 2019). A recent study in which multiple species of mammals were detected and classified using images from a crewed aircraft achieved high accuracies for three species (Eikelboom et al., 2019). Even though most studies have focused on using the visual spectrum, several have explored the use of both visual and thermal images (Kellenberger et al., 2018; Corcoran et al., 2019). At present, the usage of promising machine learning methods such as deep learning is still difficult for non-programmers and requires quite substantial investments in hardware. Large companies such as Google and Microsoft are making these methods more accessible through their cloud services, and we expect that there will be an enormous amount of progress made in the next 5 years or so.

3.3.3 Poaching

Together with habitat loss, poaching is one of the main threats to wildlife (Fa & Brown, 2009; Benítez-López et al., 2019). As referred to in this chapter, poaching refers to poaching for bushmeat and animal parts such as rhino horn, elephant tusks, and pangolin scales. Poaching occurs inside and outside of protected areas and rangers are usually insufficiently resourced to reduce or halt it. Besides, there is considerable risk for conservation staff during anti-poaching missions (Olivares-Mendez et al., 2013). There is, therefore, a strong interest in determining whether technology can facilitate the detection of poachers and whether such detections can be efficiently shared with rangers so that poaching can be reduced. Drones are one of the novel technologies being explored for poacher detection (reviewed in Wich & Koh, 2018). Ideally, poachers would be detected before they reach their target animals. This is by no means an easy task in the large landscapes where poachers often operate. Most effort to date has been focused on using drones to detect poachers of rhinos in Africa. Organizations such as WWF and Air Shepherd[15] have been using drones with thermal sensors for this purpose. There are very few details available on the relative success of these operations. Understandably, the ability of drones to facilitate poacher detection during these missions is not in the public domain. There are also reports of an anti-poaching mission using a thermal sensor-equipped drone in the Greater Kruger Area in South Africa, during which poachers were successfully detected on the drone's live feed and detained.[16]

It thus seems that poachers can be detected using thermal sensor data obtained with drones. This is promising, but much work remains to be conducted on how poachers can be differentiated from other objects that might look similar—such as similar size animals and rocks that have been warmed up by the sun (Burke et al., 2018). Also, it is important to determine the false-negative rate in such data, that is, whether there are instances in which poachers were present but were not detected on the thermal drone footage. Such false negatives are extremely important as they could lead to incorrect prioritization of poacher patrols. A recent study examined this issue in western Tanzania (Hambrecht et al., 2019). A number of human subjects, posing as poachers, were positioned in various locations in the study area to determine the influence of factors such as vegetation cover, distance from the drone flight path and flight height on detection probability on the visual spectrum and thermal cameras. The study showed that the detection probability with the thermal camera was higher than with the visual spectrum camera at dawn and dusk. It was also found that the detection probability for

thermal data decreased with an increase in canopy density and a larger distance from the flight path. Detection probability also varied between image analysts, highlighting the importance of automated detection of poachers using machine learning approaches to ensure detection is standardized. Several research groups are developing methods to automate the detection of poachers using machine learning. Other methods are also being developed (Longmore et al., 2017; Bondi et al., 2018; Bondi et al., 2019). Although numerous challenges such as differentiating poachers from other objects and data transmission over long distance remain, it is likely that, within a few years, there will be automated poacher detection from drones and other sensor platforms such as camera traps.

3.4 Case study

Mapping of the Alaotran gentle lemur (*Hapalemur alaotrensis*) habitat in Madagascar

Lac Alaotra is the largest lake in Madagascar situated in the north-east of the country (17°2ʹ to 17°6ʹS, and 48°1ʹ to 48°4ʹE). The lake has over 20,000 ha of open water and is surrounded by a large, shallow, and highly productive wetland containing both marsh and rice fields (Andrianandrasana et al., 2005). The marsh is currently estimated to cover approximately 20,000 hectares, and the dominant vegetation is reeds (*Phragmites communis*) and papyrus (*Cyperus madagascariensis*). Lac Alaotra is also Madagascar's largest inland fishery (Wallace et al., 2016). It has been referred to as Madagascar's rice bowl as it produces approximately one-third of the country's total annual crop (Penot et al., 2014). Alongside the direct and ecosystem services the lake provides to humans, the area is considered an important area for biodiversity, supporting eight endemic water birds and five endemic fish species (Andrianandrasana et al., 2005) alongside the Critically Endangered Alaotran gentle lemur. As a result, the site was recognized as a RAMSAR site in 2003 and was recently awarded the status of Nouvelle Aire Protégée ('New Protected Area').

The Alaotran gentle lemur, known locally as the Bandro, is a small, critically endangered primate that exclusively inhabits the papyrus and reed beds surrounding Lac Alaotra in Madagascar (Garbutt,

2007). It is the only primate known to live exclusively in marsh habitat (Andriaholinirina et al., 2014). The lemur feeds mainly on papyrus, reeds, and two grasses (*Echinochloa crusgalli* and *Leersia hexandra*) (Mutschler, 1999). The species is small-bodied with adults weighing between 1–1.6 kg and lives in small family groups of a male, a female, and their offspring, normally resulting in an average group size of four (Nievergelt et al., 2002).

The Alaotran gentle lemur is threatened predominantly by habitat destruction (Andriaholinirina et al., 2014). Fishermen and rice farmers regularly burn the marsh to get access to new areas. These fires are often uncontrolled and can rapidly spread, resulting in burns covering half of the lemur's habitat inside the protected area (Ralainasolo et al., 2006). Invasive species have become an increasingly important threat to lemurs in the last decade, with water hyacinth (*Eichhornia crassipes*) and *Salvinia* spp. inundating any open areas of water which occur after burns, preventing regeneration of the marsh (Andrianandrasana et al., 2005). Finally, despite a ban, hunting is thought to be a threat to the lemurs, both as pets and as a cheap source of meat.

Aerial mapping of gentle lemur habitat is critical to designing and implementing conservation strategies for the species for several reasons. Firstly, it allows us to understand the distribution and quality of habitat across the marsh and to target population surveys. Secondly, to understand the amount of marsh which has been burned in a given year and the likely impact of the gentle lemur population. This understanding of the spatial pattern of burning may also help spatially target interventions to encourage environmentally friendly behaviours in local communities.

Without aerial techniques, surveys must be conducted by canoe and are intrinsically limited to the small number of accessible canals used by fisherman. These channels are not representative habitat and, as such, are not useful for conservation planning. Satellite data have been used to successfully detect larger fires and burned areas across the Lac Alaotra marsh but are biased towards only large, easily detected burns meaning many are missed. In addition, the relatively low spatial resolution of affordable satellite data has precluded detailed mapping of habitat quality at a resolution that would enable fine-scale conservation decision-making. High-resolution mapping would, for example, enable spatial planning of habitat restoration, such as re-planting in areas that have lost the required density to support lemurs, reconnecting marsh blocks separated by burns or illegal fishing lakes, or eradication programmes for the invasive plants clogging the marsh.

To obtain a high-resolution map of the area, we flew a DJI Mavic 2 Pro with the 35 mm Hasselblad camera. Three missions were conducted to cover the

Figure 3.2 Flying the DJI Mavic on the shores of Lake Alaotra in Madagascar. © Mike Hudson.

Figure 3.3 The location of Lake Alaotra in Madagascar (left upper panel), the location of the study site on the lake (left lower panel), and the orthomosaic overlaid on satellite images. © Mike Hudson.

whole area, with a flying height 100 m above ground level. Missions were planned using the Pix4D capture app on an iPad mini 2 to plan the missions with 70% sidelap. A total of 999 images were processed into one georeferenced orthomosaic using the Pix4D cloud service (Figure. 3.2).

This orthomosaic covers the whole range of the Alaotran gentle lemur in this part of the lake. The images are of such high resolution that it allows for distinguishing the reeds and papyrus vegetation in the area and enables detailed vegetation analyses and monitoring of change (Figure. 3.3).

3.5 Limitations/constraints of conservation drones

Although drones offer promising opportunities for conservation, there are many challenges. One of these challenges is the lack of data on the durability of the drone systems used by conservation workers. There are few or no details available on, for instance, how often a motor or electronic speed controller or other part of a drone should be replaced, or how often parts should undergo maintenance. Nor are there clear data on, for example, how many belly landings a fixed-wing drone can undergo before the fuselage needs to be replaced or undergo standard maintenance. This impacts how well we can predict the number of flights or hours of flight a drone might be capable of, which has safety implications and hampers a proper comparison of costs between drones and alternative data collection methods. Since potential cost reductions are an important reason for conservation organizations to invest in drones, there is a strong need to improve our understanding of drone system durability. This is also important so that operators can develop a maintenance plan for their drones, including whether they need to build maintenance capacity themselves or whether to outsource this to third parties, and whether bespoke maintenance is required for specific drones in their fleet.

At present, there are very little data available providing cost comparisons for the use of drones versus alternative methods. Although there are studies such as those by Vermeulen et al. (2013) that provide some comparison between the survey costs with drones and those of a crewed aircraft, these generally do not cover all the costs involved such as training, computers, analyses time, and so forth. A somewhat more elaborate, albeit somewhat hypothetical, comparison was provided for orangutan surveys with drones and on the ground surveys (Wich & Koh, 2018). There is, however, a shortage of examples from the real world, even though there is a clear need for this. We encourage organizations using drones to conduct a detailed cost assessment of the various survey methods they use and put those into the public domain so that such knowledge can be shared. Only when such comparisons are available can organizations evaluate various methods and choose the most appropriate method in terms of costs and data quality.

Another challenge is that the areas that conservationists would like to cover with drone surveys are normally much larger than the areas off-the-shelf and affordable drones can currently cover. Even though method tests and development are being conducted on smaller areas (less than 10km^2), there is a need to transition to covering larger areas once methods have matured sufficiently. With the increasing capacity of batteries and the potential of combining batteries with energy from solar cells on the wings of fixed-wing systems, there are promising developments that indicate that much longer flight durations will become possible at costs within the realm of conservation projects.[17]

Training can also pose challenges, particularly related to fixed-wing systems. Although multirotors are relatively easy to fly in modes where GNSS is controlling flight altitude and position, remote pilots must also be able to fly the drones in modes that do not have GNSS enabled to ensure safe operation in case of GNSS failure. Similarly, many fixed-wing systems are becoming easier to fly with largely automated take-offs and landings. However, when GNSS fails or automated flying is interrupted, it would take considerable training to be able to continue to fly and safely land the drone. Although such training is perhaps not strictly required for largely automated systems, it would build in another layer of safety which would reduce risk to people, property, animals, the drone itself, which could also reduce costs.

Another, perhaps underappreciated, aspect of using drones is the requirement for safe storage of the large quantities of data generated from high-resolution stills and videos. Once the data is being used to process orthomosaics, the data volume increases even further. Even though the cost of storage has decreased enormously, the data volumes associated with a high number of flights might still mean that this is a considerable cost, particularly when back-ups are needed.

It is also important to consider the costs of data analyses in terms of hardware, software, and costs for the time people put in collecting and analysing the data. Particularly high costs can be associated with the hardware (e.g. >€2000 for a desktop computer) and software (e.g. >€3000 for one of the main software packages[18]) to process orthomosaics depending on the software used and the hardware requirements of that software. The costs of animal detection from large numbers of stills or videos might require numerous days for image analysts to examine images, leading to considerable costs. The development of automatic detection of animals on images will hopefully reduce those costs; however, no comparisons are yet available to evaluate this. Automated animal detection on high volumes of images will need powerful local severs with high up-front costs, while cloud computing options could also be quite costly.

The final issue to address is that the use of drones can lead to disturbance (due to e.g. noise, shape, flight pattern) of animals (Hodgson & Koh, 2016; Mulero-Pázmány et al., 2017). The relative

[17] https://sunbirds.aero/

[18] https://www.pix4d.com/ and https://www.agisoft.com/

disturbance caused by the various survey methods available should be compared before deciding on the survey method is selected. Even though it is well-established that researchers in a rainforest will disturb animals (Schaik et al., 1983) and that even camera trap surveys can influence animals (Meek et al., 2014; Meek et al., 2016), there have been virtually no studies conducted that compare the disturbance of the various survey methods (e.g. Scholten et al., 2019). It is, therefore, impossible at this stage to properly evaluate which method is best able to minimize disturbance. In relation specifically to drones, there are various reviews of the literature on the disturbance of drones on animal behaviour (Mulero-Pázmány et al., 2017; Wich & Koh, 2018), indicating that disturbance is not consistent. Some studies did not report any visual evidence of disturbance in animals' behaviour, whereas others found that animals fled (in some cases only temporarily) or produced alarm calls (see Table 6.2 in Wich & Koh, 2018). There is still a need to further refine our knowledge of the drone characteristics that influence disturbance and which factors of the animals themselves lead to them being disturbed, but a recent review (Mulero-Pázmány et al., 2017) assessed this issue in a systematic way. This study showed that animals' reactions are influenced by animal characteristics such as the level of aggregation, species, life history, and whether they are breeding or not. It also found that drone characteristics such as whether it is a fixed-wing or multirotor (data on hybrid systems are generally lacking), the system by which the drone is powered, and its flight pattern also influence animals' reaction. Based on the literature on animal disturbance caused by drones, wildlife biologists have been developing guiding principles to minimize drones' disturbance on animals (Hodgson & Koh, 2016; Wich & Koh, 2018). Although the focus on animals' visible reactions to drones is an obvious first step, and a relatively simple one, there is also a need to consider and measure animals' physiological reactions. This is a very underexplored field of research with very few studies (e.g. Ditmer et al., 2015). As much as there is a need for additional studies that investigate the physiological responses of animals to drones, there is also a need to conduct similar studies for other survey methods so that these responses can be compared.

3.6 Social impact/privacy

As the earlier sections indicate, there are several benefits from the usage of drones for conservation. Still, there have also been concerns about their potential negative social impact, for example, in issues surrounding the perception of drones by local communities regarding privacy and data security (Sandbrook, 2015). These have been highlighted often in the context of anti-poaching usage of drones (Humle et al., 2014). Particularly in situations where the usage of drones is within a law enforcement context, it has been suggested that caution is required as drones might be perceived as being part of a conservation approach that supports protected areas as 'fortresses' from which local communities are excluded (Humle et al., 2014). A similar argument could be made for other technologies commonly used in conservation (e.g. satellites, camera traps, and autonomous acoustic recording units). It might, therefore, be more beneficial if all these technologies were considered in a larger framework with respect to social and privacy implications (Wich et al., 2016b). Using such an approach could lead to the development of a general set of best practices for conservationists to adhere to when using technology. These could encompass several key aspects, such as when to involve local communities, the need to adhere to country-specific privacy regulations, and using technology within various cultural differences where there could be differences in concerns about information privacy (Bellman et al., 2004). Further discussion is needed about such issues (see Chapter 12) regarding these technologies, which currently remain underexplored. On the other side, it is important to realize that drones can have a positive social impact through the images they provide, potentially connecting people to the areas they use and changing people's attitudes and behaviours in those areas.

3.7 Future directions

The development of drones has been rapid during the last decade. Drones have become more affordable, easier to use, have longer flight durations, have better sensors, have more varied sensors, and are moving towards VTOL systems. Drones will likely continue to be improved in these aspects.

Meanwhile, there will also be more focus on using photovoltaic cells to increase flight duration and use multiple drones in swarms to coordinate coverage of larger areas for longer durations (Wich & Koh, 2018). At the same time, the role of technology in conservation is growing more generally (Allan et al., 2018). Drones, other pre-programmable or autonomous vehicles, satellites, GNSS tags on animals, camera traps, autonomous acoustic units, vibration sensors, biologgers on animals, and others, are all being increasingly used in conservation. An important next step is to integrate these various sensors into one system in which the various sensors feed into the same decision-making process that will further facilitate conservation management (Wich & Koh, 2018). In such a system, drones can be used to collect data with their own onboard sensors but can also be data mules to wirelessly relay data from other sensors such as camera traps or acoustic sensors. All these data would then ideally be analysed automatically in a software platform such as SMART[19] and inform the conservation managers, in near real-time, which next steps are needed for effective conservation.

3.8 Acknowledgements

The author would like to thank the following people and organizations: Fidimalala Ralainasolo and Richard Lewis for facilitating and supporting the work in Madagascar; the people of Andreba Village in Alaotra, and Wildlife Conservation Madagascar for permission to conduct the study at this site; the Aviation Civile de Madagascar for providing a permit for the surveys and guiding us through the relevant local legislation. Thanks to Ian Thomson for his help with processing the orthomosaic used in Figure 3.3. The map in this chapter was created using ArcGIS® software by Esri. ArcGIS® and ArcMap™ are the intellectual property of Esri and are used herein under licence. Copyright © Esri. All rights reserved. Thanks also go to the Science and Technology Facilities Council (STFC: ST/R002673/1) and UK Research and Innovation (UKRI: EP/T015403/1) for funding this work.

[19] https://smartconservationtools.org/

References

Alexander, C., Korstjens, A. H., Hankinson, E., et al. (2018). Locating emergent trees in a tropical rainforest using data from an unmanned aerial vehicle (UAV). *International Journal of Applied Earth Observation and Geoinformation*, 72, 86–90.

Allan, B. M., Nimmo, D. G., Ierodiaconou, D., Vanderwal, J., Koh, L. P., & Ritchie, E. G. (2018). Futurecasting ecological research: the rise of technoecology. *Ecosphere*, 9, e02163.

Anderson, K., & Gaston, K. J. (2013). Lightweight unmanned aerial vehicles will revolutionize spatial ecology. *Frontiers in Ecology and the Environment*, 11, 138–146.

Andriaholinirina, N., Baden, A., Blanco, M., et al. (2014). Alaotra *Reed Lemur* [Online]. Available at: http://dx.doi.org/10.2305/IUCN.UK.2014-1.RLTS.T9673A16119642.en

Andrianandrasana, H. T., Randriamahefasoa, J., Durbin, J., Lewis, R. E., & Ratsimbazafy, J. H. (2005). Participatory ecological monitoring of the Alaotra Wetlands in Madagascar. *Biodiversity & Conservation*, 14, 2757–2774.

Assmann, J. J., Kerby, J. T., Cunliffe, A. M., & Myers-Smith, I. H. (2019). Vegetation monitoring using multispectral sensors – best practices and lessons learned from high latitudes. *Journal of Unmanned Vehicle Systems*, 7, 54–75.

Bellman, S., Johnson, E. J., Kobrin, S. J., & Lohse, G. L. (2004). International differences in information privacy concerns: a global survey of consumers. *The Information Society*, 20, 313–324.

Benítez-López, A., Santini, L., Schipper, A. M., Busana, M., & Huijbregts, M. A. J. (2019). Intact but empty forests? Patterns of hunting-induced mammal defaunation in the tropics. *Plos Biology*, 17, e3000247.

Bondi, E., Fang, F., Hamilton, M., et al. (2018). SPOT poachers in action: augmenting conservation drones with automatic detection in near real time. In Thirty-Second AAAI Conference on Artificial Intelligence. Available at: https://www.aaai.org/ocs/index.php/AAAI/AAAI18/paper/download/16282/16380

Bondi, E., Fang, F., Hamilton, M., et al. (2019). Automatic detection of poachers and wildlife with UAVs. In Fang, F., Tambe, M., Dilkina, B., & Plumptre, A. (eds.). *Artificial Intelligence and Conservation (Artificial Intelligence for Social Good)*. Cambridge University Press, Cambridge, UK (pp. 77–100).

Bonnin, N., Van Andel, A. C., Kerby, J. T., Piel, A. K., Pintea, L., & Wich, S. A. (2018). Assessment of chimpanzee nest detectability in drone-acquired images. *Drones*, 2, 17.

Buckland, S. T., Anderson, D. R., Burnham, K. P., Laake, J. L., Borchers, D. L., & Thomas, L. (2001). *Introduction to Distance Sampling*. Oxford University Press, Oxford, UK.

Burke, C., Rashman, M. F., Longmore, S. N., et al. (2019). Successful observation of orangutans in the wild with thermal-equipped drones. *Journal of Unmanned Vehicle Systems*. Available at: https://doi.org/10.1139/juvs-2018-0035

Burke, C., Rashman, M. F., Mcaree, O., et al. (2018). Addressing environmental and atmospheric challenges for capturing high-precision thermal infrared data in the field of astro-ecology. In Proceedings of SPIE, Astronomical Telescopes and Instrumentation, 10–15 June 2018, Austin, Texas, USA.

Chabot, D., & Bird, D. M. (2012). Evaluation of an off-the-shelf unmanned aircraft system for surveying flocks of geese. *Waterbirds*, 35, 170–174.

Chabot, D., & Bird, D. M. (2015). Wildlife research and management methods in the 21st century: where do unmanned aircraft fit in? *Journal of Unmanned Vehicle Systems*, 3, 137–155.

Chrétien, L. P., Théau, J., & Ménard, P. (2016). Visible and thermal infrared remote sensing for the detection of white-tailed deer using an unmanned aerial system. *Wildlife Society Bulletin*, 40, 181–191.

Christie, K. S., Gilbert, S. L., Brown, C. L., Hatfield, M., & Hanson, L. (2016). Unmanned aircraft systems in wildlife research: current and future applications of a transformative technology. *Frontiers in Ecology and the Environment*, 14, 241–251.

Colefax, A. P., Butcher, P. A., Pagendam, D. E., & Kelaher, B. P. (2019). Reliability of marine faunal detections in drone-based monitoring. *Ocean & Coastal Management*, 174, 108–115.

Corcoran, E., Denman, S., Hanger, J., Wilson, B., & Hamilton, G. (2019). Automated detection of koalas using low-level aerial surveillance and machine learning. *Scientific Reports*, 9, 3208.

Cunliffe, A. M., Brazier, R. E., & Anderson, K. (2016). Ultra-fine grain landscape-scale quantification of dryland vegetation structure with drone-acquired structure-from-motion photogrammetry. *Remote Sensing of Environment*, 183, 129–143.

Ditmer, M. A., Vincent, J. B., Werden, L. K., et al. (2015). Bears show a physiological but limited behavioral response to unmanned aerial vehicles. *Current Biology*, 25, 2278–2283.

Duffy, J. P., Cunliffe, A. M., Debell, L., et al. (2017). Location, location, location: considerations when using lighweight drones in challenging environments. *Remote Sensing in Ecology and Conservation*. Available at: https://doi.org/10.1002/rse2.58

Eikelboom, J. A. J., Wind, J., Van De Ven, E., et al. (2019). Improving the precision and accuracy of animal population estimates with aerial image object detection. *Methods in Ecology and Evolution*, 10(11), 1875–1887.

Fa, J. E., & Brown, D. (2009). Impacts of hunting on mammals in African tropical moist forests: a review and synthesis. *Mammal Review*, 39, 231–264.

Fraser, R., Olthof, I., Lantz, T. C., & Schmitt, C. (2016). UAV photogrammetry for mapping vegetation in the low-arctic. *Arctic Science*, 2(3), 72–102.

Garbutt, N. (2007). *Mammals of Madagascar: A Complete Guide*. Yale University Press, New Haven, CT.

Gonzalez, L. F., Montes, G. A., Puig, E., Johnson, S., Mengersen, K., & Gaston, K. J. (2016). Unmanned aerial vehicles (UAVs) and artificial intelligence revolutionizing wildlife monitoring and conservation. *Sensors*, 16, 97.

Gooday, O. J., Key, N., Goldstien, S. & Zawar-reza, P. (2018). An assessment of thermal-image acquisition with an unmanned aerial vehicle (UAV) for direct counts of coastal marine mammals ashore. *Journal of Unmanned Vehicle Systems*, 6, 100–108.

Grooten, M., & Almond, R. E. A. (2018). *Living Planet Report—2018: Aiming Higher*. Available at: https://www.wwf.ch/sites/default/files/doc-2018-10/LPR 2018_Full%20Report%20Pages_22.10.2018_0.pdf

Groves, P. A., Alcorn, B., Wiest, M. M., Maselko, J. M., & Connor, W. P. (2016). Testing unmanned aircraft systems for salmon spawning surveys. *FACETS*, 1, 187.

Guo, X., Shao, Q., Li, Y., Wang, Y., Wang, D., Liu, J., Fan, J., & Yang, F. (2018). Application of UAV remote sensing for a population census of large wild herbivores—taking the headwater region of the Yellow River as an example. *Remote Sensing*, 10, 1041.

Hallmann, C. A., Sorg, M., Jongejans, E., et al. (2017). More than 75 percent decline over 27 years in total flying insect biomass in protected areas. *PLoS One*, 12, e0185809.

Hambrecht, L., Brown, R. P., Piel, A. K., & Wich, S. A. (2019). Detecting 'poachers' with drones: factors influencing the probability of detection with TIR and RGB imaging in miombo woodlands, Tanzania. *Biological Conservation*, 233, 109–117.

Hodgson, A., Kelly, N., & Peel, D. (2013). Unmanned aerial vehicles (UAVs) for surveying marine fauna: a dugong case study. *PLoS One*, 8, e79556.

Hodgson, J. C., & Koh, L. P. (2016). Best practice for minimising unmanned aerial vehicle disturbance to wildlife in biological field research. *Current Biology*, 26, R404–R405.

Hodgson, J. C., Mott, R., Baylis, S. M., et al. (2018). Drones count wildlife more accurately and precisely than humans. *Methods in Ecology and Evolution*, 9, 1160–1167.

Horning, N., Robinson, J., Sterling, E., Turner, W., & Spector, S. (2010). *Remote Sensing for Ecology and Conservation*. Oxford University Press, Oxford, UK.

Hu, J., Wu, X., & Dai, M. (2018). Estimating the population size of migrating Tibetan antelopes *Pantholops hodgsonii* with unmanned aerial vehicles. *Oryx*, 54, 1–9.

Humle, T., Duffy, R., Roberts, D. L., Sandbrook, C., St John, F. A. & Smith, R. J. (2014). Biology's drones: undermined by fear. *Science*, 344, 1351–1351.

Israel, M., & Reinhard, A. (2017). Detecting nests of lapwing birds with the aid of a small unmanned aerial vehicle with thermal camera. In 2017 International Conference on Unmanned Aircraft Systems (ICUAS). IEEE (pp. 1199–1207).

Kays, R., Sheppard, J., Mclean, K., et al. (2019). Hot monkey, cold reality: surveying rainforest canopy mammals using drone-mounted thermal infrared sensors. *International Journal of Remote Sensing*, 40, 407–419.

Kellenberger, B., Marcos, D., & Tuia, D. (2018). Detecting mammals in UAV images: best practices to address a substantially imbalanced dataset with deep learning. *Remote Sensing of Environment*, 216, 139–153.

Kiszka, J. J., Mourier, J., Gastrich, K., & Heithaus, M. R. (2016). Using unmanned aerial vehicles (UAVs) to investigate shark and ray densities in a shallow coral lagoon. *Marine Ecology Progress Series*, 560, 237–242.

Koski, W. R., Gamage, G., Davis, A. R., Mathews, T., Leblanc, B. & Ferguson, S. H. 2015. Evaluation of UAS for photographic re-identification of bowhead whales, Balaena mysticetus. *Journal of Unmanned Vehicle Systems*, 3, 22–9.

Laliberte, A. S., & Rango, A. (2009). Texture and scale in object-based analysis of subdecimeter resolution unmanned aerial vehicle (UAV) imagery. *IEEE Transactions on Geoscience and Remote Sensing*, 47, 761–770.

Lamba, A., Cassey, P., Segaran, R. R., & Koh, L. P. (2019). Deep learning for environmental conservation. *Current Biology*, 29, R977–R982.

Lanz, B., Dietz, S., & Swanson, T. (2018). The expansion of modern agriculture and global biodiversity decline: an integrated assessment. *Ecological Economics*, 144, 260–277.

Lhoest, S., Linchant, J., Quevauvillers, S., Vermeulen, C., & Lejeune, P. 2015. How Many HIPPOS (HOMHIP): algorithm for automatic counts of animals with infrared thermal imagery from UAV. *The International Archives of Photogrammetry, Remote Sensing and Spatial Information Sciences*, 40, 355.

Lindsey, P. A., Balme, G., Becker, M., et al. (2013). The bushmeat trade in African savannas: impacts, drivers, and possible solutions. *Biological Conservation*, 160, 80–96.

Longmore, S., Collins, R., Pfeifer, S., et al. (2017). Adapting astronomical source detection software to help detect animals in thermal images obtained by unmanned aerial systems. *International Journal of Remote Sensing*, 38, 2623–2638.

Mackenzie, D. I. (2006). *Occupancy Estimation and Modeling: Inferring Patterns and Dynamics of Species Occurrence*. Academic Press, New York, NY.

Maire, F., Alvarez, L. M., & Hodgson, A. (2015). Automating marine mammal detection in aerial images captured during wildlife surveys: a deep learning approach. In Australasian Joint Conference on Artificial Intelligence, Springer (pp. 379–385).

Maxwell, S. L., Fuller, R. A., Brooks, T. M., & Watson, J. E. (2016). Biodiversity: the ravages of guns, nets and bulldozers. *Nature*, 536, 143–145.

Meek, P., Ballard, G., Fleming, P., & Falzon, G. (2016). Are we getting the full picture? Animal responses to camera traps and implications for predator studies. *Ecology and Evolution*, 6, 3216–3225.

Meek, P. D., Ballard, G.-A., Fleming, P. J. S., Schaefer, M., Williams, W., & Falzon, G. (2014). Camera traps can be heard and seen by animals. *PLoS One*, 9, e110832.

Michez, A., Piégay, H., Lisein, J., Claessens, H., & Lejeune, P. (2016). Classification of riparian forest species and health condition using multi-temporal and hyperspatial imagery from unmanned aerial system. *Environmental Monitoring and Assessment*, 188, 1–19.

Mitchell, J., Glenn, N., Anderson, M. & Hruska, R. (2016). Flight considerations and hyperspectral image classifications for dryland vegetation management from a fixed-wing UAS. *Environmental Management and Sustainable Development*, 5(2), 17.

Mitchell, J. J., Glenn, N. F., Anderson, M. O., et al. (2012). Unmanned aerial vehicle (UAV) hyperspectral remote sensing for dryland vegetation monitoring. In 2012 4th Workshop on Hyperspectral Image and Signal Processing: Evolution in Remote Sensing (WHISPERS). IEEE (pp. 1–10).

Mulero-Pázmány, M., Jenni-Eiermann, S., Strebel, N., Sattler, T., Negro, J. J., & Tablado, Z. (2017). Unmanned aircraft systems as a new source of disturbance for wildlife: a systematic review. *PLoS One*, 12, e0178448.

Mutschler, T. (1999). Folivory in a small-bodied lemur. In Rakotosamimanana, B., Rasamimanana, H., Ganzhorn, J. U. & Goodman, S. M. (eds.). *New Directions in Lemur Studies*. Springer US, Boston, MA (pp. 221–239).

Newbold, T., Hudson, L. N., Arnell, A. P., et al. (2016). Has land use pushed terrestrial biodiversity beyond the planetary boundary? A global assessment. *Science*, 353, 288–291.

Nievergelt, C. M., Mutschler, T., Feistner, A. T., & Woodruff, D. S. (2002). Social system of the Alaotran gentle lemur (*Hapalemur griseus alaotrensis*): genetic characterization of group composition and mating system. *American Journal of Primatology: Official Journal of the American Society of Primatologists*, 57, 157–76.

Olivares-Mendez, M. A., Bissyandé, T. F., Somasundar, K., Klein, J., Voos, H., & Le Traon, Y. M. (2013). The NOAH project: giving a chance to threatened species in Africa with UAVs. In International Conference on e-Infrastructure and e-Services for Developing Countries (pp. 198–208).

Penot, E., David-Benz, H., & Bar, M. (2014). Utilisation d'indicateurs économiques pertinents pour l'évaluation des systèmes de production agricoles en termes de résilience, vulnérabilité et durabilité: le cas de la région du lac Alaotra à Madagascar. Ethique et *Économique*, 11(1), 41–61.

Pfeifer, C., Barbosa, A., Mustafa, O., Peter, H.-U., Brenning, A., & Rümmler, M. C. (2019). Using fixed-wing UAV for detecting and mapping the distribution and abundance of penguins on the South Shetlands Islands, Antarctica. *Drones*, 3, 39.

Plumptre, A. J., Baisero, D., Grantham, H., et al. (2019). Are we capturing faunal intactness? A comparison of intact forest landscapes and a first scoping of Key Biodiversity Areas of Ecological Integrity. *Frontiers in Forests and Global Change*, 2, 24.

Ralainasolo, F., Waeber, P., Ratsimbazafy, J., Joanna, D., & Richard, L. (2006). Short communication on the Alaotra gentle lemur: population estimation and subsequent implications. *Madagascar Conservation & Development*, 1, 9–10.

Rashman, M. F., Steele, I. A., Burke, C., Longmore, S. N., & Wich, S. (2018). Adapting thermal-infrared technology and astronomical techniques for use in conservation biology. In SPIE Astronomical Telescopes + Instrumentation. SPIE (p. 10).

Rees, A. F., Avens, L., Ballorain, K., et al. (2018). The potential of unmanned aerial systems for sea turtle research and conservation: a review and future directions. *Endangered Species Research*, 35, 81–100.

Reid, A., Ramos, F., & Sukkarieh, S. (2011). Multi-class classification of vegetation in natural environments using an unmanned aerial system. Robotics and Automation (ICRA), 2011 IEEE International Conference. IEEE (pp. 2953–2959).

Rieucau, G., Kiszka, J. J., Castillo, J. C., Mourier, J., Boswell, K. M., & Heithaus, M. R. (2018). Using unmanned aerial vehicle (UAV) surveys and image analysis in the study of large surface-associated marine species: a case study on reef sharks Carcharhinus melanopterus shoaling behaviour. *Journal of Fish Biology*, 93, 119–127.

Ripple, W. J., Wolf, C., Newsome, T. M., et al. (2019). Are we eating the world's megafauna to extinction? *Conservation Letters*, 12(3), e12627.

Sandbrook, C. (2015). The social implications of using drones for biodiversity conservation. *Ambio*, 44, 636–647.

Schaik, C. P. V., Noordwijk, M. A. V., Warsono, B., & Sutriono., E. (1983). Party size and early detection of predators in Sumatran forest primates. *Primates*, 24, 211–221.

Scholten, C., Kamphuis, A., Vredevoogd, K., et al. (2019). Real-time thermal imagery from an unmanned aerial vehicle can locate ground nests of a grassland songbird at rates similar to traditional methods. *Biological Conservation*, 233, 241–246.

Spaan, D., Burke, C., Mcaree, O., et al. (2019). Thermal infrared imaging from drones offers a major advance for spider monkey surveys. *Drones*, 3, 34.

Spillmann, B., Van Noordwijk, M. A., Willems, E. P., Mitra Setia, T., Wipfli, U., & Van Schaik, C. P. (2015). Validation of an acoustic location system to monitor Bornean orangutan (*Pongo pygmaeus wurmbii*) long calls. *American Journal of Primatology*, 77(7), 767–776.

Tilman, D., Clark, M., Williams, D. R., Kimmel, K., Polasky, S., & Packer, C. (2017). Future threats to biodiversity and pathways to their prevention. *Nature*, 54(6), 22900.

Tomlins, G., & Lee, Y. (1983). Remotely piloted aircraft—an inexpensive option for large-scale aerial photography in forestry applications. *Canadian Journal of Remote Sensing*, 9, 76–85.

Vermeulen, C., Lejeune, P., Lisein, J., Sawadogo, P., & Bouché, P. (2013). Unmanned aerial survey of elephants. *PLoS One*, 8, e54700.

Wallace, A. P. C., Jones, J. P. G., Milner-Gulland, E. J., Wallace, G. E., Young, R. & Nicholson, E. (2016). Drivers of the distribution of fisher effort at Lake Alaotra, Madagascar. *Human Ecology*, 44, 105–117.

Ward, S., Hensler, J., Alsalam, B., & Gonzalez, L. F. (2016). Autonomous UAVs wildlife detection using thermal imaging, predictive navigation and computer vision. In 2016 IEEE Aerospace Conference, 5–12 March 2016 (pp. 1–8).

Wearn, O. R., & Glover-Kapfer, P. (2019). Snap happy: camera traps are an effective sampling tool when compared with alternative methods. *Royal Society Open Science*, 6, 181748.

Weinstein, B. G. (2018). A computer vision for animal ecology. *Journal of Animal Ecology*, 87, 533–545.

Westoby, M., Brasington, J., Glasser, N., Hambrey, M., & Reynolds, J. (2012). 'Structure-from-motion' photogrammetry: a low-cost, effective tool for geoscience applications. *Geomorphology*, 179, 300–314.

White, E. R. (2019). Minimum time required to detect population trends: the need for long-term monitoring programs. *BioScience*, 69, 40–46.

Wich, S., Dellatore, D., Houghton, M., Ardi, R., & Koh, L. P. (2016a). A preliminary assessment of using conservation drones for Sumatran orangutan (*Pongo abelii*) distribution and density. *Journal of Unmanned Vehicle Systems*, 4, 45–52.

Wich, S., Koh, L. P., & Szantoi, Z. (2018). Classifying land cover on very high resolution drone-acquired orthomosaics. In Anemone, R. & Conroy, G. C. (eds.). *New Geospatial Approaches to the Anthropological Sciences.* School of Advances Research Press, Santa Fe, New Mexico (pp. 121–136).

Wich, S., Scott, L., & Koh, L. P. (2016b). Wings for wildlife: the use of conservation drones, challenges and opportunities. In Sandvik, K. B. & Jumbert, M. G. (eds.). *The Good Drone.* Routledge, Oxon, UK (pp. 153–167).

Wich, S. A., & Koh, L. P. (2018). *Conservation Drones.* Oxford University Press, Oxford, UK.

Wolf, P. R., Dewitt, B. A., & Wilkinson, B. E. (2014). *Elements of Photogrammetry: With Applications in GIS.* McGraw-Hill, New York, NY.

CHAPTER 4

Acoustic sensors

Anne-Sophie Crunchant, Chanakya Dev Nakka, Jason T. Isaacs, and Alex K. Piel

4.1 Introduction

4.1.1 What questions are we asking?

Animals share acoustic space to communicate vocally. Similar to food, mates, and territory, acoustic space can also be a scarce resource for which animals compete, with callers adjusting spatial, temporal, and frequency patterns in response to both abiotic and biotic factors, especially the sounds of sympatric fauna (Araya-Salas et al., 2017). Acoustic sensors offer an important method of monitoring animal behaviour by not requiring humans to be in the field with the animals. Moreover, when visual detection is limited (e.g. at night, in lush tropical forests, or when weather conditions are poor; see also Verfuss et al., 2018), the employment of acoustic sensors is an important tool to reveal and monitor the composition of acoustic communities as well as assess biodiversity, habitat integrity, and even threats. More so, as a non-invasive and remote method applicable for any taxa that produce acoustic signals, from insects to fish, herpetofauna to birds and mammals (Sugai et al., 2019), passive acoustic monitoring (PAM) is an increasingly common technique integrated into species conservation (Sugai et al., 2019).

Compared to alternative methods of biomonitoring such as point counts or drones (Chapter 3) and similarly to camera traps (Chapter 5), acoustic sensors can be deployed for long periods in the field (months or years) and simultaneously at multiple locations and, because of their large detection range, can monitor places that may be difficult to access by researchers. Moreover, they are non-invasive and require low researcher-hours for deployment compared to alternative devices that demand frequent visits or maintenance. Sensors can also be used in adverse weather conditions (Philpott et al., 2007; Elliott et al., 2011). The first device capable of listening to sounds underwater was developed during World War I, before being used for marine sciences (reviewed in Sousa-Lima et al., 2013). Recent advances in bioacoustics have expanded the applications of acoustic sensors for terrestrial species (Blumstein et al., 2011; Kalan et al., 2015; Wrege et al., 2017). Like with nearly all technology, devices have become more affordable and smaller with technological advances, from where systems began (portable handheld tape recorders) to digital audio recorders and, currently, autonomous recorders. However, the non-standardization of monitoring protocols, the labour-intensive acoustic analyses, and the limited data curation make PAM a tool that requires careful consideration prior to deployment (Sugai et al., 2019). The increasing interdisciplinary collaboration between engineers and field ecologists is driving new, affordable, and effective biomonitoring methods with reduced size and weight but increased applicability.

PAM offers a wide variety of applications for the study of wildlife ecology, behaviour, and conservation (Figure 4.1). It was initially developed for marine species (Spiesberger & Fristrup, 1990; Marques et al., 2009; 2013; Tavolga, 2012) before being adopted for birds (e.g. Efford et al., 2009; Bardeli et al., 2010; Leach et al., 2016) and more recently terrestrial species (e.g. reviewed in Huetz & Aubin, 2002; Blumstein et al., 2011; Spillmann et al., 2010; Kalan et al., 2015), insects (Penone et al., 2013), and anurans (e.g. Stevenson et al., 2015). As

Anne-Sophie Crunchant et al., *Acoustic sensors*. In: *Conservation Technology*. Edited by Serge A. Wich and Alex K. Piel, Oxford University Press.
© Oxford University Press 2021. DOI: 10.1093/oso/9780198850243.003.0004

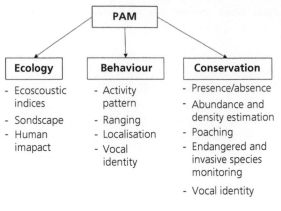

Figure 4.1 Different applications of PAM for ecology, behaviour, and conservation.

with many of the other topics discussed in this volume, understanding the relationship between the physical environment and here, animal acoustic communication, is highly interdisciplinary. That is, data collection (sound production, propagation, measurement) and analyses (and machine learning for detection and classification) are largely challenges that physicists and computer scientists are trained to address. Meanwhile, biologists tend to focus on animal vocal anatomy, call type, flexibility, and behaviour (Erbe et al., 2019).

In this chapter, we present PAM's current status as a tool for conservation biologists, using examples from both terrestrial and marine systems, focusing on recent developments and applications. Besides PAM, there are other systems that integrate acoustic sensors, namely telemetry and biologgers. These systems are often invasive techniques, such as collar-mounted acoustic sensors (e.g. Lynch et al., 2013; Yan et al., 2019; Wijers et al., 2021); we focus instead here on PAM, given its historic and increasing use in conservation. Next, we review how PAM has been used to address related topics across taxonomic groups under these themes (behaviour, ecology, and conservation).

a Behaviour

Ranging, territoriality, and activity patterns

Acoustic sensors are powerful tools to reveal elements of animal behaviour, especially when subjects cannot be observed visually by researchers. Historically (Watkins and Schevill, 1972) but also

more recently (Elliott et al., 2011), PAM has been the primary means of revealing cetacean habitat use and behaviour. It is beyond the scope of this chapter to comprehensively review this body of work. However, PAM has been used to reveal seasonal and daily temporal movement patterns of harbour porpoises (*Phocoena phocoena*), common (*Delphinus delphis*) and bottlenose dolphins (*Tursiops truncatus*) (Dede et al., 2014), or night-time movements of Yangtze finless porpoises (*Neophocaena phocaenoides asiaeorientalis*) (Akamatsu et al., 2008), seasonal migrations of mink whales (*Balaenoptera acutorostrata*) (Risch et al., 2014), and simultaneous ranging behaviour of seven adjacent killer whale (*Orcinus orca*) pods (Yurk et al., 2010). It has also been used to study right whale (*Eubalaena glacialis*) use of corridors in urbanized coastal regions (Morano et al., 2012). The study of marine mammal behaviour using PAM is usually based on either stationary underwater hydrophones or towing multiple hydrophone arrays behind sea vessels.

There is far less, but nonetheless growing application of PAM to terrestrial species. These studies involve attaching acoustic sensors to stationary locations (e.g. trees usually) and setting recording schedules (i.e. duty cycling, e.g. continuous, 10 minutes every hour, to save battery life and memory). Some of the earliest work assessed wildlife behavioural responses to human activity (see below). Wrege and colleagues (2010) investigated whether forest elephants (*Loxodonta cyclotis*) avoided areas of higher oil exploration activity in Loango National Park, Gabon. To their surprise, elephants did not, and in fact, neither dynamite blast intensity nor explosion frequency affected elephant presence. In fact, elephant abundance figures followed patterns of other, non-disturbed areas. However, temporal (activity) patterns of behaviour changed in the explored areas, with elephants exhibiting more nocturnal calling behaviour the closer they were to areas of oil exploration (Wrege et al., 2010).

While PAM has rarely been used to assess territoriality, strategic acoustic sensor deployments across territorial boundaries can reveal how vocalization patterns change with caller location. Chimpanzees (*Pan troglodytes*) are highly xenophobic and change their behaviour near territorial boundaries. In Tai National Park, Ivory Coast, Kalan and colleagues

(2016) used PAM and showed that chimpanzee drumming events were more frequent near territorial boundaries, supporting earlier observational results of vocalizations by Wilson et al., (2007). PAM has been used to assess similar questions about vocalizations and territoriality in birds. Male and female wren (*Thryothorus rufalbus*) co-defend territories and were hypothesized to have a territory size and overlap comparable between sexes. Researchers found that both sexes exhibited congruent patterns of singing behaviour within their territory (Osmun & Mennill, 2011). Additional acoustic monitoring of other terrestrial taxa from numerous other groups, including insects, amphibians, and mammals, has provided data on breeding partner coordination in calling, either by overlapping or alternating call elements (e.g. Geissmann, 2002; Bryant et al., 2016). This is also the case for more than 220 species of birds (Mennill et al., 2012). Given the difficulty of not only following wide-ranging species like chimpanzees and elephants, but also monitoring their behaviour at key areas like boundaries, PAM is well poised to reveal location-specific calling patterns.

In species that produce context-specific vocalizations, recording and identifying call types inform on animal activity. For example, chimpanzees (e.g. Crockford & Boesch, 2003), bottlenose dolphins (Janik & Slater, 1998), chipmunks (*Tamias striatus*) (Burke da Silva et al., 1994), humpback whales (*Megaptera novaeangliae*) (Mercado III et al., 2007), short-finned pilot whales (*Globicephala macrorhynchus*) (Jensen et al., 2011), and white-bearded wildebeest (*Connochaetes taurinus mearnsi*) (Calabrese et al., 2018) exhibit call types that reveal behaviour such as travelling, foraging, greeting, cohesion maintenance, or rutting. Thus, we can remotely infer intersite and interseasonal social context, behaviour, and habitat use across diel phases by knowing how animals are calling. For example, in bottlenose dolphins, this approach was used to identify areas of higher feeding activity (Elliott et al., 2011), whereas in Serengeti wildebeests, call types revealed a highly synchronized reproduction (Calabrese et al., 2018). There are obvious limitations to this employment of PAM, however. Silent individuals go undetected, and so presence/behaviour is not always revealed. Moreover, call types are not always discretely different, but instead grading into

each other, making the relationship between call type and behaviour inherently difficult to decipher.

Localization

Using PAM to study more specific social dynamics like subgroup coordination and reunions or territory boundaries requires knowing caller locations, not just caller presence. By localizing callers on the landscape through acoustic triangulation (Klimley et al., 2001; Blumstein et al., 2011) we can also reconstruct animal movements if individual identity of callers is known. Acoustic triangulation uses the time difference of arrival of sounds to multiple sensors to estimate the sound origin. Besides these behaviour purposes, localization also improves algorithms to automatically detect a species by separating animal sounds from background noise and estimate density (see next), and quantify sound amplitude or directionality (Rhinehart et al., 2020). Initial systems exploited the speed and distance that sounds propagate underwater, with marine vessels toeing hydrophones only centimetres apart (e.g. Filatova et al., 2006; reviewed in Marques et al., 2013). Studies of birds then followed (e.g. Wang et al., 2005a; Mennill et al., 2006). An early prototype system used an eight-microphone cable-based array with sensors placed approximately 75 m apart to study duetting rufous-and-white wrens (*Thryothorus rufalbus*) (Mennill et al., 2006). This system consistently and accurately localized callers to within 3 m. A later prototype with portable and wireless sensors, including GNSS (Global Navigation Satellite System) for time synchronization, showed similar accuracy localization and offered increased deployment flexibility (Mennill et al., 2012). Location accuracy varies as a function of intercaller and intersensor distance (as well as wind, humidity, and vegetation), which affects the sound transmission, signal-to-noise ratio, sensor synchronization accuracy, and recording sample rate (reviewed in Rhinehart et al., 2020).

b Ecology

Ecoacoustic indices

Biodiversity assessments can be very labour-intensive and time-consuming, often requiring detailed taxonomic identification of obscure animal species (Sueur et al., 2012). However, acoustic approaches to biodiversity have the advantage that they can inform on a community level (group of different species) without species identification and document biodiversity using various acoustic diversity indices (Depraetere et al., 2012). Over 60 indices have been developed to address different research questions, such as biodiversity, temporal patterns, and habitat and site effect, among others (Buxton et al., 2018b). However, ecoacoustic (or acoustic ecology) indices are often site-specific (e.g. Gasc et al., 2015; Harris et al., 2016). Moreover, for estimates to be reliable, data should be collected—and indices tested—at a large scale, whereas in reality, that is rarely done and thus sample size is often small. Thus, although generalizing across taxa and ecosystems has not yet been well demonstrated (Buxton et al., 2018b), ecoacoustic indices can be used in longitudinal studies for specific sites. This approach has recently been manifested in soundscape analyses, which include both animal's use of acoustic space and their interactions with all other sources of abiotic and biotic sound sources, including how they partition acoustic space to increase the likelihood of sound transmission to conspecifics (e.g. Ruppé et al., 2015). This field has received increased attention recently, given the increasing global anthropogenic footprint and associated animal vocal and non-vocal responses (behavioural adaptations) to disturbance (Rendell and Gordon, 1999; Brumm et al., 2004; Holt et al., 2015).

Soundscapes

Pijanowski et al.'s (2011a; 2011b) early descriptions of soundscapes integrated biophony (biological sounds), geophony (environmental sounds), and anthrophony (anthropogenic sounds) to characterize the sound environment in a landscape context. They are grounded in the interaction of bioacoustics, landscape ecology, community ecology, and engineering (Gasc et al., 2017). PAM is best placed to document these interactions given the need for continuous recordings of multiple frequencies and often, at times of the day not easy for recording, for example, pre-dawn. In contrast to species-specific PAM studies tuned to the frequency of the target vocalizations, soundscape recordings for biodiversity are more general and serve

to detect audible signals ranging from 20 Hz to 20 kHz (Pijanowski et al., 2011a; 2011b). Analyses of entire soundscapes—versus specific sounds—have the potential to allow us to examine the biodiversity and interspecific dynamics of vocally conspicuous species, especially where communities are disturbed by logging, urban and agricultural expansion, or energy development (Deichmann et al., 2017). However, extensive preliminary work before the disturbance of a system is needed to have a comparative baseline necessary to estimate any subsequent change (Deichmann et al., 2018).

Historically, questions that sought to understand the complex sound environment focused on birds, which were shown to reduce singing competition by acoustically partitioning the timing of overlapping frequencies (Cody & Brown, 1969; Popp et al., 1985). In a simulation of the effect that call length would have on this phenomenon, Suzuki et al., (2012) found that callers with longer songs were least likely to make timing adjustments to avoid overlap, and instead, those species with shorter songs modified behaviour to fit into the available soundscape space. Similarly, acoustic niche partitioning has been demonstrated in primates as well. Schneider et al. (2008) examined the calling pattern of Kloss gibbons (*Hylobates klossii*), Mentawai macaques (*Macaca siberu*), pig-tailed langurs (*Simias concolor*), and Mentawai leaf monkeys (*Presbytis potenziani*) in Siberut, SE Asia. All four species vocalized mostly early morning when background noise is low, but callers showed temporal offsets of calling patterns, likely to reduce acoustic competition.

Human impact

Similar to how acoustic sensors capture entire communities of animal acoustic activity, they also capture the human acoustic element. Human-generated noise can impact the basic ecological functions of a biological community, especially acoustic communication processes (Slabbekoorn & Ripmeester, 2008). The anthropogenic effects of noisy soundscapes have been largely investigated in two domains: the influence of (1) urbanization on birdsong and (2) shipping and natural resource exploration on marine mammals. We discuss each here briefly.

The rapid expansion of urban areas results in a rise in ambient noise levels, generated by ever-growing transportation networks and human activities. Early studies showed the impact of anthropogenic sounds on the timing of the dawn chorus (Arroyo-Solís et al., 2013), with those bird species that are the least variable in their timing being the most affected. Species richness and species abundance also decline significantly with increasing noise (Laiolo, 2010; Proppe et al., 2013). A recent study from LaZerte et al. (2017) demonstrated that male mountain chickadees (*Poecile gambeli*) adjust songs, calls, and chorus composition with increasing ambient and experimental anthropogenic noise. In contrast to calls, songs are longer vocalizations and are more complex in terms of structure and thus, carry more information. In their study, LaZerte et al. (2017) showed that males increase the frequency of their calls in noisy areas and also use more songs than calls.

PAM has also been a key method in evaluating marine mammal responses to drilling, sonar testing by military vessels, and general movements of commercial shipping (e.g. Pirotta et al., 2014; Dyndo et al., 2015; Wisniewska et al., 2018). Noise created by marine vessels is not isotropic (Arveson & Vendittis, 2000); instead, vessel direction, speed, size, cargo type, and weight influence the resulting pattern of underwater disturbance (Erbe et al., 2019). As a result, animals vocally respond in different ways to different types of disturbance. For example, harp seals (*Pagophilus groenlandicus*) reduced the loudness of their calls when travelling near vessels in the Gulf of St. Lawrence, Canada (Terhune et al., 1979), while Florida manatees (*Trichechus manatus latirostris*) shift their orientation, depth, travelling speed, or diving behaviour as a response to approaching vessels (Rycyk et al., 2018). However, the diversity in manatee response behaviour represents the interaction of physics and behaviour, as mentioned earlier. Initial suggestions that manatees were unable to detect oncoming vessels because of manatee hearing were later unsupported (Gerstein, 2002). Instead, the more likely explanation of these accidents concerns manatee preferences to travel near the ocean surface (where noise levels

are low compared to lower depths) and their travelling speed (Gerstein et al., 1999). This example symbolizes the complex interaction between animal, sound, and human behaviour when trying to not only identify the sources of human–wildlife conflict, but also develop conservation strategies. The extensive acoustic frequency and geographic spectrum of data that PAM collects allow researchers to tackle these issues.

c Conservation

While also providing critical data that reveal changes in species abundance and distribution over time, systematic monitoring is also necessary to assess the impacts of management decisions and evaluate wildlife status (Akçakaya et al., 2018; Martin et al., 2018). PAM can inform on abundance, distribution, and wildlife population trends—critical information for conservation scientists.

Occupancy

Occupancy is the proportion of an area used by a species (MacKenzie et al., 2017) and the study relative to the presence/absence of a species. Occupancy statistical models use detection/non-detection data from multiple visits of a given area to infer the probability of species presence. Occupancy modelling provides a useful tool to assess the population status (i.e. declining, stable or increasing) of any species. It can be applied to numerous marine (e.g. whales: Miller & Miller, 2018) and terrestrial species (e.g. birds: Campos-Cerqueira & Aide, 2016; chimpanzees: Kalan et al., 2015). The increased survey coverage of PAM compared to camera trapping sometimes results in better detection rates (Rayment et al., 2018; Enari et al., 2019; Crunchant et al., 2020). However, some species change the frequency of vocalizing in response to external contexts, such as human hunting pressure, or vocalizations can be sex-specific, and thus, other census methods are preferable in these scenarios. Studies from people acting as sensors to monitor animal calls have revealed that low caller activity may result from human disturbance (e.g. Kone & Refisch, 2007; Hicks & Roessingh, 2010). In these cases, PAM should be complemented with other methods to validate results.

Abundance and density

Species abundance and density estimates are two important measures for species monitoring, especially to evaluate extinction risk and to assess the efficacy of conservation policy and practice. PAM was first used to reveal animal density for marine species (reviewed in McDonald & Fox, 1999; Marques et al., 2013) and only more recently for terrestrial species (Stevenson et al., 2015; Measey et al., 2017; Sebastián-González et al., 2018). Current methods to estimate abundance or density have been adapted from direct (visual) observation methods. They include capture-recapture (CR), distance sampling (DS), and via a spatially explicit capture-recapture framework (SECR), reviewed in more detail in Marques et al. (2013) and the study examples provided next:

  CR: when individual identification is possible (e.g. Norwegian ortolan—*Emberiza hortulana* abundance estimation with automatic song-type and individual identity recognition; Adi et al., 2010).

  DS: when the distance of the animal from the sensors are known, point transect and cue counting methods can be used (e.g. North Pacific right whale—*Eubalaena japonica* density using passive acoustic cue counting and sound propagation model Marques et al., 2011).

  SECR: when distances to detected animals are not known, but the same sound can be detected and localized across multiple sensors and when individual recognitions is not possible (e.g. ovenbird—*Seiurus aurocapilla* density estimation with multiple four-microphones arrays, Dawson & Efford, 2009; minke whale—*Balaenoptera acutorostrata* density from multiple hydrophones, Marques et al., 2012; Cape peninsula moss frog *Arthroleptella lightfooti* density from a six microphones array, Stevenson et al., 2015; Measey et al., 2017).

Call rate

To convert calling patterns to animal abundance if methods are based on acoustic cues, information on call rate is necessary, a parameter that is usually collected during focal follows of individuals. Call rate is enormously complex to quantify, as it changes with caller age or sex class, group size and composition, social context, and environmental surroundings, among others (Pérez-Granados et al., 2019). Knowing how often calls are produced can also reveal aggregation patterns of fission-fusion species for whom party composition is ephemeral. Payne et al. (2003), for example, showed that the rate of three different call types varied predictably with herd size in savanna elephants (*L. africana*) from Amboseli, Kenya. Similar relationships have been shown with porpoises (Wang et al., 2005b) and beaked whales (*Ziphiidae* spp.) (Dimarzio et al., 2008) among marine mammals.

Vocal identity

Individual recognition in many species can be achieved with acoustic features of their vocalizations (e.g. giant pandas (*Ailuropoda melanoleuca*—Charlton et al., 2009); four passerines: Gansu leaf warbler (*Phylloscopus kansuensis*), Chinese leafwarbler (*Phylloscopus yunnanensis*), Hume's warbler (*Phylloscopus humei*), and Chinese bulbul (*Pycnonotus sinensis*—Cheng et al., 2010); eagle owl (*Bubo bubo*—Grava et al., 2008); bottlenose dolphins (*Tursiops truncatus*—Kershenbaum et al., 2013), and these features will play the same role as physical marks (tags, scares, patterns) used in camera trapping for individual identification (Terry et al., 2005; Laiolo, 2010). These acoustic features allow us to study, for example, survival rate based on non-invasive acoustic mark-capture-recapture methods as, for instance, with great bittern males (*Botaurus stellaris*) (Gilbert et al., 2002) or Dupont's lark passerine males (*Chersophilus duponti*) (Vögeli et al., 2008) that have distinctive vocalizations, and also estimate site fidelity with the same principle (Grava et al., 2008). Neural networks used to determine call similarities among killer whales (*Orcinus orca*) have shown that calls from individuals from the same matriline were more similar than those from different matrilines (Nousek et al., 2006). This demonstrates that social affiliations also affect vocal identity. Lastly, and especially important for nearly all conservation scientists, vocal individuality can be integrated into census methods to count individuals within a population (e.g. Terry & McGregor, 2002; Hoodless et al., 2008).

Poaching

The illegal killing of wildlife is a widespread and pervasive threat. In addition to revealing the aforementioned parameters of calling behaviour, PAM can also be used as a law enforcement tool to assist conservationists in combatting poaching by localizing gunshots, for instance. Over 2 years, acoustic data have been continuously recorded within Cameroon's Korup National Park. Spatiotemporal gun hunting patterns have been derived from a gunshot detection algorithm, allowing adaptation of anti-poaching patrol activities (Astaras et al., 2017). Multiple platforms have been field-tested and have shown great promise in this context. CARACAL, a low-cost hardware and system, can locate gunshots with an accuracy of under 35 m within an array of seven sensors 500 m apart (Wijers et al., 2021). AudioMoth, another low-cost hardware, can detect gunshots up to 500 m (Hill et al., 2018). Rainforest Connection[1] is a non-profit group that transforms old smartphones into autonomous solar-powered acoustic sensors. Sensors are deployed in the rainforest, and sounds of gunshots or chainsaw are picked up, recorded, and sent in real-time to a server in the cloud via a GSM connection. With this system, they are trying to prevent animal poaching by detecting patterns of activity in major roads used by poachers and illegal deforestation operations in rainforests around the world.

Endangered and invasive species monitoring

The use of acoustic sensors to detect and monitor invasive species, such as freshwater drum (*Aplodinotus grunniens*) (Rountree & Juanes, 2017), Red-billed Leiothrix (*Leiothrix lutea*) (Farina et al., 2013), and pest insects (Mankin et al., 2011) is promising to allow control and eventual eradication (Juanes, 2018). Acoustic monitoring reveals the presence of new or cryptic species as well. In the Gulf of Trieste, Italy, visual surveys have been carried out for decades and failed to detect the presence of the cusk-eel (*Ophidion rochei*), an uncharismatic

[1] https://rfcx.org/

nocturnal predator species classified as Data Deficient in the IUCN Red List. However, after only ten surveys using hydrophones to assess fish acoustic signals, researchers detected its presence off the coast of the Adriatic Sea (Picciulin et al., 2019).

4.1.2 Traditional methods and how technologies overcome limitations

Point counts, especially for bird surveys, or dung counts for elephants (Jones et al., 2012) have long been the established method for collecting data on animal presence and abundance (e.g. Scott et al., 1981; Sedláček et al., 2015). Point counts consist of recording all animals of interest visually or aurally detected during a time period at a given point and certain species-specific radius. However, this technique is limited by the need for trained observers (Hobson et al., 2002) and typically restricted in spatiotemporal coverage. One way to overcome spatial limitations is to have multiple observers working simultaneously, as has been successfully demonstrated in northern yellow-cheeked gibbon (*Nomascus annamensis*) density estimation (Kidney et al., 2016). Call-response surveys, which involve broadcasting conspecific vocalizations to elicit responses, have been used to monitor different species, such as different waterbird species (pied-billed grebe (*Podilymbus podiceps*), American bittern (*Botaurus lentiginosus*), least bittern (*Ixobrychus exilis*), Virginia rail (*Rallus limicola*), and sora (*Porzana carolina*) (Gibbs & Melvin, 1993), and coyotes (*Canis latrans*) (Hansen et al., 2015). However, limitations on the accessibility to the survey area exist and bias can manifest when detectability is low. Leach et al. (2016) conducted a direct comparison between point counts and automated acoustic monitoring in an Australian rainforest and found that point counts detected significantly more species across an elevational gradient. They concluded that 'quiet' species (those that did not vocalize but were detected from visual cues) at least partially explained the variation. Nonetheless, they reported no differences in community-level patterns (e.g. turnover in species composition) across elevation from the two methods (Leach et al., 2016).

In the case of larger terrestrial mammals, dung counts have traditionally been used to assess species abundance and distribution (Kuehl et al., 2007; Jones et al., 2012), but like point counts, these can be time-intensive and spatially limiting. PAM offers a reliable alternative. Acoustic surveys for land mammals have successfully estimated elephant abundance, with confidence intervals half as wide as traditional dung count methods (Thompson et al., 2009). For chimpanzees, they revealed their presence across an unusually large home range across seasons (Crunchant et al., 2020). In comparing visual and acoustic methods for studying anurans in an Afromontane wetland, visual surveys revealed more species during the day, while acoustic methods revealed more at night (Sinsch et al., 2012). Alas, combining methods may inevitably be the ideal solution in these scenarios.

4.2 PAM: from data collection to data analyses

4.2.1 Data collection

An acoustic sensor is composed of any sound recorder and a microphone/hydrophone. The choice of the acoustic sensor is made as a function of the different parameters necessary to monitor, such as the species studied and the frequency range of its vocalizations, the environment (marine or terrestrial), the design of the study (study length, area covered, etc.), and the budget available, among others. Table 4.1 lists some of the most common bioacoustic sensor manufacturers.

Sounds can be recorded at different sampling rates, which is the number of times per second a sound is sampled, measured in hertz (Hz). Sensors can cover different spectra, from infrasound (typically sounds below a frequency of 20 Hz, e.g. whales, elephants), audible sounds (e.g. birds, mammals) to ultrasound (typically sounds above a frequency of 20 kHz, e.g. bats), and recordings depend on the targeted species. The sampling rate must be at least twice as high as the targeted sound's highest frequency to contain necessary acoustic information (namely Nyquist frequency). Therefore, the sampling rate for ultrasounds is much higher than for audible sounds (i.e. typically

44.1 kHz) and is usually between 200 and 400 kHz, which generate bigger files requiring larger data storage.

4.2.2 Analysis

Historically, acoustic data were analysed by visually sifting through hundreds, if not thousands, of spectrograms and aurally verifying sounds of interest. A spectrogram is a visual representation of a sound recording in the time-frequency domain. Time is represented on the x axis, frequency on the y axis, and the signal amplitude is shown as colour density (Figure 4.2). One needs to set the frequency window according to the species of interest and identify patterns corresponding to the targeted call of interest.

This visualization allows analysts to identify and annotate calls manually or automatically through machine learning-based algorithms, themselves based on the acoustic signal itself or on its spectrograms (e.g. Digby et al., 2013; Helble et al., 2015; Clink et al., 2019).

Increasingly and because of the growing data set size, automatic detection is becoming more common (see Chapter 11). Different pipelines exist to process sound files and create algorithms to detect, classify, and then automate sound identification based directly on acoustic signal or the derived spectrogram. These include supervised and unsupervised methods, with thresholding (e.g. Digby et al., 2013), spectrogram cross-correlation (e.g. Aide et al., 2013), random forest (e.g. Ross & Allen, 2014), hidden Markov models (e.g. Kogan & Margoliash, 1998; Enari et al., 2019), convolutional neural networks (e.g. Mac Aodha et al., 2018), support vector machines, and others. Next, we describe the design of a chimpanzee detector using convolutional neural networks, a supervised method without the need for a signal pre-extraction.

Each method has advantages and disadvantages, however. For instance, some methods do not require large training data sets but are sensitive to background noise and overlapping signals that will mask the signal of interest (e.g. Digby et al., 2013). Some methods do not require training data sets, but subsequent identification is necessary (e.g. unsupervised clustering algorithms, Frasier et al., 2016).

Table 4.1 Principal bioacoustic sensor manufacturers for terrestrial and marine environments (adapted from Browning et al., 2017)

Company	Summary	Species/habitat	Study example	Price/unit (US$)	Website
AudioMoth (UK)	Low-cost and open-source device	Terrestrial, audible range and ultrasonic	Hill et al., 2018	60–75	https://www.openacousticdevices.info/
Chelonia (UK)	C-POD and DeepC-POD	Marine	Brandt, et al., 2011	3800–3900	https://www.chelonia.co.uk/
Dodotronic (UK)	USB, parabolic, and analogue microphones; hydrophones	Terrestrial, marine, audible range and ultrasonic	Kloepper et al., 2016	133–1111	https://www.dodotronic.com/
Elekon (Switzerland)	Bat recorders and detectors	Terrestrial, bats	Weier, et al., 2018	1192–4726	https://www.batlogger.com/en/
Frontier Labs (Australia)	Bioacoustic Audio Recorder, with omnidirectional microphone, integrated GPS unit and sampling rate up to 96 kHz	Terrestrial, audible range	Metcalf et al., 2020	538–740	https://frontierlabs.com.au/
Ocean Instruments (New Zealand)	Self-contained sound recorder	Marine	Lillis et al., 2018	2850–6300	http://www.oceaninstruments.co.nz/
Petterson Elektronik (Sweden)	Ultrasonic bat detectors, ultrasound USB microphones	Terrestrial, bats	Nakano & Mason, 2018	173–4090	https://batsound.com/
Solo (UK)	Open-source, low-cost, customizable, Raspberry Pi	Terrestrial, audible range	Whytock & Christie, 2017	108	http://solo-system.github.io home.html
Titley Scientific (UK)	Bat detectors	Terrestrial, bats	Teets et al., 2019	891–2590	https://www.titley-scientific.com/uk/
Wildlife Acoustics (USA)	Song Meter SM4, SM4Bat, SM Mini, SM Mini Bat, handheld bat detector Echo Meter Touch 2	Terrestrial, audible range, and ultrasonic	Hagens et al., 2018	179–899	https://www.wildlifeacoustics.com/

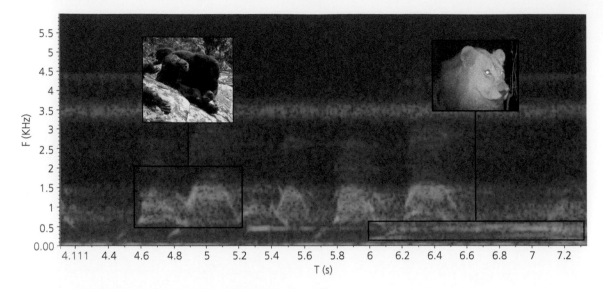

Figure 4.2 Spectrogram of a lion and chimpanzees calling (credit: AS Crunchant).

For a more detailed review of the common methods, see Gibb et al. (2019) and Bianco et al. (2019). While the growing demand and application of artificial intelligence and especially machine learning (see Chapter 11) is applicable to PAM data as well, a major issue with automated detection is the generation of false positives (incorrect identification) and false negatives (failure to identify signals of interest). When the researcher assigns the confidence score that a call has been matched with a known call, the threshold can impact false positive and false negative rates. If the threshold is low, for example, the number of false negatives decreases, but the number of false positives may also increase. On the contrary, if the threshold is high, the number of false negatives might be high. A high rate of false positives could result in an inflated population size estimate, while the opposite could underestimate the number of individuals in a population, which could impact decision-makers. False negatives occur when, for instance, the signal is too faint to be recognized, or when the signal-to-noise ratio is low. It can also happen when the signal is masked by other sounds, such as other species (e.g. dawn and dusk choruses) or abiotic noises (wind). This can change seasonally; for example, detectability might decrease during times of high rain or wind (e.g. Kalan et al., 2015). Automated

detection performance can also vary as a function of site-specific characteristics and quality of recording. Ideally, training sounds are recorded under different conditions and from different locations to improve the detector's generalizability. A manual step is always necessary to validate or reject the classifications, at least for part of the results, as the accuracy is not 100% (e.g. Heinicke et al., 2015). We also need to bear in mind that algorithm development is often laborious and complex, requires programming and/or signal-processing expertise, and typically requires a large, annotated data set for training.

Numerous software packages are available for acoustic analyses, from free and open-source software to proprietary ones. Table 4.2 lists some of the most common platforms.

4.3 Case study: detecting wild chimpanzees using PAM

With the sixth extinction crisis ongoing (Barnosky et al., 2011; Ceballos et al., 2015; Johnson et al., 2017), one million species are at risk of extinction. For instance, numerous studies report the dramatic and global decline of chimpanzees over the past decades (e.g. Campbell et al., 2008; Junker et al., 2012; Kühl et al., 2017). We need reliable,

Table 4.2 Software available for analyses (adapted from Browning et al., 2017)

Software	Availability	Summary	Website/Reference
Anabat Insight	Free and paid versions	Bat call detection, annotation, mapping	https://www.titley-scientific.com/
ARBIMON II	Free initially, charges apply for larger quantities of data	Web-based analysis platform: store, visualize, annotate audio recordings; pattern matching detectors; soundscape analyses	https://arbimon.sieve-analytics.com/
AviaNZ	Free, open source	Manual annotation, automated detection	http://www.avianz.net/
Audacity	Free, open source	Multitrack audio editor, to listen, visualize, subset, and annotate files	https://www.audacityteam.org/
AudioTagger	Free	Listen, visualize, annotate audio files	https://github.com/groakat/AudioTagger
AviSoft	Avisoft Lite (free) Avisoft Pro (licence needed)	Visualization, annotation, spectrogram cross-correlation (find TDOAs), geo-referencing tools, and noise analyses	http://www.avisoft.com/
BatScope4	Free	Visualization, analyse, classification of bat recordings	https://www.wsl.ch/en/services-and-products /software-websites-and-apps/batscope-4.html
CARACAL	Free	Acoustic localization	https://github.com/OpenWild/caracal
CPOD.exe	Free	Analyse data collected by T-PODs and C-PODs	https://www.chelonia.co.uk/cpod_downloads.htm
gibbonR (R package)	Free, open source	Classification, detection, and visualization using machine learning	https://github.com/DenaJGibbon/gibbonR-package
iBatsID	Free	Classification of European bat call recordings to genus and species; requires call parameters extracted by SonoBat	http://ibatsid.eu-west-1.elasticbeanstalk.com/
Ishmael	Free	Visualization, annotations tools, sound localization, and automated call recognition	http://www.bioacoustics.us/ishmael.html
Kaleidoscope	Free trial, licence needed	Visualization, annotation of recordings, automated identification of bats, cluster analysis, noise analysis, batch processing	https://www.wildlifeacoustics.com/products /kaleidoscope-pro
LibROSA (python package)	Free	Visualization	https://librosa.org/librosa/
OpenSoundscape	Free	Preprocessing audio data, machine learning models training, spatial localization of sounds	https://github.com/kitzeslab/opensoundscape

continued

Table 4.2 *Continued*

Software	Availability	Summary	Website/Reference
PAMGUARD	Free, open source	Developed for marine mammals, detection, classification, localization	https://www.pamguard.org/
Pumilio	Free, open source	Web-based management system for ecological recordings, with visualization and manipulation of sound files	http://ljvillanueva.github.io/pumilio/
Raven	Raven Lite (free), Raven Pro (licence needed)	Visualization, annotation, call detection, and spectrogram correlation	https://ravensoundsoftware.com/
scikit-maad (Python package)	Free, open source	Compute ecoacoustic indices	https://github.com/scikit-maad/scikit-maad
Seewave (R package)	Free, open source	Sound analyses and synthesis with acoustic indices calculations	http://rug.mnhn.fr/seewave/
Songscope	Free	Spectrogram visualization	https://www.wildlifeacoustics.com/download/200-song-scope-software
Sonobat	Proprietary	Visualization, call detection, parameter extraction, and species classification	https://sonobat.com/
SonoChiro	Proprietary	Automated bat identifications	http://sonochiro.biotope.fr/
Soundecology (R package)	Free, open source	Functions to calculate indices for soundscape ecology	https://cran.r-project.org/web/packages/soundecology/
SoundFinder (R package)	Free, open source	Position estimation	(Wilson et al., 2014)
Tadarida	Free	Developing and applying an acoustic classifier	https://github.com/YvesBas
WarbleR (R package)	Free, open source	Spectrogram visualization, feature extraction, cross-correlation functions, batch processing	https://cran.r-project.org/web/packages/warbleR/index.html

efficient, and affordable methods to prioritize conservation actions to monitor and ultimately mediate species loss. Chimpanzees are elusive and only a few communities are habituated to humans; thus, most chimpanzees are difficult to observe in the wild. Chimpanzees have large ranges for a terrestrial mammal and rely on loud calls to communicate. PAM is a useful way to detect them.

For three months, we deployed a PAM system that enabled the localization of loud chimpanzee calls from the Issa Valley, Tanzania. The acoustic array consisting of four audio recorders was deployed around the perimeter and on both sides of a single valley known to be an important part of the Issa chimpanzee community's territory. Each acoustic sensor was composed of a microphone unit integrated with a nano-computer Raspberry Pi (Raspberry Pi 3 Model B Motherboard), a GPS unit and three 10 W solar panels and two 44 V batteries (Voltaic systems) and was protected in a Pelicase (Pelican 1170 Case) (Figure 4.3). Sounds were recorded continuously, saved as 1 h audio files at 11,025 Hz sampling rate in the .flac format and stored in a 64 GB SD card. Each sensor was placed at a maximum distance of 500 m from each other to maximize the likelihood of triangulation while simultaneously minimizing the likelihood of missing calls, as calls can carry up to 1 km.

Our goal in designing a chimpanzee detector was twofold. First, we needed a system for processing and analysing many hours of existing data recordings. Our second and ultimate goal was to make detections in real-time on the PAM system, thus enabling real-time localization and more efficient use of data storage on these remote devices. Since the detector had to be designed to run on the Raspberry Pi and ultimately perform near real-time inference, it was required to have a low memory footprint. However, the design's memory constraint could not come at the expense of achieving good classification accuracy.

The detection system was trained with an existing data set of audio recordings that were recorded at a sampling rate of 11,025 Hz, 16-bit depth, and a mono channel. These 1-hour recordings were split into single input audio files with a length of 4 seconds. The raw audio files were then converted to a feature vector consisting of Mel Frequency Cepstral Coefficients (MFCCs) using a Hann window of hop length 512. The windowed signals were then padded with zeros to form consistent vectors of length 2048. This corresponds to a 186-millisecond recording at the 11,025 Hz sampling rate. The 4-second audio clip was converted to a two-dimensional feature vector of size 40×87, which represents 40 MFCC channels. By organizing the feature vectors into this image-like representation, we could utilize convolutional neural networks (CNNs), which have been well studied in the image detection domain.

The high-level system architecture for the CNN can be seen in Figure 4.4. The initial feature vector is first passed through a convolutional layer containing 16 filters to produce 16 feature maps. The purpose of this first convolutional layer is to extract simple patterns such as lines. Here, each pixel in each feature map represents the output of a convolutional layer. The second level of the system is another convolutional layer containing 32 filters to extract higher complexity structures. The outputs from this convolutional layer are then passed through a max-pooling layer to produce space-invariant, low-level features. The purpose of the pooling layer is to reduce the overall size of the representation and reduce the total number of parameters to compute in the network. Next, a fully connected layer is used to obtain the non-linear relationships between these features. This dense layer uses 128 hidden nodes and is followed by a final fully connected layer which gives the probabilities associated with each of the 12 possible classes under consideration. Additionally, the activation function is the Exponential Linear Units (Clevert et al., 2015), and dropout is performed after each max-pooling layer at a rate of 0.3.

When training and testing a classifier, it is generally better to include more classes than just a single positive (chimpanzee vocalizations) and a single negative (anything but chimpanzee vocalizations). In particular, the resulting multiclass classifier is generally more robust to false positives and has improved sensitivity. Ideally, the additional classes would come from sounds commonly heard in the deployment area if such a data set were readily available. Part of the ongoing work for this study is to build such a data set, but at the time

(a) (b)

Figure 4.3 Acoustic unit as described in the text. (a) Sensor and three solar panels deployed at the top of a valley, (b) Raspberry Pi 3 model B motherboard with GPS unit and two 44 V batteries (Voltaic Systems) protected in a Pelicase (Pelican 1170 Case).

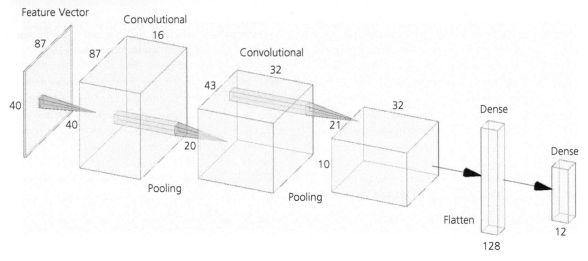

Figure 4.4 Convolutional neural network architecture.

of writing, this data set is not of sufficient size for use. The proposed multiclass detection system was trained and tested on a data set using sounds from the Urbansound8K (public acoustic data set) along with two distinct chimpanzee vocalizations. The resulting 12 possible classes included an air conditioner, car horn, children playing, a dog barking, drilling, an engine idling, a gunshot, a jackhammer, a siren, street music, a chimpanzee shriek, and a chimpanzee pant hoot. This particular data set was chosen for two reasons. First, as described earlier, it allows the development of a more robust multiclass classifier. Second, it is a well-studied data set, thus allowing the comparison of the proposed

classifier's performance to other benchmarked classifiers on the non-chimpanzee classes as a sanity check.

While accuracy is a very intuitive measure of performance, it is generally not enough to evaluate model performance effectively. Accuracy is simply the ratio of observations that are predicted correctly to the total number of observations. This metric works best when the data set is mostly balanced and the classes are evenly distributed. Optimizing purely for accuracy makes the model sensitive and much more prone to detecting falsely labelled examples as positive. False positives are cases the model incorrectly labels as positive that are actually negative.

Precision is the ratio of true positive predictions to the total number of positive predictions. Recall is the ratio of true positive predictions to the total number of positive occurrences. To balance out the true positive rate (precision) and sensitivity (recall) together, we use a metric called the F1 score. The F1 score is the harmonic mean of precision and recall, where a score of 1 is considered perfect. The resulting F1 scores for chimpanzee shrieks and pant hoots were 0.89 and 0.78, respectively, which signify accurate and reliable classification results.

The resulting post-training size of the model described earlier was suitable for operation on a Raspberry Pi 3 platform. The neural network just described is designed to have 4 seconds of audio as input to the system. Each 1-second chunk of audio data is appended to the previous 3 seconds to ensure that calls lasting longer than 1 second are detected. If a chimpanzee call is detected above a prescribed probability threshold, the respective audio clip is saved to file on the Raspberry Pi. By only saving audio signals which are predicted to contain useful information, the limited storage space of the Raspberry Pi can be more efficiently utilized.

4.4 Limitations/constraints

Despite the diverse conservation applications for PAM as an instrument to reveal animal presence and movement, the approach has limitations.

Power limitation—For continuous recordings, PAM is particularly energy-consuming and supplying continuous power is challenging. The incorporation of radio transmission into the system is also energy-consuming, as well as on-board detection algorithms. For areas with sufficient sun exposure, sensors can be recharged with solar panels, potentially then offering unlimited deployment durations in terms of energy (Beason et al., 2018; Hill et al., 2018; Sethi et al., 2018), but with a limited period of time due to data storage if files are stored on-board. For areas without sufficient solar exposure, alkaline or rechargeable batteries are options, but those can be costly to import, store, and dispose of properly.

Data storage—Acoustic data generate large data sets, and data storage can be problematic. File size depends on sample rate, number of channels, duration, and file format. Whether recordings are in mono or stereo depends on the study design. Often mono recordings are sufficient, but stereo recordings can be useful if one of the microphones fails in the field. If data are stored on the recording devices, researchers must visit sensors regularly (every few weeks or months, depending on the recording schedule and how the audio files are recorded and on which format) to retrieve and download the audio data with final data sets quickly approaching terabytes of audio files. Furthermore, data backup is necessary, and archival storage can be problematic in many countries (uploading speed to cloud, storage cost, storage environment of the data). One way to overcome this problem is to store audio data in compressed file formats without compromising sound quality, such as the lossless format .flac, for instance. Alternatively, some systems transmit data from sensors to a central station via wireless networks, for example, via a 3G, radio-antenna system, Iridium satellite system, or satellite internet, which can also be costly (Aide et al., 2013; Saito et al., 2015; Sethi et al., 2018; Baumgartner et al., 2019) or even lightweight aircraft to receive transmission (see Chapter 3). However, transmission of raw data is virtually impossible due to large sizes. If only calls detected with an onboard processor are transmitted, the detector's performance must be determined a priori and additional validation with raw data would not be possible. This can decrease the quality of the data and generate false negatives and false positives errors.

Data processing—As a result of the enormous data sets generated by continuous acoustic recordings, sophisticated, big data processing and analyses are required to post-process (e.g. filter) sounds of interest (Knight et al., 2017). Without automated detection, PAM data analyses are extremely time-consuming and therefore not advisable for regular monitoring via manual analyses. However, in the past few years, major improvements in automated detection (species-specific identification and call types for examining vocal repertoire) (Knight et al., 2017, see next) have transformed this process. Such algorithms have been successfully demonstrated to detect many acoustically active taxa, including elephants (Zeppelzauer et al., 2015), bats (Mac Aodha et al., 2018), humpback whales

(Helble et al., 2015), sika deer (*Cervus nippon*), and Japanese macaques (*Macaca fuscata*) (Enari et al., 2019), manatees (Castro et al., 2016), Diana (*Cercopithecus diana*), and king colobus (*Colobus polykomos*) monkeys (Heinicke et al., 2015), among others. However, for species that exhibit high intra- and extra-individual variation in acoustic structure, like chimpanzees (Clark & Wrangham, 1993; Riede et al., 2004), developing an algorithm is more challenging.

Despite these improvements, manual validation to clean false positives remains compulsory and involves listening to a subsample of pre-identified calls to confirm their identity. This validation is necessary over time and must be done repetitively to ensure against performance decline. Signal degradation and often structural changes to the primary sound are a result of propagation through the environment. They vary due to fluctuating environmental variables such as temperature, humidity, and wind, and the distance and orientation of the caller to the microphone (Schrader & Hammerschmidt, 1997). To compensate for this, intermediate processing steps, such as noise reduction, would increase the recording quality and benefit detecting sounds of interest. However, this is costly in time and money for analysis and algorithm development.

Costs—Besides data and power capacity, another principal constraint of acoustic sensors is the high cost of the on-the-shelf devices and accessories (i.e. microphones, batteries, etc.). For instance, the SM4 from Wildlife Acoustics costs US$849 per unit (www.wildlifeacoustics.com) and requires high capacity batteries (D batteries at 10,000 mAh) that are expensive (i.e. US$76 for a 4-pack). However, custom-made units can be much cheaper (see Table 4.1, e.g. Solo or AudioMoth—US$60–108) but not often made in mass quantities and thus risk structural interunit variability. Software to analyse the data can also be expensive (e.g. US400$ for a non-profit Raven licence or US$680 for Sonobat 4 Universal), which can limit the use of PAM as a biomonitoring tool for small NGOs.

4.5 Social impact/privacy

Privacy—Like drones (Chapter 3) and camera traps (Chapter 5), acoustic sensors can reveal information on third parties that threaten personal privacy,

especially when sensors are deployed in urban areas. Human voices could be recorded without the speaker's knowledge and sensitive information overheard. There is currently little discussion on the regulation of acoustic sensors and nothing governmentally instituted (as compared to drones, see Chapter 3). In order to mitigate these concerns, sensor locations could be published on a website for the public to access, including duration, study objectives, and contact information, among other details; however, with that information revealed, sabotage and theft could easily result, complicating how scientists protect privacy without risking the loss of valuable equipment. Another solution could be to include band filters to remove frequencies with human voices; the trade-off here would be the loss of animal vocalizations that overlap in these frequencies. Data sharing could also be restricted, and not allowed if human voices are heard. More simply, a general warning (as with CCTV cameras) could be placed on the boundaries of a protected area.

Social impact—Over the last few years, citizen science projects have been increasingly used to help collect and analyse large data sets and to promote the importance of biodiversity to the general public. Citizen science is a collaborative approach to research projects conducted by amateur and professional scientists, from data collection to data analysis. Some of the most well-known projects involving citizen science include, for example, Chimp&See,[2] Snapshot Serengeti (Swanson et al., 2015),[3] iNaturalist,[4] and solicit help with images and videos. Citizen science projects with acoustic data are rare, but some exist. For instance, in Australia, Rowley and colleagues (2019) have developed 'FrogID', a database of frog occurrences based on acoustic validation. The platform relies on participants using their smartphones to record frog sounds and submit them for subsequent identification. This has allowed scientists to build a database that includes rare and threatened species, document the decline of native frog species from parts of their range, and detect invasive species. In only one year, over 66,000

---

[2] https://www.chimpandsee.org
[3] https://www.snapshotserengeti.org
[4] https://www.inaturalist.org

frog observations have been made, representing 13% of the total number of previous records made in Australia (Rowley et al., 2019).

Similarly, in England, the Norfolk Bat Survey was launched by the British Trust for Ornithology in April 2013 to conduct a large-scale survey on bat activity and distribution (Newson et al., 2015) and also indirectly on bush-crickets (Orthoptera of the family Tettigoniidae) (Newson et al., 2017). The public can borrow acoustic sensors to record bat calls following a determined protocol (more information can be found on their website).[5] After 2 years, the project generated over 600 million audio recordings (Newson et al., 2015) and, after 4 years, 1.9 million bat recordings have been analysed. The data set is one of the most extensive in the world for bats. Finally, the large ultrasonic audio data set collected along road-transects across Europe and labelled by citizen scientists has allowed scientists to use deep learning to detect bat species (Mac Aodha et al., 2018). However, many of the calls have been misidentified due to the difficulty of recognizing bat calls and the inexperience of some of the citizen scientists.

Similarly, acoustic monitoring and citizen science offer several advantages for assessing insect biodiversity at large spatial and temporal extents (Penone et al., 2013). The authors found that urbanization has a negative effect on the average mass of Orthoptera communities. In summary, PAM has both advantages and disadvantages (i.e. extensive audio data sets capture diverse acoustic soundscapes full of biodiversity and human activity); broadly, the increasingly large data sets require machine learning or armies of citizen scientists to help sieve through the noise. We see real potential for societal impact with these citizen science platforms and the broader interest in what especially soundscapes provide. That is, already related sounds are incorporated into mobile phone applications used to calm tension (e.g. 'Calm'—Huberty et al., 2019), wake people, or provide background noise. Soundscapes from different parts of the world are also made publicly available (e.g. Nature Sound Map, Safe Project's Acoustics

team)[6] and sometimes in real-time (Sethi et al., 2020).

4.6 Future directions

As is the case of nearly every technological approach described in this book, the ongoing development of new technologies and the increasingly interdisciplinary nature of field ecologists, computer scientists, engineers, and bioinformaticians are driving new affordable and effective acoustic biomonitoring methods. We close this chapter with where and how we see the future use of acoustic sensors in conservation science.

Combining new technologies—Similar to PAM, drones can provide real-time feedback for rapid surveys and offer an aerial perspective (cf. Chapter 3). By combining aerial and acoustic technologies, otherwise labour and time-intensive species monitoring is being revolutionized by remotely recording sounds with drone-mounted microphones. Drones have already been integrated with acoustic sensors that have captured bat and bird sounds (Wilson et al., 2017; August & Moore, 2019), and there are plans to diversify applications to terrestrial species. The excessive drone noise recorded by mounted sensors was initially an impediment to this work, but new signal-processing algorithms and drone architectures that reduce this noise are promising (Hioka et al., 2019).

As stated earlier, camera traps and acoustic sensors provide complementary information on wildlife behaviour. By combining these two sensor types into a single system, we capture complementary extents of human disturbance on wildlife, study biotic interactions and animal behaviour at multiple scales, and provide a more thorough picture of animal presence, movement, and communication (Buxton et al., 2018a).

AI, real-time monitoring, and edge computing—Given the desire and usefulness of identifying acoustic events in real-time (e.g. gunshots), acoustic monitoring systems are now being developed with on-board signal processing for event detection (e.g. Hill et al., 2018). Klinck et al. (2012) were one of

---

[5] https://www.batsurvey.org/

[6] https://www.naturesoundmap.com/ and http://acoustics.safeproject.net/06:00/10/51503

the first to demonstrate this for marine mammals with an underwater vehicle (SeaGlider), equipped with an acoustic sensor and on-board data processing capabilities to passively scan for marine mammals. Methods with edge computing are in development (Sheng et al., 2019). Edge computing, or local processing, uses data analysis at or near the source, on edge devices, and thus does not require transmitting large audio data volumes (Chapter 13; Sheng et al., 2019). The advantage of on-board processing is the potential to detect in real-time and record only sounds of interest, dramatically reducing data set sizes, for example, AudioMoth (Hill et al., 2018). The cost, however, is the loss of broader acoustic data that may provide researchers with future questions to pursue (e.g. acoustic biodiversity, non-focal species presence) and to control for quality. The reality is that acoustic data sets can be mined for years to come to examine patterns and metrics that we may not yet realize are important. As always, then, the trade-off of collecting large data sets against the costs to manage them is one that any researcher must carefully consider. New projects looking at providing internet to remote areas, such as the project Loon (Nagpal & Samdani, 2017) and its use of helium balloons launched in the stratosphere, are making acoustic real-time monitoring possible worldwide.

4.7 Acknowledgements

We thank Ammie Kalan and Peter Wrege for their helpful comments on previous versions of the chapter.

References

Adi, K., Johnson, M. T., & Osiejuk, T. S. (2010). Acoustic censusing using automatic vocalization classification and identity recognition. *The Journal of the Acoustical Society of America*, 1272, 874–883.

Aide, T. M., Corrada-Bravo, C., Campos-Cerqueira, M., Milan, C., Vega, G., & Alvarez, R. (2013). Real-time bioacoustics monitoring and automated species identification. *PeerJ*, 1, e103.

Akamatsu, T., Nakazawa, I., Tsuchiyama, T., & Kimura, N. (2008). Evidence of nighttime movement of finless porpoises through Kanmon Strait monitored using a stationary acoustic recording device. *Fisheries Science*, 745, 970–975.

Akçakaya, H. R., Bennett, E. L., Brooks, T. M., et al. (2018). Quantifying species recovery and conservation success to develop an IUCN Green List of Species. *Conservation Biology*, 325, 1128–1138.

Araya-Salas, M., Wojczulanis-Jakubas, K., Phillips, E. M., Mennill, D. J., & Wright, T. F. (2017). To overlap or not to overlap: context-dependent coordinated singing in lekking long-billed hermits. *Animal Behaviour*, 124, 57–64.

Arroyo-Solís, A., Castillo, J. M., Figueroa, E., López-Sánchez, J. L., & Slabbekoorn, H. (2013). Experimental evidence for an impact of anthropogenic noise on dawn chorus timing in urban birds. *Journal of Avian Biology*, 443, 288–296.

Arveson, P. T., & Vendittis, D. J. (2000). Radiated noise characteristics of a modern cargo ship. *The Journal of the Acoustical Society of America*, 1071, 118–129.

Astaras, C., Linder, J. M., Wrege, P., Orume, R. D., & Macdonald, D. W. (2017). Passive acoustic monitoring as a law enforcement tool for Afrotropical rainforests. *Frontiers in Ecology and the Environment*, 155, 233–234.

August, T., & Moore, T. (2019). Autonomous drones are a viable tool for acoustic bat surveys. *BioRvix*, 673772. Available at: https://www.biorxiv.org/content/10.1101/673772v1.full.pdf

Bardeli, R., Wolff, D., Kurth, F., Koch, M., Tauchert, K.-H., & Frommolt, K.-H. (2010). Detecting bird sounds in a complex acoustic environment and application to bioacoustic monitoring. *Pattern Recognition Letters*, 3112, 1524–1534.

Barnosky, A.D., Matzke, N., Tomiya, S., et al. (2011). Has the Earth's sixth mass extinction already arrived? *Nature*, 4717336, 51–57.

Baumgartner, M. F., Bonnell, J., Van Parijs, S. M., et al. (2019). Persistent near real-time passive acoustic monitoring for baleen whales from a moored buoy: system description and evaluation. *Methods in Ecology and Evolution*, 10(9), 1476–1489.

Beason, R. D., Riesch, R., & Koricheva, J. (2018). AURITA: an affordable, autonomous recording device for acoustic monitoring of audible and ultrasonic frequencies. *Bioacoustics*, 284, 381–396.

Bianco, M. J., Gerstoft, P., Traer, J., Ozanich, E., Roch, M. A., Gannot, S., & Deledalle, C.-A. (2019). Machine learning in acoustics: theory and applications. *The Journal of the Acoustical Society of America*, 1465, 3590–3628.

Blumstein, D. T., Mennill, D. J., Clemins, P., et al. (2011). Acoustic monitoring in terrestrial environments using microphone arrays: applications, technological considerations and prospectus. *Journal of Applied Ecology*, 483, 758–767.

Brandt, M. J., Diederichs, A., Betke, K., & Nehls, G. (2011). Responses of harbour porpoises to pile driving at the

Horns Rev II offshore wind farm in the Danish North Sea. *Marine Ecology Progress Series*, 421, 205–216.

Browning, E., Gib, R., Glover-Kapfer, P., & Jones, K. E. (2017). Passive acoustic monitoring in ecology and conservation. WWF Conservation Technology Series, 12, 75.

Brumm, H., Voss, K., Köllmer, I., & Todt, D. (2004). Acoustic communication in noise: regulation of call characteristics in a New World monkey. *Journal of Experimental Biology*, 2073, 443–448.

Bryant, J. V., Brulé, A., Wong, M. H. G., et al. (2016). Detection of a new Hainan gibbon (Nomascus hainanus) group using acoustic call playback. *International Journal of Primatology*, 37, 534–547.

Burke Da Silva, D. L., & Weary, D. M. (1994). Context-specific alarm calls of the eastern chipmunk, Tamias striatus. *Canadian Journal of Zoology*, 726, 1087–1092.

Buxton, R. T., Lendrum, P. E., Crooks, K. R., & Wittemyer, G. (2018a). Pairing camera traps and acoustic recorders to monitor the ecological impact of human disturbance. *Global Ecology and Conservation*, 16, e00493.

Buxton, R. T., McKenna, M. F., Clapp, M., et al. (2018b). Efficacy of extracting indices from large-scale acoustic recordings to monitor biodiversity. *Conservation Biology*, 325, 1174–1184.

Calabrese, J. M., Moss Clay, A., Estes, R. D., Thompson, K. V., & Monfort, S. L. (2018). Male rutting calls synchronize reproduction in Serengeti wildebeest. *Scientific Reports*, 81, 1–9.

Campbell, G., Kuehl, H., N'Goran Kouamé, P., & Boesch, C. (2008). Alarming decline of West African chimpanzees in Côte d'Ivoire. *Current Biology*, 1819, 903–904.

Campos-Cerqueira, M., & Aide, T. M. (2016). Improving distribution data of threatened species by combining acoustic monitoring and occupancy modelling. *Methods in Ecology and Evolution*, 711, 1340–1348.

Castro, J. M., Rivera, M., & Camacho, A. (2016). Automatic manatee count using passive acoustics. Proceedings of Meetings on Acoustics, 23, 010001.

Ceballos, G., Ehrlich, P. R., Barnosky, A. D., García, A., Pringle, R. M., & Palmer, T. M. (2015). Accelerated modern human–induced species losses: entering the sixth mass extinction. *Science Advances*, 15, e1400253.

Charlton, B. D., Zhihe, Z., & Snyder, R. J. (2009). Vocal cues to identity and relatedness in giant pandas (Ailuropoda melanoleuca). *The Journal of the Acoustical Society of America*, 1265, 2721–2732.

Cheng, J., Sun, Y., & Ji, L. (2010). A call-independent and automatic acoustic system for the individual recognition of animals: a novel model using four passerines. *Pattern Recognition*, 4311, 3846–3852.

Clark, A. P., & Wrangham, R. W. (1993). Acoustic analysis of wild chimpanzee pant hoots: do Kibale Forest chimpanzees have an acoustically distinct food arrival pant hoot? *American Journal of Primatology*, 312, 99–109.

Clevert, D. A., Unterthiner, T., & Hochreiter, S. (2015). Fast and accurate deep network learning by exponential linear units (ELUs). *arXiv preprint*, p.arXiv:1511.07289.

Clink, D. J., Crofoot, M. C., & Marshall, A. J. (2019). Application of a semi-automated vocal fingerprinting approach to monitor Bornean gibbon females in an experimentally fragmented landscape in Sabah, Malaysia. *Bioacoustics*, 283, 193–209.

Cody, M. L., & Brown, J. H. (1969). Song asynchrony in neighbouring bird species. *Nature*, 2225195, 778.

Crockford, C., & Boesch, C. (2003). Context-specific calls in wild chimpanzees, Pan troglodytes verus: analysis of barks. *Animal Behaviour*, 661, 115–125.

Crunchant, A.-S., Borchers, D., Kühl, H., & Piel, A. (2020). Listening and watching: do camera traps or acoustic sensors more efficiently detect wild chimpanzees in an open habitat? *Methods in Ecology and Evolution*, 114, 1–11.

Dawson, D. K., & Efford, M. G. (2009). Bird population density estimated from acoustic signals. *Journal of Applied Ecology*, 466, 1201–1209.

Dede, A., Öztürk, A. A., Akamatsu, T., Tonay, A. M., & Öztürk, B. (2014). Long-term passive acoustic monitoring revealed seasonal and diel patterns of cetacean presence in the Istanbul Strait. *Journal of the Marine Biological Association of the United Kingdom*, 946, 1195–1202.

Deichmann, J. L., Acevedo-Charry, O., Barclay, L., et al. (2018). It's time to listen: there is much to be learned from the sounds of tropical ecosystems. *Biotropica*, 505, 713–718.

Deichmann, J. L., Hernández-Serna, A., Delgado, C., Campos-Cerqueira, M., & Aide, T. M. (2017). Soundscape analysis and acoustic monitoring document impacts of natural gas exploration on biodiversity in a tropical forest. *Ecological Indicators*, 74, 39–48.

Depraetere, M., Pavoine, S., Jiguet, F., Gasc, A., Duvail, S., & Sueur, J. (2012). Monitoring animal diversity using acoustic indices: implementation in a temperate woodland. *Ecological Indicators*, 131, 46–54.

Digby, A., Towsey, M., Bell, B. D., & Teal, P. D. (2013). A practical comparison of manual and autonomous methods for acoustic monitoring. *Methods in Ecology and Evolution*, 47, 675–683.

Dimarzio, N., Moretti, D., Ward, J., et al. (2008). Passive acoustic measurement of dive vocal behavior and group size of Blainville's beaked whale (*Mesoplodon densirostris*) in the Tonge of the ocean (TOTO). *Canadian Acoustics*, 361, 166–172.

Dyndo, M., Wiśniewska, D. M., Rojano-Doñate, L., & Madsen, P. T. (2015). Harbour porpoises react to low levels of high frequency vessel noise. *Scientific Reports*, 5, 1–9.

Efford, M. G., Dawson, D. K., & Borchers, D. L. (2009). Population density estimated from locations of individuals on a passive detector array. *Ecology*, 9010, 2676–2682.

Elliott, R. G., Dawson, S. M., & Henderson, S. (2011). Acoustic monitoring of habitat use by bottlenose dolphins in Doubtful Sound, New Zealand. *New Zealand Journal of Marine and Freshwater Research*, 454, 637–649.

Enari, H., Enari, H. S., Okuda, K., Maruyama, T., & Okuda, K. N. (2019). An evaluation of the efficiency of passive acoustic monitoring in detecting deer and primates in comparison with camera traps. *Ecological Indicators*, 98, 753–762.

Erbe, C., Marley, S. A., Schoeman, R. P., Smith, J. N., Trigg, L. E., & Embling, C. B. (2019). The effects of ship noise on marine mammals—a review. *Frontiers in Marine Science*, 6, 606.

Farina, A., Pieretti, N., & Morganti, N. (2013). Acoustic patterns of an invasive species: the Red-billed Leiothrix (Leiothrix lutea Scopoli 1786) in a Mediterranean shrubland. *Bioacoustics*, 223, 175–194.

Filatova, O. A., Fedutin, I. D., Burdin, A. M., & Hoyt, E. (2006). Using a mobile hydrophone stereo system for real-time acoustic localization of killer whales (Orcinus orca). *Applied Acoustics*, 6711–6712, 1243–1248.

Frasier, K. E., Henderson, E., Bassett, H. R., & Roch, M. A. (2016). Automated identification and clustering of subunits within delphinid vocalizations. *Marine Mammal Science*, 323, 911–930.

Gasc, A., Francomano, D., Dunning, J. B., & Pijanowski, B. C. (2017). Future directions for soundscape ecology: the importance of ornithological contributions. *The Auk*, 1341, 215–228.

Gasc, A., Pavoine, S., Lellouch, L., Grandcolas, P., & Sueur, J. (2015). Acoustic indices for biodiversity assessments: analyses of bias based on simulated bird assemblages and recommendations for field surveys. *Biological Conservation*, 191, 306–312.

Geissmann, T. (2002). Duet-splitting and the evolution of gibbon songs. *Biological Reviews*, 771, 57–76.

Gerstein, E. (2002). Manatees, bioacoustics and boats. *American Scientist*, 902, 154.

Gerstein, E. R., Gerstein, L., Forsythe, S. E., & Blue, J. E. (1999). The underwater audiogram of the West Indian manatee (*Trichechus manatus*). *The Journal of the Acoustical Society of America*, 1056, 3575–3583.

Gibb, R., Browning, E., Glover-Kapfer, P., & Jones, K. E. (2019). Emerging opportunities and challenges for passive acoustics in ecological assessment and monitoring. *Methods in Ecology and Evolution*, 10, 169–185.

Gibbs, J. P., & Melvin, S. M. (1993). Call-response surveys for monitoring breeding waterbirds. *The Journal of Wildlife Management*, 571, 27–34.

Gilbert, G., Tyler, G. A., & Smith, K. W. (2002). Local annual survival of booming male Great Bittern Botaurus stellaris in Britain, in the period 1990–1999. *Ibis*, 1441, 51–61.

Grava, T., Mathevon, N., Place, E., & Balluet, P. (2008). Individual acoustic monitoring of the European eagle owl Bubo. *Ibis*, 1502, 279–287.

Hagens, S. V, Rendall, A. R., & Whisson, D. A. (2018). Passive acoustic surveys for predicting species ' distributions: optimising detection probability. *PLoS One*, 13(7), e0199396.

Hansen, S. J. K., Frair, J. L., Underwood, H. B., & Gibbs, J. P. (2015). Pairing call-response surveys and distance sampling for a mammalian carnivore. *Journal of Wildlife Management*, 794, 662–671.

Harris, S. A., Shears, N. T., Radford, C. A., & Reynolds, J. (2016). Ecoacoustic indices as proxies for biodiversity on temperate reefs. *Methods in Ecology and Evolution*, 76, 713–724.

Heinicke, S., Kalan, A. K., Wagner, O. J. J. J., et al. (2015). Assessing the performance of a semi-automated acoustic monitoring system for primates. *Methods in Ecology and Evolution*, 67, 753–763.

Hicks, T. C., & Roessingh, P. (2010). Chimpanzees (Pan troglodytes schweinfurthii) in the *Northern Democratic Republic* of Congo *Adapt Their Long-Distance Communication Behavior* to *Human Hunting Pressure*. Institute for Biodiversity and Ecosystem Dynamics (IBED).

Helble, T. A., Ierley, G. R., D'Spain, G. L., & Martin, S. W. (2015). Automated acoustic localization and call association for vocalizing humpback whales on the Navy's Pacific Missile Range Facility. *The Journal of the Acoustical Society of America*, 1371, 11–21.

Hill, A. P., Prince, P., Piña Covarrubias, E., Doncaster, C. P., Snaddon, J. L., & Rogers, A. (2018). AudioMoth: evaluation of a smart open acoustic device for monitoring biodiversity and the environment. *Methods in Ecology and Evolution*, 9(5), 1199–1211.

Hioka, Y., Kingan, M., Schmid, G., McKay, R., & Stol, K. A. (2019). Design of an unmanned aerial vehicle mounted system for quiet audio recording. *Applied Acoustics*, 155, 423–427.

Hobson, K. A., Rempel, R. S., Greenwood, H., Turnbull, B., & Van Wilgenburg, S. L. (2002). Acoustic surveys of birds using electronic recordings: new potential from an omnidirectional microphone system. *Wildlife Society Bulletin*, 303, 709–720.

Holt, M. M., Noren, D. P., Dunkin, R. C., & Williams, T. M. (2015). Vocal performance affects metabolic rate in dolphins: implications for animals communicating in noisy environments. *Journal of Experimental Biology*, 21811, 1647–1654.

Hoodless, A. N., Inglis, J. G., Doucet, J. P., & Aebischer, N. J. (2008). Vocal individuality in the roding calls of Woodcock Scolopax rusticola and their use to validate a survey method. *Ibis*, 1501, 80–89.

Huberty, J., Green, J., Glissmann, C., Larkey, L., Puzia, M., & Lee, C. (2019). Efficacy of the mindfulness meditation mobile app 'calm' to reduce stress among college students: randomized controlled trial. *Journal of Medical Internet Research*, 7(6), e14273.

Huetz, C., & Aubin, T. (2002). Bioacoustics approaches to locate and identify animals in terrestrial environments. In Le Galliard, J., Guarini, J., & Gaill, F. (eds.). *Sensors for Ecology, Towards Integrated Knowledge of Ecosystems*. CNRS, Paris, France (pp. 83–96).

Janik, V. M., & Slater, P. J. B. (1998). Context-specific use suggests that bottlenose dolphin signature whistles are cohesion calls. *Animal Behaviour*, 564, 829–838.

Jensen, F. H., Perez, J. M., Johnson, M., Soto, N. A., & Madsen, P. T. (2011). Calling under pressure: short-finned pilot whales make social calls during deep foraging dives. *Proceedings of the Royal Society B: Biological Sciences*, 2781721, 3017–3025.

Johnson, C. N., Balmford, A., Brook, B. W., et al. (2017). Biodiversity losses and conservation responses in the Anthropocene. *Science*, 3566335, 270–275.

Jones, T., Bamford, A. J., Ferrol, D., Hieronimo, P., Mcwilliam, N., & Rovero, F. (2012). Vanishing wildlife corridors and options for restoration: a case study from Tanzania. *Tropical Conservation Science*, 54, 463–474.

Juanes, F. (2018). Visual and acoustic sensors for early detection of biological invasions: current uses and future potential. *Journal for Nature Conservation*, 42, 7–11.

Junker, J., Blake, S., Boesch, C., et al. (2012). Recent decline in suitable environmental conditions for African great apes. *Diversity and Distributions*, 1811, 1077–1091.

Kalan, A. K., Mundry, R., Wagner, O., Heinicke, S., Boesch, C., & Kühl, H. S. (2015). Towards the automated detection and occupancy estimation of primates using passive acoustic monitoring. *Ecological Indicators*, 54, 217–226.

Kalan, A. K., Piel, A. K., Mundry, R., et al. (2016). Passive acoustic monitoring reveals group ranging and territory use: a case study of wild chimpanzees (Pan troglodytes). *Frontiers of Zoology*, 131, 34.

Kershenbaum, A., Sayigh, L. S., & Janik, V. M. (2013). The encoding of individual identity in dolphin signature whistles: how much information is needed? *PLoS One*, 810, 1–7.

Kidney, D., Rawson, B. M., Borchers, D. L., Stevenson, B. C., Marques, T. A., & Thomas, L. (2016). An efficient acoustic density estimation method with human detectors applied to gibbons in Cambodia. *PLoS One*, 115, e0155066.

Klimley, A. P., Le Boeuf, B. J., Cantara, K. M., Richert, J. E., Davis, S. F., & Sommeran, S. Van (2001). Radio-acoustic positioning as a tool for studying site-specific behavior of the white shark and other large marine species. *Marine Biology*, 138, 429–446.

Klinck, H., Mellinger, D.K., Klinck, K., et al. (2012). Near-real-time acoustic monitoring of beaked whales and other cetaceans using a Seaglider™. *PLoS One*, 75, 1–8.

Kloepper, L. N., Linnenschmidt, M., Blowers, Z., Branstetter, B., Ralston, J., & Simmons, J. A. (2016). Estimating colony sizes of emerging bats using acoustic recordings. *Royal Society Open Science*, 33, 1–5.

Knight, E. C., Hannah, K. C., Foley, G. J., Scott, C. D., Brigham, R. M., & Bayne, E. (2017). Recommendations for acoustic recognizer performance assessment with application to five common automated signal recognition programs. *Avian Conservation and Ecology*, 122, 14.

Kogan, J. A., & Margoliash, D. (1998). Automated recognition of bird song elements from continuous recordings using dynamic time warping and hidden Markov models: a comparative study. *Journal of Acoustical Society of America*, 1034, 2185–2196.

Kone, I., & Refisch J. (2007). Can monkey behavior be used as an indicator for poaching pressure? A case study of the Diana guenon (*Cercopithecus diana*) and the western red colobus (*Procolobus badius*) in the Tai National Park, Côte d'Ivoire. *Cambridge Studies in Biological and Evolutionary Anthropology*, 151, 257–289.

Kuehl, H. S., Todd, A., Boesch, C., & Walsh, P. D. (2007). Manipulating decay time for efficient large-mammal density estimation: gorillas and dung height. *Ecological Applications*, 178, 2403–2414.

Kühl, H. S., Sop, T., Williamson, E. A., et al. (2017). The critically endangered western chimpanzee declines by 80%. *American Journal of Primatology*, 799. doi: 10.1002/ajp.22681.

Laiolo, P. (2010). The emerging significance of bioacoustics in animal species conservation. *Biological Conservation*, 1437, 1635–1645.

LaZerte, S. E., Otter, K. A., & Slabbekoorn, H. (2017). Mountain chickadees adjust songs, calls and chorus composition with increasing ambient and experimental anthropogenic noise. *Urban Ecosystems*, 20(5), 989–1000.

Leach, E. C., Burwell, C. J., Ashton, L. A., Jones, D. N., & Kitching, R. L. (2016). Comparison of point counts and automated acoustic monitoring: detecting birds in a rainforest biodiversity survey. *Emu*, 116(3), 305–309.

Lillis, A., Apprill, A., Suca, J. J., Becker, C., Llopiz, J. K., & Mooney, T. A. (2018). Soundscapes influence the settlement of the common Caribbean coral Porites astreoides irrespective of light conditions. *Royal Society Open Science*, 5(12), 181358.

Lynch, E., Angeloni, L., Fristrup, K., Joyce, D., & Wittemyer, G. (2013). The use of on-animal acoustical recording devices for studying animal behavior. *Ecology and Evolution*, 37, 30–2037.

Mac Aodha, O., Gibb, R., Barlow, K. E., et al. (2018). Bat detective—deep learning tools for bat acoustic signal detection. *PLoS Computational Biology*, 143, 1–19.

MacKenzie, D. I., Nichols, J. D., Royle, J. A., Pollock, K. H., Bailey, L., & Hines, J. E. (2017). *Occupancy Estimation and Modeling: Inferring Patterns and Dynamics of Species Occurrence*. Elsevier, Burlington, MA.

Mankin, R. W., Hagstrum, D. W., Smith, M. T., Roda, A. L., & Kairo, M. T. K. (2011). Perspective and promise: a century of insect acoustic detection and monitoring. *American Entomologist*, 57, 30–44.

Marques, T. A., Munger, L., Thomas, L., Wiggins, S., & Hildebrand, J. A. (2011). Estimating North Pacific right whale Eubalaena japonica density using passive acoustic cue counting. *Endangered Species Research*, 133, 163–172.

Marques, T. A., Thomas, L., Martin, S. W., Mellinger, D. K., Jarvis, S., Morrissey, R. P., Ciminello, C. A., & DiMarzio, N. (2012). Spatially explicit capture-recapture methods to estimate minke whale density from data collected at bottom-mounted hydrophones. *Journal of Ornithology*, 152 (Suppl. 2), 445–455.

Marques, T. A., Thomas, L., Martin, S. W., et al. (2013). Estimating animal population density using passive acoustics. *Biological Reviews*, 882, 287–309.

Marques, T. A., Thomas, L., Ward, J., DiMarzio, N., & Tyack, P. L. (2009). Estimating cetacean population density using fixed passive acoustic sensors: an example with Blainville's beaked whales. *The Journal of the Acoustical Society of America*, 125(4), 1982–1994.

Martin, T. G., Kehoe, L., Mantyka-Pringle, C., et al. (2018). Prioritizing recovery funding to maximize conservation of endangered species. *Conservation Letters*, 11(6), e12604.

McDonald, M. A., & Fox, C. G. (1999). Passive acoustic methods applied to fin whale population density estimation. *The Journal of the Acoustical Society of America*, 105(5), 2643–2651.

Measey, G. J., Stevenson, B. C., Scott, T., Altwegg, R., & Borchers, D. L. (2017). Counting chirps: acoustic monitoring of cryptic frogs. *Journal of Applied Ecology*, 54(3), 894–902.

Mennill, D. J., Battiston, M., Wilson, D. R., Foote, J. R., & Doucet, S. M. (2012). Field test of an affordable, portable, wireless microphone array for spatial monitoring of animal ecology and behaviour. *Methods in Ecology and Evolution*, 3(4), 704–712.

Mennill, D. J., Burt, J. M., Fristrup, K. M., & Vehrencamp, S. L. (2006). Accuracy of an acoustic location system for monitoring the position of duetting songbirds in tropical forest. *The Journal of the Acoustical Society of America*, 119(5 Pt 1), 2832–2839.

Mercado III, E., Schneider, J. N., Green, S. R., Wang, C., Rubin, R. D., & Banks, P. N. (2007). Acoustic cues available for ranging by humpback whales. *Journal of the Acoustic Society of America*, 121, 2499.

Metcalf, O. C., Lees, A. C., Barlow, J., Marsden, S. J., & Devenish, C. (2020). hardRain: an R package for quick, automated rainfall detection in ecoacoustic datasets using a threshold-based approach. *Ecological Indicators*, 109, 105793.

Miller, B. S., & Miller, E. J. (2018). The seasonal occupancy and diel behaviour of Antarctic sperm whales revealed by acoustic monitoring. *Scientific Reports*, 8(1), 5429.

Morano, J. L., Rice, A. N., Tielens, J. T., Estabrook, B. J., Murray, A., Roberts, B. L., & Clark, C. W. (2012). Acoustically detected year-round presence of right whales in an urbanized migration corridor. *Conservation Biology*, 26(4), 698–707.

Nagpal, L., & Samdani, K. (2017). Project Loon: innovating the connectivity worldwide. In RTEICT 2017—2nd IEEE International Conference on Recent Trends in Electronics, Information and Communication Technology, Proceedings, 2018 (pp. 1778–1784).

Nakano, R., & Mason, A. C. (2018). Early erratic flight response of the lucerne moth to the quiet echolocation calls of distant bats. *PLoS One*, 13(8), 1–12.

Newson, S. E., Bas, Y., Murray, A., & Gillings, S. (2017). Potential for coupling the monitoring of bush-crickets with established large-scale acoustic monitoring of bats. *Methods in Ecology and Evolution*, 8(9), 1051–1062.

Newson, S. E., Evans, H. E., & Gillings, S. (2015). A novel citizen science approach for large-scale standardised monitoring of bat activity and distribution, evaluated in eastern England. *Biological Conservation*, 191, 38–49.

Nousek, A. E., Slater, P. J. B., Wang, C., & Miller, P. J. O. (2006). The influence of social affiliation on individual vocal signatures of northern resident killer whales (Orcinus orca). *Biology Letters*, 2(4), 481–484.

Osmun, A. E., & Mennill, D. J. (2011). Acoustic monitoring reveals congruent patterns of territorial singing behaviour in male and female tropical wrens. *Ethology*, 117(5), 385–394.

Payne, K. B., Thompson, M., & Kramer, L. (2003). Elephant calling patterns as indicators of group size and composition: the basis for an acoustic monitoring system. *African Journal of Ecology*, 41(1), 99–107.

Penone, C., Le Viol, I., Pellissier, V., Julien, J. F., Bas, Y., & Kerbiriou, C. (2013). Use of large-scale acoustic monitoring to assess anthropogenic pressures on orthoptera communities. *Conservation Biology*, 27(5), 979–987.

Pérez-Granados, C., Bota, G., Giralt, D., et al. (2019). Vocal activity rate index: a useful method to infer terrestrial bird abundance with acoustic monitoring. *Ibis*, 161(4), 901–907.

Philpott, E., Englund, A., Ingram, S., & Rogan, E. (2007). Using T-PODs to investigate the echolocation of coastal bottlenose dolphins. *Journal of the Marine Biological Association of the United Kingdom*, 87(1), 11–17.

Picciulin, M., Kéver, L., Parmentier, E., & Bolgan, M. (2019). Listening to the unseen: passive acoustic monitoring reveals the presence of a cryptic fish species. *Aquatic Conservation: Marine and Freshwater Ecosystems*, 29(2), 202–210.

Pijanowski, B. C., Farina, A., Gage, S. H., Dumyahn, S. L., & Krause, B. L. (2011a). What is soundscape ecology? An introduction and overview of an emerging new science. *Landscape Ecology*, 26(9), 1213–1232.

Pijanowski, B. C., Villanueva-Rivera, L. J., Dumyahn, S. L., et al. (2011b). Soundscape ecology: the science of sound in the landscape. *BioScience*, 61(3), 203–216.

Pirotta, E., Brookes, K. L., Graham, I. M., & Thompson, P. M. (2014). Variation in harbour porpoise activity in response to seismic survey noise. *Biology Letters*, 10(5), 20131090.

Popp, J. W., Ficken, R. W., & Reinartz, J. A. (1985). Short-term temporal avoidance of interspecific acoustic interference among forest birds. *Auk*, 102(4), 744–748.

Proppe, D. S., Sturdy, C. B., & St. Clair, C. C. (2013). Anthropogenic noise decreases urban songbird diversity and may contribute to homogenization. *Global Change Biology*, 19(4), 1075–1084.

Rayment, W., Webster, T., Brough, T., Jowett, T., & Dawson, S. (2018). Seen or heard? A comparison of visual and acoustic autonomous monitoring methods for investigating temporal variation in occurrence of southern right whales. *Marine Biology*, 165, 12.

Rendell, L. E., & Gordon, J. C. D. (1999). Vocal response of long-finned pilot whales (*Globicephala melas*) to military sonar in the Ligurian Sea. *Marine Mammal Science*, 15(1), 198–204.

Rhinehart, T. A., Chronister, L. M., Devlin, T., & Kitzes, J. (2020). Acoustic localization of terrestrial wildlife: current practices and future opportunities. *Ecology and Evolution*, 10(13), 6794–6818.

Riede, T., Owren, M. J., & Arcadi, A. C. (2004). Nonlinear acoustics in pant hoots of common chimpanzees (*Pan troglodytes*): frequency jumps, subharmonics, biphonation, and deterministic chaos. *American Journal of Primatology*, 64(3), 277–291.

Risch, D., Castellote, M., Clark, C. W., et al. (2014). Seasonal migrations of North Atlantic minke whales: novel insights from large-scale passive acoustic monitoring networks. *Movement Ecology*, 2(1), 1–17.

Ross, J. C., & Allen, P. E. (2014). Random forest for improved analysis efficiency in passive acoustic monitoring. *Ecological Informatics*, 21, 34–39.

Rountree, R. A., & Juanes, F. (2017). Potential of passive acoustic recording for monitoring invasive species: freshwater drum invasion of the Hudson River via the New York canal system. *Biological Invasions*, 19(7), 2075–2088.

Rowley, J. J. L., Callaghan, C. T., Cutajar, T., et al. (2019). FrogID: citizen scientists provide validated biodiversity data on frogs of Australia. *Herpetological Conservation and Biology*, 14(1), 155–170.

Ruppé, L., Clément, G., Herrel, A., Ballesta, L., Décamps, T., Kéver, L., and Parmentier, E. (2015). Environmental constraints drive the partitioning of the soundscape in fishes. *Proceedings of the National Academy of Sciences of the United States of America*, 112(19), 6092–6097.

Rycyk, A. M., Deutsch, C. J., Barlas, M. E., et al. (2018). Manatee behavioral response to boats. *Marine Mammal Science*, 34(4), 924–962.

Saito, K., Nakamura, K., Ueta, M., et al. (2015). Utilizing the Cyberforest live sound system with social media to remotely conduct woodland bird censuses in Central Japan. *Ambio*, 44(4), 572–583.

Schneider, C., Hodges, K., Fischer, J., & Hammerschmidt, K. (2008). Acoustic niches of Siberut primates. *International Journal of Primatology*, 29(3), 601–613.

Schrader, L., & Hammerschmidt, K. (1997). Computer-aided analysis of acoustic parameters in animal vocalisations: a multi-parametric approach. *Bioacoustics*, 7(4), 247–265.

Scott, J. M., Ramsey, F. L., & Kepler, C. B. (1981). Distance estimation as a variable in estimating bird numbers. *Estimating Numbers of Terrestrial Birds*, 6, 334–340.

Sebastián-González, E., Camp, R. J., Tanimoto, A. M., et al. (2018). Density estimation of sound-producing terrestrial animals using single automatic acoustic recorders and distance sampling. *Avian Conservation and Ecology*, 132.

Sedláček, O., Vokurková, J., Ferenc, M., Djomo, E. N., Albrecht, T., & Hořák, D. (2015). A comparison of point counts with a new acoustic sampling method: a case study of a bird community from the montane forests of Mount Cameroon. *Ostrich*, 86(3), 213–220.

Sethi, S. S., Ewers, R. M., Jones, N. S., Orme, C. D. L., & Picinali, L. (2018). Robust, real-time and autonomous monitoring of ecosystems with an open, low-cost, networked device. *Methods in Ecology and Evolution*, 9(12), 2383–2387.

Sethi, S. S., Ewers, R. M., Jones, N. S., Signorelli, A., Picinali, L., & Orme, C. D. L. (2020). SAFE Acoustics: an open-source, real-time eco-acoustic monitoring networks in the tropical rainforests of Borneo. *Methods in Ecology and Evolution*, 11(10), 1182–1185.

Sheng, Z., Pfersich, S., Eldridge, A., Zhou, J., Tian, D., & Leung, V. C. M. (2019). Wireless acoustic sensor networks and edge computing for rapid acoustic monitoring. *IEEE/CAA Journal of Automatica Sinica*, 6(1), 64–74.

Sinsch, U., Lümkemann, K., Rosar, K., Schwarz, C., & Dehling, J. M. (2012). Acoustic niche partitioning in an anuran community inhabiting an afromontane wetland (Butare, Rwanda). *African Zoology*, 47(1), 60–73.

Slabbekoorn, H., & Ripmeester, E. A. P. (2008). Birdsong and anthropogenic noise: implications and applications for conservation. *Molecular Ecology*, 72–83.

Sousa-Lima, R. S., Norris, T. F., Oswald, J. N., & Fernandes, D. P. (2013). A review and inventory of fixed autonomous recorders for passive acoustic monitoring of marine mammals. *Aquatic Mammals*, 39(1), 23–53.

Spiesberger, J. L., & Fristrup, K. M. (1990). Passive localization of calling animals and sensing of their acoustic environment using acoustic tomography. *The American Naturalist*, 135(1), 107–153.

Spillmann, B., Dunkel, L. P., Van Noordwijk, M. A., et al. (2010). Acoustic properties of long calls given by flanged male orang-utans (*Pongo pygmaeus wurmbii*) reflect both individual identity and context. *Ethology*, 116(5), 385–395.

Stevenson, B. C., Borchers, D. L., Altwegg, R., Swift, R. J., Gillespie, D. M., & Measey, G. J. (2015). A general framework for animal density estimation from acoustic detections across a fixed microphone array. *Methods in Ecology and Evolution*, 6(1), 38–48.

Sueur, J., Gasc, A., Grandcolas, P., & Pavoine, S. (2012). Global estimation of animal diversity using automatic acoustic sensors. In: Gasc, A., Sueur, J., Pavoine, S., Pellens, R., & Grandcolas, P. (eds.). *Sensors for Ecology*. CNRS, Paris, France (pp. 99–117).

Sugai, L. S. M., Silva, T. S. F., Ribeiro Jr., J. W., & Llusia, D. (2019). Terrestrial passive acoustic monitoring: review and perspectives. *BioScience*, 69(1), 5–11.

Suzuki, R., Taylor, C. E., & Cody, M. L. (2012). Soundspace partitioning to increase communication efficiency in bird communities. *Artificial Life and Robotics*, 17(1), 30–34.

Swanson, A., Kosmala, M., Lintott, C., Simpson, R., Smith, A., & Packer, C. (2015). Snapshot Serengeti, high-frequency annotated camera trap images of 40 mammalian species in an African savanna. *Scientific Data*, 2, 1–14.

Tavolga, W. (2012). Listening backward: early days of marine bioacoustics. In: Popper, A. N., & Hawkins, A. D. (eds.). *The Effects of Noise on Aquatic Life*. Springer, New York, NY (pp. 11–14).

Teets, K. D., Loeb, S. C., & Jachowski, D. S. (2019). Detection probability of bats using active versus passive monitoring. *Acta Chiropterologica*, 21(1), 205.

Terhune, J. M., Stewart, R. E. A., & Ronald, K. (1979). Influence of vessel noises on underwater vocal activity of harp seals. *Canadian Journal of Zoology*, 57(6), 1337–1338.

Terry, A., & McGregor, P. (2002). Census and monitoring based on individually identifiable vocalizations: the role of neural networks. *Animal Conservation*, 5(2), 103–111.

Terry, A., Peake, T., & McGregor, P. (2005). The role of vocal individuality in conservation. *Frontiers in Zoology*, 2(10), 1–16.

Thompson, M. E., Schwager, S. J., & Payne, K. B. (2009). Heard but not seen: an acoustic survey of the African forest elephant population at Kakum Conservation Area, *Ghana*, 1, 1–8.

Verfuss, U. K., Gillespie, D., Gordon, J., et al. (2018). Comparing methods suitable for monitoring marine mammals in low visibility conditions during seismic surveys. *Marine Pollution Bulletin*, 126, 1–18.

Vögeli, M., Laiolo, P., Serrano, D., & Tella, J. L. (2008). Who are we sampling? Apparent survival differs between methods in a secretive species. *Oikos*, 117(12), 1816–1823.

Wang, H., Chen, C. E., Ali, A. M., Asgari, S., Hudson, R. E., Yao, K., Estrin, D., & Taylor, C. (2005a). Acoustic sensor networks for woodpecker localization. *Advanced Signal Processing Algorithms, Architectures, and Implementations XV*, 5910, 591009 1–591009 12. Available at: http://proceedings.spiedigitallibrary.org/proceeding.aspx?articleid=870182.

Wang, K., Wang, D., Akamatsu, T., Li, S., & Xiao, J. (2005b). A passive acoustic monitoring method applied to observation and group size estimation of finless porpoises. *The Journal of the Acoustical Society of America*, 118(2), 1180–1185.

Watkins, W. A., & Schevill, W. E. (1972). Sound source location with a three-dimensional hydrophone array. *Deep Sea Research*, 19, 691–706.

Weier, S. M., Grass, I., Linden, V. M. G., Tscharntke, T., & Taylor, P. J. (2018). Natural vegetation and bug abundance promote insectivorous bat activity in macadamia orchards, South Africa. *Biological Conservation*, 226, 16–23.

Whytock, R. C., & Christie, J. (2017). Solo: an open source, customizable and inexpensive audio recorder for bioacoustic research. *Methods in Ecology and Evolution*, 8(3), 308–312.

Wijers, M., Loveridge, A., Macdonald, D. W., & Markham, A. (2021). CARACAL: a versatile passive acoustic monitoring tool for wildlife research and conservation. *Bioacoustics*, 30(1), 1–17.

Wijers, M., Trethowan, P., Markham, A., et al. (2018). Listening to lions: animal-borne acoustic sensors improve bio-logger calibration and behaviour classification performance. *Frontiers in Ecology and Evolution*, 6, 1–8.

Wilson, A. M., Barr, J., & Zagorski, M. (2017). The feasibility of counting songbirds using unmanned aerial vehicles. *The Auk: Ornithological Advances*, 134(2), 350–362.

Wilson, D. R., Battiston, M., Brzustowski, J., & Mennill, D. J. (2014). Sound Finder: a new software approach for localizing animals recorded with a microphone array. *Bioacoustics*, 23(2), 99–112.

Wilson, M. L., Hauser, M. D., & Wrangham, R. W. (2007). Chimpanzees (*Pan troglodytes*) modify grouping and vocal behaviour in response to location-specific risk. *Behaviour*, 144(12), 1621–1653.

Wisniewska, D. M., Johnson, M., Teilmann, J., Siebert, U., Galatius, A., Dietz, R., & Madsen, P. T. (2018). High rates of vessel noise disrupt foraging in wild harbour porpoises (*Phocoena phocoena*). *Proceedings of the Royal Society B: Biological Sciences*, 285(1872), 20172314.

Wrege, P. H., Rowland, E. D., Keen, S., & Shiu, Y. (2017). Acoustic monitoring for conservation in tropical forests: examples from forest elephants. *Methods in Ecology and Evolution*, 8(10), 1292–1301.

Wrege, P. H., Rowland, E. D., Thompson, B. G., & Batruch, N. (2010). Use of acoustic tools to reveal otherwise cryptic responses of forest elephants to oil exploration. *Conservation Biology*, 24, 1578–1585.

Yan, X., Zhang, H., Li, D., et al. (2019). Acoustic recordings provide detailed information regarding the behavior of cryptic wildlife to support conservation translocations. *Scientific Reports*, 9(1), 1–11.

Yurk, H., Filatova, O., Matkin, C. O., Barrett-Lennard, L. G., & Brittain, M. (2010). Sequential habitat use by two resident killer whale (*Orcinus orca*) clans in Resurrection bay, Alaska, as determined by remote acoustic monitoring. *Aquatic Mammals*, 36(1), 67–78.

Zeppelzauer, M., Hensman, S., & Stoeger, A. S. (2015). Towards an automated acoustic detection system for free ranging elephants. *Bioacoustics*, 24(1), 13–29.

CHAPTER 5

Camera trapping for conservation

Francesco Rovero and Roland Kays

5.1 Introduction

5.1.1 What questions are being asked?

Camera trapping is the use of motion-sensitive cameras to record images or videos of animals passing in front of them. Camera trapping has advanced our knowledge of elusive and rare fauna across the planet thanks to the parallel development of better camera technology and new analytical approaches to address ecological questions (reviews in Rovero & Zimmermann, 2016; Wearn & Glover-Kapfer, 2017). Although scientists have been using camera traps for a century, the last 15 years have seen the technology improve to a point that makes it the preferred method for many types of wildlife studies, and nowadays thousands of research projects across diverse terrestrial habitats deploy camera traps, with tens of millions of images being accumulated (Steenweg et al., 2017). Such diffusion of camera trapping worldwide has been accompanied by an increase in its relevance and impact on wildlife conservation (Figure 5.1).

This chapter reviews the use and potential of camera trapping in conservation science. In particular, (1) it introduces the importance and challenges of studying reclusive wildlife; (2) introduces the key aspects of camera traps that make them efficient and widely used; (3) the core chapter section reviews the variety of ways camera trapping has informed conservation, first presenting which wildlife metrics are typically used with camera trap data, from species presence to abundance estimation and community metrics, and then reviewing the wealth of applications that directly or indirectly contribute to conservation, such as habitat preference and range models, threat assessments, general monitoring, and evaluations of conservation interventions; (4) we present case studies showing how camera trapping can effectively link ecological monitoring to conservation, including how data and images can be used to engage the public and policymakers with conservation issues, and how this work is being scaled up through citizen science and networks of standardized data collection coupled with cyber-infrastructures for automatized analyses. The chapter ends by reviewing possible technological improvements for camera traps and how they would aid conservation in the future.

5.1.2 Traditional methods and limitations

Camera traps represent a method to record the presence of an animal at a particular place and time, including information about their behaviour. This is most similar to traditional methods of direct observation or visual censuses. However, given that most animals are highly mobile and afraid of people, camera traps have a distinct advantage over manual surveys for many species. Indeed, camera traps generally are more efficient and collect better data than visual censuses in direct comparisons (Wearn & Glover-Kapfer, 2019), even for arboreal primates, where significant extra effort is necessary to climb and set the camera traps (Whitworth et al., 2016). Camera traps are also non-invasive, making them preferred to live-trapping animals unless this is necessary to identify species, take samples, or attach tracking tags. Recording animal signs such as

Francesco Rovero and Roland Kays, *Camera trapping for conservation*. In: *Conservation Technology*.
Edited by Serge A. Wich and Alex K. Piel, Oxford University Press.
© Oxford University Press 2021. DOI: 10.1093/oso/9780198850243.003.0005

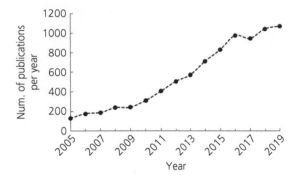

Figure 5.1 The number of publications per year from studies related to applications of camera trapping in conservation (data from Google Scholar search using 'camera trapping AND conservation', F. Rovero, R. Kays).

tracks and faeces is another useful survey method that produces similar data to camera traps, although it can be dependent on appropriate weather conditions (Gompper et al., 2006), as well as complicate analyses (e.g. calculating defecation rates for censuses, etc.). Acoustic sensors also provide wildlife presence data akin to camera traps in many ways, but appropriate for a different suite of more vocal species and require more complicated methods to identify species (see discussion on localization in Chapter 4).

Animal tracking is an alternative approach to record an animal's location and movements by using satellite systems or radio signals (Millspaugh et al., 2012). However, tracking is usually limited to only a few individuals of one or a few species, providing less information about the wider animal community. Camera trapping represents a complementary approach, monitoring a single site and sampling all medium-to-large mammals and birds that pass through it. The method covers the wider animal community but does not provide information on where animals move once they leave the field of view and does not distinguish between individuals (excepting those with unique markings).

5.1.3 How technology addresses this

Although there are a variety of sensors that could be used to detect motion and trigger a camera, most camera traps today use a passive infrared (PIR) sensor. The PIR detects a change in the surface temperature of objects in the sensor detection zone relative to the background objects, such as when an animal moves in front of the camera (see Welbourne et al., 2016 for details). The target area is illuminated by a flash at night, with the large majority of camera models using an infrared LED (light-emitting diode) flash that is detectable by digital cameras as a black-and-white image, without blinding an animal with a bright white-light flash. Some of these LEDs can be seen as a faint red glow when they fire, although there are also 'black', or 'no-glow' versions of LED flashes that emit light in the 940 and 850 nanometre spectrum, respectively, that are presumed to be invisible to animals (Figure 5.2; see

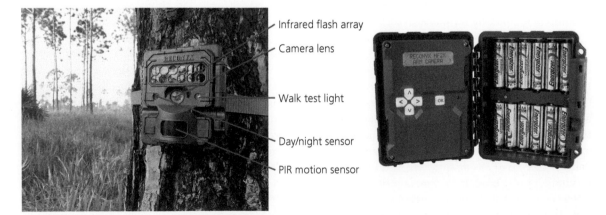

Figure 5.2 Main components of a Passive InfraRed (PIR) camera trap with infrared flash (left; credit R. Kays), and internal view with loaded AA batteries and commands for settings (right; credit Reconyx Inc.).

Rovero & Zimmermann, 2016 for more details on current camera technology).

While this system applies to most commercially available camera models, cameras with different but basic components are also available. These include cameras with active infrared sensors (AIR), triggered when passing subjects break an infrared beam, and cameras with white LED or incandescent xenon flashes, enabling them to take colour images at night. Digital videos with audio are increasingly common. While most researchers today store images on memory cards, some camera trap models can transmit images via GSM (Global System for Mobile Communication) connections. The use of these networked cameras is limited, notably by their high cost, and because networks typically are slow and have limited coverage.

The rapid technological improvements of camera traps have generally improved this tool's efficiency for the study of elusive fauna. This is fundamentally ensured by a fast trigger speed (i.e. the time elapsed between the animal entering the detection zone of the PIR and the camera triggering, typically <0.5 sec), their capacity to detect small passing animals (i.e. >50 g), their compact dimensions, camouflaged colours, and invisible flash. The low power requirements, and their resistance to extreme climates, enable these cameras to function continuously over several months. This in turn allows researchers to collect large amounts of data (in the order of thousand detections per camera), potentially over vast areas and for a large variety of species.

5.2 New Technology: hardware, software, and data analysis

Over the last 20 years, improvements in camera trap technology have mostly come as the result of general improvements in cameras and trigger sensitivity (Kays, 2016). Scientists have benefitted from improvements driven by the photography industry that resulted in better film, lenses, and flashes, and more recently, higher-resolution digital photographs, video (including audio), and better infrared (IR) night images. Triggers have evolved from string or pressure pads to highly sensitive AIR and PIR sensors. We expect these improvements

to continue to help camera trapping in the future (e.g. 360° cameras) but predict the most impactful changes will come from the addition of other sensors to complement the visual image. At the end of the chapter (Section 5.7), we review the future technological developments in camera traps we think have the most potential to aid conservation.

Given the power, cost, and bandwidth limitations of transmitting thousands of pictures wirelessly, most camera trapping data is moved by retrieving memory cards from digital cameras into laptop or desktop computers. There are a variety of software solutions developed in the last few years to process the images (Young et al., 2018), including stand-alone desktop tools (CPW Photo Warehouse[1]), cloud-based solutions (eMammal[2], Wildlife Insights[3], Conservation AI[4]), and those that use crowdsourcing to identify animals (Zooniverse[5]). There has been a surge in research into using AI tools to automatically identify species in images (He et al., 2016; Norouzzadeh et al., 2018), and these are now being integrated into data management tools (e.g. Wildlife Insights; see Figure 5.5D).

Analysis of camera trapping data has primarily advanced through statistical packages developed through R (Niedballa et al., 2016; Rovero & Zimmerman, 2016), with more automated reporting being restricted to a few simple statistics (eMammal), or packages of limited release, such as those developed by the Zoological Society of London[6] and by Wildlife Insights.

5.3 Review of camera trapping conservation applications

Medium-to-large ground-dwelling species of wildlife around the world are threatened by illegal hunting and habitat loss, with additional threats

---

[1] https://cpw.state.co.us/learn/Pages/ResearchMammals Software.aspx
[2] https://emammal.si.edu/
[3] https://www.wildlifeinsights.org/
[4] https://conservationai.co.uk/
[5] https://www.zooniverse.org/
[6] https://www.zsl.org/conservation/how-we-work/conservation-technology/zsl-camera-trap-data-management-and-analysis

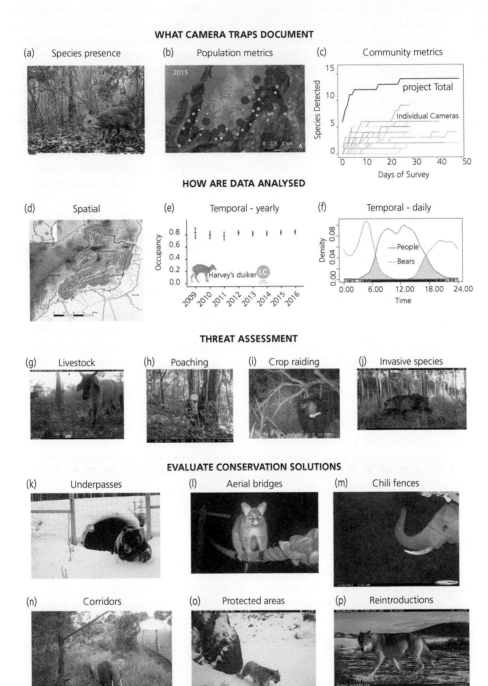

WHAT CAMERA TRAPS DOCUMENT

(a) Species presence

(b) Population metrics

(c) Community metrics

HOW ARE DATA ANALYSED

(d) Spatial

(e) Temporal - yearly

(f) Temporal - daily

THREAT ASSESSMENT

(g) Livestock

(h) Poaching

(i) Crop raiding

(j) Invasive species

EVALUATE CONSERVATION SOLUTIONS

(k) Underpasses

(l) Aerial bridges

(m) Chili fences

(n) Corridors

(o) Protected areas

(p) Reintroductions

associated with climate change, invasive species, and pollution (Dirzo et al., 2014). Camera traps have become one of the primary tools for studying this group of species, leading to a wealth of new information documenting patterns of distribution and abundance that have been critical for establishing conservation status and developing solutions. In this section, we review this rapidly growing body of work (Figure 5.3). First, we show examples of how camera traps have been used to document what species live in an area and quantify how common they are. We then show how researchers use this distribution information to build spatial models that deduce what factors might be limiting a species' range and abundance and create maps that predict these over larger areas. While studies have typically targeted one or a few species of particular interest, we also review more recent studies that focus on species richness and spatiotemporal patterns of the broader community of larger mammals detected by camera traps within an area. These studies are often designed to evaluate the impact of a particular problem on wildlife, and therefore we can then review how this research has detailed specific conservation problems created by threats ranging from hunting and logging to invasive species and land-use changes.

5.3.1 What cameras are good at documenting

5.3.1.1 Species presence

The earliest and most basic contribution of camera trapping to wildlife conservation is to simply document the presence of a species in a given place and time. Resulting data have revealed numerous range extensions and even the discovery, or re-discovery, of particularly rare species that had gone undetected for extensive periods. For example, the saola (*Pseudoryx nghetinhensis*), a large antelope found in mountain forests of Laos and Vietnam, was described in 1992 from skulls, teeth, and skins found in local inhabitants' houses as hunting trophies, was photographed for the first time in 1998 with camera traps (Whitfield, 1998). The grey-faced sengi or elephant-shrew (*Rhynchocyon udzungwensis*) lives in two forest fragments in Tanzania and belongs to an ancient lineage of African mammals; its first ever detection by camera traps in 2005 (Rovero & Rathbun, 2006) led to the subsequent description as a new species in 2008. The silver-backed chevrotain (*Tragulus versicolor*) was first described in 1910, but since then only one verifiable record existed from the early 1990s; this tiny forest antelope was recently camera-trapped in Vietnam (Nguyen et al., 2019).

Recent, camera-trap-based range extensions of rare or little known species include the first record of the endangered Arabian Tahr (*Arabitragus jayakari*) in the Hatta Mountain, Emirates (Aguhob et al., 2018), and the first record of the marbled cat (*Pardofelis marmorata*) in Nepal (Lama et al., 2019). Similarly, camera trapping has documented the re-establishment and expansion of large carnivores, such as the golden jackal (*Canis aureus*) in Europe (Ivanov et al., 2016), and the persistence of rare species in human-dominated landscapes, such as the bushdog (*Speothus venaticus*) in Brazil (Azevedo et al., 2016). Camera traps have even proven useful to study rare birds; for example, the black rail (*Laterallus jamaicensis*), one of the most elusive birds in North America, was documented breeding in South Carolina for the first time in more than a century using camera traps (Hand et al. 2019).

5.3.1.2 Community diversity

These significant single-species discoveries are often part of general biodiversity assessments that also produce inventories of the larger community of

---

Figure 5.3 Examples of key uses of camera trapping in conservation. (a) Silver-backed chevrotain (*Tragulus versicolor*) re-discovered by scientists in Vietnam after 25 years since the last record (credit: GWC/SIE/Leibniz-IZW/NCPC); (b) Map with brown bear (*Ursus arctos*) occupancy estimates in northern Italy (credit: F. Rovero and M. Rinaldi); (c) Species accumulation curves (credit: H. Boone); (d). Map of density estimates of leopard (*Panthera pardus*) in Tanzania's Udzungwa Mountains (credit: Havmøller et al., 2019); (e) Estimated occupancy of Harvey's duiker (*Cephalophus harveyi*) during a 8-year period in Tanzania (credit: Oberosler et al., 2020b); (f) Diel activity pattern and overlap of people and brown bears in northern Italy (credit: F. Rovero); (g–j) Examples of threat assessments (credits left to right: R. Kays, F. Rovero, S. Krief/Sebitoli Chimpanzee Project, R. Kays); (k–p) Examples of evaluations of conservation interventions (credits left to right: Highwaywilding.org, B. Taylor and R. Goldingay. Southern Cross University, Southern Tanzania Elephant Program, S. Green, F. Rovero, C. Hansen).

medium-to-large, ground-dwelling mammals and birds. For this purpose camera traps are unmatched in terms of conducting efficient (data/effort) and effective (high detectability) wildlife surveys (Wearn & Glover-Kapfer, 2019), as they are relatively easy to deploy and allow researchers to detect a variety of species. For example, camera trap surveys have recently revealed the conservation value of previously neglected areas, such as the northern coastal forests in Kenya (Amin et al., 2018), and previously unsurveyed or inaccessible areas in Myanmar (Moo et al., 2018) and Cambodia (Whitworth et al., 2019). Researchers have also recently explored innovative ways to use camera traps beyond sampling large terrestrial species, such as their deployment on forest canopies (Bowler et al., 2017), showing that arboreal species may be more impacted by anthropogenic disturbances than their ground-dwelling counterparts (Whitworth et al., 2019).

5.3.1.3 Population metrics: occupancy, abundance, density

Once a species is documented at a site, the next obvious question is 'how common is it?' Camera traps can provide answers to this question through a variety of population metrics: occupancy, relative abundance, and density. These measures can be used in monitoring schemes to plot spatiotemporal variation. Occupancy is the percentage of sites where a species is present. It is typically estimated with models that correct for imperfect detection (i.e. the probability that a species is detected when present at a site) and include habitat covariates (MacKenzie et al., 2017). Several of the studies we review hereafter used single-species occupancy modelling as the analytic approach of choice, and their sampling design typically consists of randomly-placed arrays of evenly spaced camera traps along a gradient of anthropogenic disturbance. Occupancy analysis has also been applied to the entire assemblage of species that camera traps typically detect, with a class of models that can estimate the richness of all species—or groups of species, grouped for example by diet or body size—as well as model the average community occupancy and its

variation with covariates (e.g. Rich et al., 2016; Cavada et al., 2019).

If camera traps are set un-baited and in a randomized, stratified, or gridded layout, the rate at which they detect species should be directly proportional to that species' local abundance. While there are several new approaches to convert these rates to real estimates of density (e.g. Howe et al., 2017; Moeller et al., 2018), these emerging methods are complex and often rely on stringent assumptions or collecting extra data such as the distance of animals from the camera, the camera field of view, or the animals' speed of movements. However, even without these additional data, several studies have shown that raw detection rate can serve as an index of abundance within a species to be compared across different areas (Rovero & Marshall, 2009; Parsons et al., 2017).

The gold standard for counting animals is capture-recapture models based on individuals that are individually recognizable. For camera traps, this works best for species with natural marks, typically felids. The seminal paper on tiger (*Panthera tigris*) density estimation in India by Karanth and Nichols (1998) triggered a wealth of studies around the world, contributing to future assessments on the conservation status of iconic and often wide-ranging and threatened carnivores with unique markings, and other species such as orangutan (*Pongo pygmaeus*; Spehar et al., 2015). We note that this approach generally requires robust sampling designs relative to those used in presence/absence surveys, as a sufficient number of individuals photographed at different sites is needed for reliable inference on density (Wearn & Glover-Kapfer, 2017). Another advantage of identifying individual animals is that, in addition to density, other demographic parameters such as recruitment and survival can be estimated, as shown for snow leopards (*Panthera uncia*) in Mongolia (Sharma et al., 2014) and tigers in China (Wang et al., 2018) and India (Sadhu et al., 2017).

Unfortunately, none of these three popular population metrics is perfect, and there is considerable interest in developing new approaches for using camera traps to estimate the abundance of wildlife populations. While occupancy modelling has the theoretical ability to account for differences in detectability among species, camera traps clearly violate the assumption of closure required for this inference, in that individual animals are moving in and out of the sample area (view of camera) within a continuous habitat, unlike classic cases where occupancy relates to occurrence across discrete habitat patches (Efford & Dawson, 2012; MacKenzie et al., 2017). This raises questions about how the temporal and spatial grain of sampling and the movement/home range patterns affect the underlying occupancy–abundance relationship (Steenweg et al., 2018; Rogan et al., 2019). Furthermore, the variance in occupancy is fixed between 0–1, and the metric often fails for very low values (e.g. rare species) and very high values (e.g. very common species that saturate sites) (Kays et al., 2017). Using detection rate as a measure of relative abundance requires standardization of how cameras are set, and has a number of assumptions and cautions that need to be considered since it can be affected by animal abundance and movement rates (Sollmann et al., 2013). Furthermore, unless extra measures are taken to account for differences in detectability between species (Rowcliffe et al., 2011; Hofmeester et al., 2019), these metrics cannot be compared among species. While capture-recapture density estimates are the most robust to detectability differences, and most comparable across sites and species, they are also the strictest in terms of required sample size and study design (see earlier), and most limited in which species they can be applied. Although some exceptional tropical sites could have multiple species with unique markings (e.g. 7 in Botswana; Rich et al., 2019), for most mammal species different individuals are not identifiable from camera trap photos. With the increasing use of camera traps, there is growing interest in deriving improved metrics through various approaches, including quantifying the area sampled by the camera trap (Howe et al., 2017) and by accounting for animal movement (Rowcliffe et al., 2016). This type of innovative thinking and statistical development is part of the maturation of camera trapping analyses into analytical methods, and can only work with detailed and standardized reporting of procedures

adopted by individual studies (Hofmeester et al., 2019).

In the next section, we review studies that use these various wildlife metrics obtained from camera trapping to inform conservation or assess conservation interventions.

5.3.2 Camera traps for conservation—spatial and temporal comparisons

Camera trap data are particularly well suited for making spatial and temporal comparisons of animal populations, which are fundamental to directing conservation action: where and when are species doing well, where are they declining, and what factors are different between these situations. The wildlife population metrics we have described can be compared across areas to determine habitat condition/quality for a species, and over time to monitor trends and assess the impact of conservation action. Some of these studies are general in scope, quantifying the niche dimensions for a species over large areas, while other studies target specific conservation threats. Importantly, however, we note that the conservation relevance of most studies we review here is inherently only indirect, and often intangible, as studies do not necessarily trigger conservation action or change in conservation practice following impact assessment. Also, any such action is often not recorded in the scientific literature. To try and address this issue, therefore, we highlight in the case studies (Section 5.4) some examples of how research triggered conservation outcomes.

5.3.2.1 Habitat preference, niche models, range models

Various statistical approaches use spatial comparisons of animal population metrics (i.e. occupancy, abundance, or density) to discover which suite of habitat and/or anthropogenic factors best explain the distribution or abundance of species. Depending on the scale or exact models, these might be referred to as species distribution models (SDM), environmental niche models (ENM), occupancy modelling, or simply habitat models (Guisan & Thuiller 2005). For our purposes, they are all similar in that they identify what factors are likely to affect a species' conservation status.

For example, habitat modelling for the endangered Arabian tahr (*Arabitragus jayakari*), a rare and little known mountain ungulate from the Middle East, estimated that occupancy was highest within protected areas and in rugged terrain, was negatively correlated with elevation and rainfall, and decreased with proximity to villages and their domestic goats (Ross et al., 2019). These results are relevant to the design of new protected areas and reintroduction programmes, and helped revise local livestock grazing rules. These models can also be applied at larger scales. For example, in a country wide analysis, researchers used habitat models for nine wildlife species to assess the function of Panama as an intercontinental land bridge. They found that most species were subject to at least one gap in habitat larger than their known dispersal distance and highlighted which protected areas are of highest priority for conservation in the region (Meyer et al., 2020).

Conservation biologists have recognized the limitations of any one data set for addressing large-scale questions, and began to work together to combine efforts and increase the scale of their models. The Borneo Carnivore Database is an outstanding example of this, collecting data on all 20 species of carnivores from the island and using them to build models predicting each species distribution based on presence-only modelling (Kramer-Schadt et al., 2016). Similarly, a recent study pooled data on all pangolin species from 103 surveys conducted in 22 countries across their range in Asia and Africa (Khwaja et al., 2019). Another consortium collected camera trap data from 19 sites in the range of the endangered Asian tapir (*Tapirus indicus*), finding them to still occupy most protected areas, and showing that tapir occurrence was negatively correlated with measures of human disturbance (Linkie et al., 2013). A similar effort for Baird's tapir (*Tapirus bairdii*) in Central America used an integrated modelling framework followed by a hotspot analysis to identify the most critical core habitat for the species (Schank et al., 2017). We expect these large-scale collaborations will increase in the future, highlighting the importance of maintaining camera trap

databases (Ahumada et al., 2019). New statistical tools enabling the integration of multiple data types into habitat models (Pacifici et al., 2017) also hold the promise of enabling a new generation of large-scale modelling by combining camera trap data with other data types (e.g. animal tracking, citizen science, museum records; Kays et al., 2020b) and providing a big improvement in our understanding of where species live and what affects their distribution.

With the recent extension of occupancy models applied to multiple species within a community, numerous studies have been able to apply habitat modelling to entire communities rather than just single species. For example, Rich et al., (2016) found that the protection of an area best predicted average occupancy of large-bodied species and herbivores in a mammalian community in Botswana, while medium-sized species increased in non-protected areas. Cavada et al. (2019) studied a community of nearly 50 species across a heterogeneous landscape in Tanzania and found that species richness did not vary among areas with different habitats, but overall community occupancy consistently decreased with proximity to human settlements.

5.3.2.2 Threat assessments

Spatial comparisons of animal populations are also useful to assess the vulnerability of wildlife to anthropogenic threats. By measuring aspects of local animal communities experiencing different levels of disturbance, camera traps can help researchers quantify the impact of poaching, habitat fragmentation, land-use changes, road construction, and other human activities. In contrast to general SDM studies, these threat assessments typically have an a priori study design that distributes camera traps among sites with varying levels of disturbance to improve statistical inference.

Ghoddousi et al. (2019), for example, related decline in population abundance of ungulate species in Iran to species-specific hunting intensity derived from hunters' interviews and trophy seizure records. Hunting was also identified as the number one threat—often acting in synergy or following habitat loss and fragmentation—in explaining the decline of large carnivores such as the jaguar (*Panthera onca*) in the Atlantic forests of South America (Paviolo et al., 2016) and the Indochinese leopard (*Panthera pardus delacouri*) across South-east Asia (Rostro-García et al., 2016). Hunting has also been correlated with declining wildlife in Latin America, with sites close to people with more signs of hunting having consistently lower detection rates of several wildlife species in studies in an Atlantic forest in Brazil (Lessa et al., 2017) and in a rainforest in Chiapas, Mexico (Porras et al., 2016). However, where managed, hunting might have a lower impact on communities. For example, Kays et al., (2017) found that managed hunting in the United States slightly reduced the abundance of harvested species (white-tailed deer *Odocoileus virginianus*, raccoons *Procyon lotor*, eastern grey squirrels *Sciurus carolinensis*), but had no impact on non-game wildlife species.

Unfortunately, in much of the world, hunting is not well regulated, and an original application of camera trapping is related to detecting hunters. For example, a year-long camera trap sampling in a protected area in Brazil's Atlantic forest revealed that the occurrence of poachers was highest in areas with higher accessibility, while their detectability was higher near waterways, forest edges, settlements, and in periods with higher lunar light intensity (Ferreguetti et al., 2018). Bisi et al. (2019) coupled camera trapping targeting wildlife with data from ranger patrols in Myanmar to model distribution of both wildlife and humans and hence identify areas where poaching activities are most intense. An inherent issue with these applications, however, is the difficulty of adequately detect both wildlife and humans on cameras, as the optimal, evenly distributed sampling and site selection chosen for a robust wildlife study may not be adequate to detecting poachers that likely use a limited set of access routes and move along established paths. On the other hand, the common setting of camera traps at approx. 50 cm from the ground along wildlife trails that may also be used by poachers will expose cameras to theft (see Section 5.5).

Threat assessment has often been addressed by comparing wildlife metrics across land-use types. Li et al. (2018) sampled across widely different habitats in southwest China, from dry hot valleys to subtropical montane and alpine areas, allowing them to map species richness across the region,

and predict occupancy of several mammal species based on habitat type. Using a multitaxa approaches also allows researchers to consider what functional traits affect how species respond to anthropogenic disturbance. For example, Rovero et al. (2019) used data from 16 tropical forests of the Tropical Ecology, Assessment and Monitoring (TEAM) Network— nearly 1000 camera trap sites—to study variation in community structure with anthropogenic disturbance and found that insectivores were the most vulnerable guild across all sites.

A number of studies have focused on landscapes with a mixture of natural, disturbed, and restored areas, revealing declining animal populations in some areas, but surprising resilience of wildlife in other contexts. Using a large array of cameras across a working landscape in Canada, Stewart et al. (2019) found that local natural features promote biodiversity while distant anthropogenic features suppress it, and therefore that protected areas, alone, were not enough to protect species. Shamoon et al. (2018) found wildlife could survive in an agricultural region of Israel, but changed their temporal activity due to perceived threat from humans during daytime. This, combined with elevated nocturnal predation risk, excluded prey species from large areas of an agricultural region designated as an ecological corridor. Similarly, chimpanzees in Sierra Leone did not avoid areas with humans and settlements, although they adjusted their diel activity pattern to avoid encountering people (Garriga et al., 2019). In the Chernobyl Exclusion Zone (CEZ), camera trapping found that introduced Przewalski's wild horses (*Equus ferus przewalski*) routinely use abandoned structures in the CEZ with visitation patterns tending to be nocturnal in winter and crepuscular in summer (Schlichting et al., 2019).

Wildlife sometimes surprises us by adapting to changes in the environment, and camera traps are useful for showing examples of animals persisting in highly developed areas where they are not expected. For example, the highest density of servals (*Leptailurus serval*) ever recorded was in an industrial area in South Africa, presumably due to a high abundance of prey and the absence of persecution and/or competitor species (Loock et al., 2018). Similarly, bobcats (*Lynx rufus*) in Texas reach their highest density in developed areas around Dallas (Young et al., 2019). To test for this phenomenon's generality, Parsons et al. (2018) worked with citizen scientists to set camera traps along the gradient of development, from urban-suburban-exurban-rural-wild, in Raleigh, North Carolina, and Washington, DC. They expected to find a steady decline in animal abundance and diversity with increasing development. Instead, they found few differences across the gradient (Raleigh) or increased populations and diversity in suburbia (DC), with declines seen only in the most developed parts of the gradient. Given that human–wildlife conflicts are increasingly triggered by the expansion of urban and developed areas, boosting this work is critical to understand if and how species can adapt to such development, and what aspects of urbanization are conducive to coexisting with wildlife species across a range of conditions and countries. The emergence of new research networks, such as the Urban Wildlife Information Network,[7] which standardizes protocols and shares data across cities, should provide a rich set of comparisons to start to address these questions (Magle et al., 2019).

Invasive species are increasingly recognized as an important threat to native wildlife, and camera trapping has proven useful in a variety of assessments, including for eradication programmes. In Australia, for example, predation by introduced feral cats (*Felis catus*) and red fox (*Vulpes vulpes*) is regarded as the greatest threat currently facing native mammals. Ruykys and Carter (2019) used camera trapping successfully in combination with other detection methods to locate these two predators and inform eradication efforts within a fenced wildlife sanctuary, while Comer et al. (2018) used camera traps and occupancy models to assess the efficacy of a feral cat eradication programme. Anton et al. (2018) found camera traps most effective at detecting multiple invasive species, including relatively small ones such as rats (*Rattus* spp.) and mice (*Mus musculus*) in New Zealand. Other studies have used co-occurrence modelling as a robust tool to determine how invasive species impact native ones. In Madagascar, for example, (Murphy et al.,

2019) found that the presence of dogs (*Canis familiaris*), cats, and small Indian civet (*Viverricula indica*) had a predominantly negative effect on native bird and small mammal occupancy and/or detection. Similarly, occupancies of four of the eight carnivores known from a study area in Ecuador were best predicted by occupancy of domestic dogs rather than measures of habitat loss and fragmentation (Zapata-Ríos & Branch, 2018).

Camera traps have been used to study wildlife behaviours that trigger human–wildlife conflicts, with livestock depredation by large carnivores being a primary one. In a large cattle ranch in the Pantanal, the declining prey base due to deforestation and ecosystem homogenization determined frequent livestock depredation by jaguars and pumas (*Puma concolor*) (de Souza et al., 2018). Camera trapping data coupled with ranch managers' recordings of depredation events allowed these authors to determine wildlife richness and distribution in relation to ranching activities. This in turn provided evidence to design more viable landscape management options able to maintain the full diversity of wildlife communities and enhance connectivity among natural areas. Smit et al. (2019) provided a novel use of camera traps to study the crop foraging behaviour of African elephants (*Loxodonta africana*) in farmland adjacent to the Udzungwa Mountains in Tanzania and determine the age/sex composition of crop-raiding elephants. They set camera traps along elephant pathways and positioned them strategically on trees at a height of 3 m and oriented downwards to capture the head, pinnae, and tusks of passing elephants allowing them for manual individual identification. Similarly, in Botswana, camera traps allowed Pozo et al. (2019) to determine how bricks made of dry chilli, elephant dung, and water effectively deter elephants' crop-raiding.

Camera trapping has been used extensively to determine how habitat conversion and resource extraction impact wildlife and help us move towards finding sustainable solutions. The typical approach is to contrast results from areas with different management regimes. Thus, to determine how logging impacts wildlife, researchers have sampled unlogged and logged forests. For example, the density of marbled cat (*Pardofelis marmorata*)

---

[7] https://urbanwildlifeinfo.org/

in Indonesia, as estimated from capture-recapture analyses of camera trapping data, dropped to 50% from primary to selectively logged areas, and no animals were found in plantations (Hearn et al., 2016). In Malaysia, Jamhuri et al. (2018) deployed 120 camera trap sites within selectively logged and unlogged forests. They found that unlogged forest had greater richness of mammals (16 versus 10 species), a result associated with greater abundance of large trees, lianas, and canopy cover in the unlogged forest. Also in Malaysia, Wearn et al. (2017) targeted small and large mammals using live traps and camera traps, and sampled a landscape in Borneo that included old-growth forest, logged forest, and oil palm plantations. For 57 species, authors found that estimated local population abundance was 28% higher in logged forest, with local animal biomass that increased even more (by 113% in herbivores), due to the abundant forage available once canopies are opened up. Notably, small mammals increased in their local abundance proportionately much more than large mammals (169% compared to 13%). In plantations, in contrast, mammalian abundance declined substantially (by 47%) compared to forest areas. The authors conclude that, in the absence of hunting, even the most intensively logged forests can conserve the abundance and functionality of mammals, and suggest that the protection of high carbon stock logged forest could therefore yield substantial conservation benefits. The validity of this recommendation is supported by a large camera trap study in Guatemala and Peru showing that responsibly managed logging concessions for long-term production—through third-party certification, low-volume timber extraction, set-aside high-value forest areas, prohibit hunting, etc.—maintain important populations of large and medium-sized mammals, including large herbivores and large carnivores (Tobler et al., 2018). Researchers set camera traps in systematic grids designed for jaguar's density estimation and used data on 27 'by-catch' species to study community metrics and species-specific responses to logging. An original study conducted in a Malaysian production forest used camera traps set to target small mammals (rodents and insectivores) belonging to 12 genera (Yamada et al., 2016). The authors found that maintaining logging residues (i.e. natural fallen trees in logged areas, woody debris, and left over upper part of logged trees), instead of removing them, promotes diversity, richness, and detection rates of small mammals.

The rapid spread and intensification of oil palm plantations, especially in South America and Southeast Asia, has become a pervasive threat to wildlife. In Colombia, occupancy modelling of 12 mammals showed variable patterns in their use of oil palm farms (Pardo et al., 2019). Generalist mesocarnivores, white-tailed deer, and giant anteater were more likely to use oil palm plantations, while the remaining species, including ocelot and lesser anteater, showed preferences for forest. Distance to the nearest forest had mixed effects on species habitat use, while understory vegetation facilitated the presence of species with oil palm plantations. At the community level, Pardo et al. (2018) found a significant decrease in total mammalian species richness in response to increasing oil palm cover, with subtle changes until oil palm cover reached 45–75%, beyond which mammalian species composition drastically changed. These studies suggest that maintaining understory vegetation and protecting or restoring forest patches and corridors within plantations is critical to maintain wildlife. These findings, however, were not confirmed by the study in Malaysia (Wearn et al., 2017), which found a sharp decline in mammalian abundance in plantations. The authors observed that increases in forest cover from 0 to 30% within plantations, the likely range which could realistically be manipulated in oil palm landscapes, resulted in very limited wildlife increases. They suggest that a land-sparing approach might better serve mammal conservation in the region, in which companies are encouraged to invest in the off-site conservation of large, contiguous forest areas rather than retaining small forest patches. Low tolerance to extensive plantations was displayed by mesocarnivores in Sumatra in a study by Jennings et al. (2015), reporting that only 3 of potentially 14 species of this guild were found to use plantations. Similarly, in Borneo, just 22 (35%) of 63 forest mammals were found to ever use plantations, and just 19 (31%) were recorded >200 m from the forest edge (Wearn et al., 2017).

Several studies have looked into the effects of oil extraction projects on wildlife. For example, in

Uganda, Fuda et al. (2018) surprisingly found that herbivore species richness was greater at sites that were within a disturbance matrix but located >1 km from drilling sites, roads, and other infrastructures, and within 500 m of restored drill pad sites, compared to more remote sites, whereas species richness did not vary across areas for other guilds. Occupancy models fitted for 15 relatively common species indicated no difference in occupancy probability across the three areas, but giraffe occupancy was higher at sites with more restored drill pads. The authors concluded that most species did not avoid areas affected by oil activities in this study, and new vegetation growth may attract some herbivores to restored oil sites; the authors could not exclude that historical or current hunting may also explain some of the results. Kolowski and Alonso (2010) assessed whether seismic operations conducted to search oil deposits within an extractive concession in the Peruvian Amazon impacted ocelot abundance and activity pattern, and found no differences in these metrics before and during operations. However, 3 years of camera trapping in Canadian oil sands revealed some species benefitted from the development, while others declined (Fisher & Burton, 2018). Statistical models showed the difference was how they responded to the linear features associated with prospecting, and the conversion of mature forest to early seral vegetation.

Livestock grazing represents another challenge to wildlife if it causes too much competition for herbivores, or too much conflict for large predators. However, there is also a chance for sustainable grazing if stocking levels are limited and herds are protected from predators. Most sites find a decline in wildlife with increasing stocking rates, including studies in Africa (Kinnaird & O'Brien, 2012), South America (Puechagut et al., 2018), and Asia (Sharma et al., 2015). However, many see the potential for the coexistence of smaller-scale operations and wildlife (Drouilly & O'Riain, 2019), especially if appropriate incentives can be introduced to improve access to ecotourism benefits, create forging agreements to maintain wildlife habitat and corridors, restore degraded rangelands, expand opportunities for grazing leases, and allow direct benefits to landowners through wildlife harvesting (Kinnaird & O'Brien, 2012).

Camera traps have also been used to assess how the ever-expanding amount of roads across the planet represents an increasingly alarming threat to wildlife. Road building triggers a cascade of processes that impact wildlife, from habitat loss and road kills to connectivity loss, increased human accessibility to people and hence poaching, construction of additional secondary roads, etc. In the Ecuadorian Amazon, Espinosa et al. (2018) used grids of camera traps in areas with contrasting accessibility by hunters, measured as the distance of camera traps to three sources of access (roads, rivers, and settlements). They found that higher accessibility to hunters was associated with lower occupancy and biomass of game species. This in turn impacted the density of these species' main predator, the jaguar, whose density estimated through capture-recapture was up to 18 times higher in the most remote site compared with the most accessible site.

5.3.2.3 Wildlife monitoring

Standardized monitoring of wildlife populations allows conservationists to spot problems quickly, and if done across a variety of sites, can be used to address several important conservation questions. A prime example is the TEAM project, a network of 17 sites of protected areas across three continents that have collected standardized camera trapping data according to a robust sampling design since the mid-to-late 2000s (TEAM Network, 2011; Rovero & Ahumada, 2017). This consists of 60 camera trap sites sampled yearly for 30 days and arranged into a regular grid with cell size of 2 km^2. Several sites in the network have already collected over 10 years of data' allowing researchers to study population trends. The geographic breadth of sites samples, the standardization in data collection, and the open-access data policy represent the main strengths of TEAM Network, providing for a data set of unprecedented value and potential. For example, Beaudrot et al. (2016) determined occupancy trends for 511 populations sampled for multiple years and found that the occupancy of 39% of these either had no change or increased, while 22% decreased and 39% did not have sufficient data for analyses. Authors also derived the Wildlife Picture

Index (WPI; O'Brien et al., 2010), an aggregate community metric signalling changes in species richness and occupancy. As an official CBD Aichi Targets indicator, specifically formulated for camera trapping data, WPI represents an excellent example of how camera trapping can inform conservation management from global to regional and national levels. A growing number of studies have used TEAM data for addressing diverse ecological and conservation questions in a monitoring framework, for example, the effects of invasive predators, human presence, and habitat quality on changes in the carnivore community in Ranomafana National Park, Madagascar (Farris et al., 2017).

5.3.2.4 Evaluations of conservation interventions

Aside from general monitoring, camera trapping has been used to evaluate conservation actions to see if they helped wildlife as intended. For example, in the Udzungwa Mountains of Tanzania, Oberosler et al. (2020a) assessed how protection affected mammal populations and communities by contrasting a forest in a national park versus a poorly protected reserve where hunting and habitat degradation are major threats. Using the TEAM Network protocol in both forests, they found declines in species richness, community structure, and species-specific occupancy in the unprotected forest. Moreover, while in the unprotected forest some mammal populations underwent declines in the last decades, in the protected forest another study that used 8 years of TEAM data revealed a general stability of temporal trends in species occupancy (Oberosler et al., 2020b), which were associated with a ban on firewood collection by local communities and a decrease in hunting pressure. A similar assessment aimed at evaluating the results of protection in Iguaçu National Park, Brazil, found that the occupancy of 11 species declined, close to the edge of the park, or close to tourism infrastructure, by 13–23% relative to maximum values in areas not affected by such threats (da Silva et al., 2018).

Chen et al. (2019) found that the local abundance of two hunted species studied across six protected areas in China was more related to local people's perception of law enforcement than to objective levels of enforcement, showing the importance of community outreach as a fundamental component of protected area management. In Bhutan, an impressive camera trap effort generating 10 million photos assessed how the richness and diversity of mammals varied across protected areas, biological corridors, and intervening non-protected areas (Dorji et al., 2019). While protected areas contained abundant wildlife, the authors also stressed the value of non-protected areas that may hold globally threatened species not found in reserves, similarly to findings from India (Velho et al., 2016) and Colombia (Boron et al., 2016).

Strategically placed camera traps are a primary method to assess use of wildlife corridors. For example, Green et al. (2018) studied how elephants use the 14 km-long Mount Kenya Elephant Corridor, linking two protected areas. The authors found that elephants use the corridor both for movement and as a habitat extension, depending on vegetation cover and human disturbance. More frequently, studies have assessed how natural strips of vegetation, such as riparian forests, are used by wildlife for dispersal or movements across suitable habitat patches, providing critical recommendations for corridor design. This is the case of linear remnants of natural riparian forests located within a large extent of tree plantations and used in Sumatra by a range of mammals (Yaap et al., 2016). This study evaluated the minimum corridor dimensions needed to ensure local movements across habitat patches, recommending 200 m width and 4 km length. In Brazil, Paolino et al. (2018) found that riverine 'Areas of Permanent Protection' prescribed by the national legislation are critical for the persistence of ocelot (*Leopardus pardalis*) within sugarcane and plantations. Understanding which types of corridors are used by which species is key for landscape planning. Wang et al. (2018) found that not all potential giant panda (*Ailuropoda melanoleuca*) corridors in central China are effective for other wildlife species, and recommend that conservation planners should prioritize corridor development based on a multispecies perspective without loss of connectivity for the priority species. On the other hand, the cameras set by LaPoint et al. (2013) found that corridors used by fishers (*Pekania pennant*) were also used by other species to navigate suburban natural areas in New York, but that these were not predicted well by traditional corridor models.

Passages that allow animals to crossroads are often the pinch-point in landscape connectivity plans, and targeted camera trapping has been critical in understanding how wildlife use these. There has been extensive work studying which passage designs are used by which species (Clevenger et al., 2011). Drainage culverts are an easy solution, typically already included in highway design to allow water to flow, but camera traps have shown that many larger species will not use these small underpasses. This work has led to the creation of wildlife-specific solutions, including expensive bridges forcing herds of migratory ungulates to pass over, rather than under, the road (Clevenger et al., 2011). However, there are also less expensive solutions for smaller species, including tall poles to help gliding animals (Goldingay et al., 2019). This type of work has also been extended to other types of movement barriers. For example, Balbuena et al. (2019) found that five species of arboreal mammals use both natural and semi-artificial canopy bridges to cross a gas pipeline right-of-way in the Peruvian Amazon. Monitoring road crossings can also double as population assessments, as seen in Croatia, where Šver et al. (2016) set camera traps along green bridges and viaducts to monitor grey wolves (*Canis lupus*) for 5 years. Results show that wolves regularly use these bridges, but there is also an apparent decrease in the number of detection events over time which may signal a population decline in the region.

5.4 Case studies

To make a difference in conservation, we have to not only document the problems but also engage local communities, provide information to other stakeholders, and take action on the ground. Here we provide examples of how camera traps can be useful in numerous dimensions of conservation, providing brief case studies around each of the five principles proposed by Robinson et al. (2018) that ensure that threatened species monitoring actually leads to threatened species conservation. These collectively show how camera trapping can effectively and directly contribute to all steps that link ecological monitoring to conservation.

5.4.1 Integrate monitoring with management

Monitoring becomes ineffective in conservation if it is not linked to management; hence, threat assessment should inform management actions whose outcomes should, in turn, be evaluated through continuous monitoring. For example, in Bwindi Impenetrable Forest, Uganda, camera trapping data showed a decline of the African golden cat (*Caracal aurata*). Upon reporting to the park authorities, managers noticed that the monitored locations were heavily used by the increasing number of tourists visiting the park for mountain gorilla watching, which subsequently redirected people to different trails (Ahumada et al., 2016). Similarly, in Costa Rica, results of alarmingly declining species (particularly the lowland paca *Cuniculus paca*, agouti *Dasyprocta punctata* and nine-banded armadillo *Dasypus novemcinctus*) found by Ahumada et al. (2013) were shared with park authorities and measures were taken to increase patrolling and control poaching. Following 10 years of monitoring, data show that two of the species that were declining the fastest (agouti and lowland paca) have at least stabilized since threat mitigation was enforced (Ahumada et al., 2016). In the Udzungwa Mountains of Tanzania, the camera trapping research on crop-raiding elephants described earlier (Smit et al., 2019) has triggered a community-based programme to mitigate human–elephant conflicts,[8] as well as guided the design of an elephant corridor, which is under implementation by the government. In the same area, the assessment of declining mammal communities in unprotected forests has motivated an anti-poaching programme in a large and recently gazetted Nature Reserve.

5.4.2 Design fit-for-purpose monitoring programmes

Threatened species monitoring programmes should be scientifically robust, which typically means that

---

[8] https://stzelephants.org/projects/human-elephants-co-existence/

they need to be targeted, question-driven, and yield statistically robust sample sizes (Kays et al., 2020a). As we have described in Section 5.3, camera traps are a powerful tool to study populations and communities, but the exact spacing of cameras and sampling duration might vary, depending on targets. For example, very rare and far-ranging such as tigers and snow leopards (e.g. Sharma et al., 2014, Wang et al., 2018) need widely spaced cameras run for long periods. Smaller-scale programmes run for shorter intervals are more realistic to maintain and can still be effective for a suite of species, as shown by the TEAM Network protocol (Beaudrot et al., 2019). The important thing is that the trade-offs of camera trapping study designs are now fairly well known (Kays et al., 2020a) so that managers can create a monitoring programme to fit local needs. Also, the success of these can be evaluated during the study; for example, a review of a vertebrate monitoring programme in Northern Australia that included camera trapping for mammals revealed its inability to detect changes in occupancy for some species, resulting in a redesign and improvement of the programme (Einoder et al., 2018).

5.4.3 Engage people and organizations

Monitoring programmes that engage stakeholders at the outset are more likely to have their results accepted and acted on (Robinson et al., 2018). Camera traps offer a good tool in this regard as citizens can be involved in the actual monitoring, and because the pictures are of broad interest. McShea et al. (2016) showed that volunteers could efficiently and accurately run camera traps, helping survey larger areas, and Schuttler et al. (2018) showed that even school student groups could produce useful data. These projects used the eMammal data management system, allowing volunteers to identify the animals in the pictures, but includes a quality control step where experts validate the identifications. Other projects with cameras set by professionals have still benefitted from engaging the public to help identify the photographs through the Zooniverse platform[9] (Swanson et al., 2015).

Engaging stakeholders through camera trapping is a relatively new idea, but few studies have quantified its effects. Forrester et al. (2017) used before-after surveys to find that volunteers who ran camera traps became advocates for conservation, being 84% more likely to discuss local mammals or local mammal conservation. Parsons et al. (2018) also reported changes in attitudes towards wildlife from participants in monitoring local nature preserves, including some who were so motivated by the experience they shared the results with self-published newsletters within their neighbourhood. Monitoring projects that work with schools to run camera traps can also have amplified impact on local communities. Prominent examples include those in India, receiving considerable attention from local press, and Mexico, where students directly met with local politicians to share their pictures (Schuttler et al., 2018). A few groups have started running camera traps with local communities, finding that sharing the pictures with local ranchers and landowners can improve their tolerance of potential conflict with wildlife, such as large cats and livestock (e.g. Yaguará Panamá[10]). The sharing of information with locals is critical, since otherwise the technology and surveillance can spark resentment, marginalize local indigenous communities, and cause a rift between conservation organizations and local communities (Shrestha & Lapeyre, 2018).

5.4.4 Ensure good data management

Camera trap monitoring can enable powerful comparisons of wildlife community composition and change across study sites or over time, but these data need to be managed and secured to meet these goals. Sharing photographs with the public and monitoring results with locals also has to be balanced against protecting the privacy of people photographed, securing data on sensitive species that might be poached, as well as preventing camera theft or damage. While small-scale monitoring programmes might be able to manage this with simple spreadsheets and folders of pictures, larger schemes need custom software. The TEAM Network is one prominent example of this, and their

---

[9] https://www.zooniverse.org/

[10] https://yaguarapanama.org/

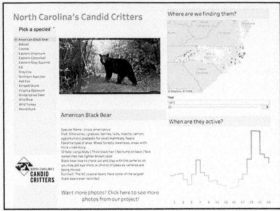

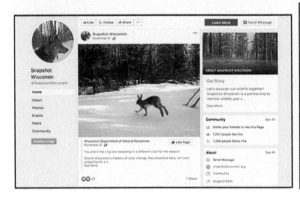

Figure 5.4 Monitoring programmes that use camera traps have the advantage of a large selection of charismatic animal photos to help communicate the value of monitoring through newsletters (top left), social media posts (bottom left), data interactives (top right), and videos (bottom right, source: Snapshot Wisconsin and North Carolina's Candid Critters).

database was used to archive >3.2 million photos obtained across nearly 1100 camera trap sites with a standardized protocol (Rovero & Ahumada, 2017). The same software architecture is now available as Wild.ID[11] for other monitoring programmes to adopt. eMammal is another cloud-based camera trap repository based at the Smithsonian Institution that hosts data from dozens of camera trap projects worldwide, managing data embargoes, protecting the location of endangered species, and facilitating basic data analysis and visualization (McShea et al., 2016). The sustainability of these software solutions is challenging due to the yearly costs of maintenance and software development (Kays et al., 2020a). These challenges led several conservation

groups to collaborate to create the Wildlife Insights platform to manage camera trap data and integrate artificial intelligence (AI) to improve efficiency and create improved automated analytics (Ahumada et al., 2019).

5.4.5 Communicate the value of monitoring

Many of the aspects that make camera traps an attractive tool to engage local stakeholders also make them a powerful tool to communicate the value of monitoring to local communities and beyond. Two large-scale, state-wide citizen science projects are good examples of this. Both the Snapshot Wisconsin[12] and North Carolina

---

[11] https://github.com/ConservationInternational/Wild.ID

[12] https://dnr.wi.gov/topic/research/projects/snapshot/

Candid Critters[13] projects have extensive communications programmes to share their projects' results through regular newsletters, social media posts, live webinars, data interactives, and feature videos (Figure 5.4).

5.5 Limitations/constraints of camera trapping

It is important to recognize some of the limitations of camera traps. First, they typically do not trigger on small species (<50 g) or poikilothermic species, although there have been some modifications in design to improve this (Glen et al., 2013; Hobbs & Brehme, 2017). Second, even larger species are not perfectly detected due to inefficient motion sensors or slow trigger times (e.g. Urbanek et al., 2019). Thus it is important to remember that camera traps capture just a sample of the animal community and miss some individuals and species. Third, some species groups can be difficult, or impossible, to accurately identify with typical camera trap images, for example, ungulates and small carnivores, depending on exactly which species overlap spatially at a given location (Gooliaff & Hodges, 2018). Finally, as with any technology in the field, many things can go wrong, including humidity damaging the electronics, elephants stomping on the camera, insects crawling into it, or thieves stealing it (Kays et al., 2011; Meek et al., 2019). Some of these risks can be minimized by using high-quality, tropicalized camera traps (humidity and insect damage) and hidden/camouflaged positioning of camera traps (wildlife damages and thefts). However, while camera damage and theft are common problems in camera trap studies, their impact is rarely documented and will likely vary across areas. Some studies have used camouflaged positioning of cameras, setting at much higher height from the floor (1.5–2 m) and targeting known sites of access by poachers, such as waterways (e.g. Hossain et al., 2016). Surprisingly, despite these precautions, the authors of this study in Bangladesh report that nearly 50% of the camera traps (20 out of 41 deployed) were lost to supposed thefts.

[13] https://www.nccandidcritters.org/

5.6 Social impact/privacy

Camera traps inevitably catch pictures of people in addition to animals, which potentially raises the issue of privacy infringement. Sandbrook et al. (2018) collated information from 253 researchers using camera traps in 65 countries: > 90% of them reported capturing images of people, in most cases unintentionally. Yet, images of people are widely used to inform conservation practice as they may provide evidence of potential anthropogenic threats to wildlife. The ethics and legality of photographing people without their permission varies across countries, cultures, and land ownership classes. It is imperative that camera trappers are sensitive to privacy concerns, aware of relevant local surveillance laws, careful about exactly where they put their cameras and how they manage any data where people can be identified. Camera trapping is a key method for conserving wildlife but permission from local authorities is needed to use the cameras, as is acceptance of the local community to avoid cultural conflicts. Therefore, a heightened respect for the privacy of the people we photograph is paramount, regardless of local laws.

To our best knowledge, only a few countries have national laws regulating the use of camera trapping, with particular regard to protecting privacy. For example, in the United States, the general rule is that taking photographs and video of things or people that are plainly visible in public spaces is allowed, unless specifically forbidden. Thus, regulations on camera trapping will be set by local public management agencies or private landowners. In Italy, a privacy policy has been recently proposed stating that the local authority should authorize use of camera trapping; however, this has not been enforced yet.

It is also important to point out that before initiating any kind of camera trapping project, researchers must verify what permissions may be required, which may often be granted by multiple authorities (e.g. national-level wildlife authorities and ministries, local authorities). Working in protected areas most often requires permission from park authorities, and permits are also generally needed to work on private lands, where owners might be particularly sensitive to the privacy of people using their land. Letting authorities and landowners know the locations of camera traps and the periods they are operative will also ensure that cameras are not damaged by scheduled management such as logging or burns.

We recommend the following best practices for dealing with human pictures and minimize chances to infringe privacy. When setting cameras in the field, we recommend: (1) Set them low (typically <0.5 m) and slightly incline them downwards to minimize chances of photographing people's faces. (2) Put labels on the camera itself with a brief description and contact phone number or website so people can know what the project is for. Interestingly, Clarin et al. (2014) found that a personal label reduced the overall number of negative interactions by 40–60%, compared with a neutral or threatening label type. (3) Some countries or protected areas might require additional signage—usually to be placed at the edges of an area where camera traps are operational—to alert people that the area is under camera surveillance.

As for use of people's pictures, we recommend: (1) Do not seek publicity on social media with funny or embarrassing pictures of people caught on camera trap. (2) Do not share pictures of people online if their faces are in view. (3) Use a face blurring tool to anonymize pictures saved in a database. (4) Pictures of people doing illegal activities can be shared with local authorities as relevant, while any other pictures of people should not be shared. For example, Wildlife Insights, an emerging camera trap data platform, does not share images of people publicly, although the images are available to project owners (Ahumada et al., 2019).

5.7 Future directions

Camera trapping science underwent a massive expansion in the last 15 years and has revolutionized the way we study wildlife, in turn offering unprecedented opportunities and insight for conservation. We have reviewed the technical aspects of camera traps that make them so efficient and widely used, the wildlife metrics obtained, and the various ways camera trap studies help conservation. Here, we conclude by reviewing what we consider the main possible technological improvements

of camera trapping data collection, management, and analysis in the future, and how they would aid conservation.

We expect that current sensor and camera technology will continue to improve, including: (1) faster trigger speeds, which will result in fewer missed detections of passing animals; notably, this was the top request from a survey of 258 camera trappers around the world (Wearn & Glover-Kapfer, 2019); (2) camera systems that also detect small mammals (i.e. <50 g), which will detect a much wider variety of species; (3) cameras with 360° detection zone, hence monitoring a wider area resulting into fewer missed detections; and (4) improved video recording relative to current standards (i.e. with faster triggers and improved routines for data management), which will result into better AI species identification, easier individual identification, and more chances to study animal behaviours. However, we predict the most impactful changes will come from the addition of other sensors to complement the visual image (Figure 5.5). The most critical improvement needed is to be able to make measurements automatically from images. For example, measuring the distance that the animal is away from the camera would allow a more precise estimate of survey area for each species (Rowcliffe et al., 2011), and density estimation using distance sampling approach for species that are not individually identifiable (Howe et al., 2017). Measurement of the paths that animals make in front of the camera could also be useful for parameterizing random encounter models (Rowcliffe et al., 2014), and in other research questions associated with animal movement (Rowcliffe et al., 2016). Finally, accurately measuring the sizes of animals could be useful for species and individual identification. Stereo cameras are the most likely solution for this challenge (e.g. Intel® RealSense™ Depth Cameras, see Figure 5.5C), but LIDAR (Laser Imaging Detection and Ranging) and polarized light sensors might also work. Thermal imagery sensors could provide new data additional variables (e.g. animal temperature) and improve species detection and identification.

The final class of hardware improvements likely to affect camera trapping in the future is the addition of onboard AI software models embedded

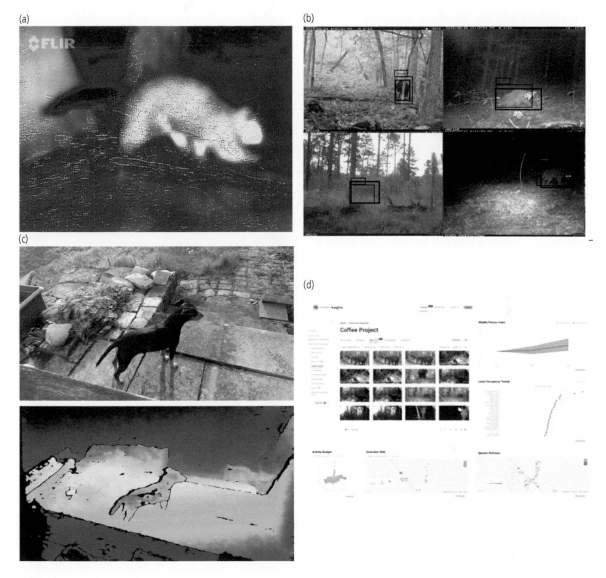

Figure 5.5 Four advances in the near horizon for camera trapping include: (a) thermal imagery (credit: R. Kays); (b) artificial intelligence identification of species (credit: Yousif, 2019); (c) automated distance measurement (coded as colour here; credit: H. Naik); (d) automated statistical reporting (source: Wildlife Insights).

into the cameras themselves (i.e. edge computing). This would require improved (faster and more efficient) processing power on the remote camera itself, and AI algorithms customized to the local animal species (He et al., 2016). This approach's main benefit is that species detection results could be transmitted wirelessly, allowing real-time monitoring. Although technology is available for wireless image

transfer, this is currently not widely used by camera trappers because of their associated high cost, high bandwidth requirements, and limited network availability in many remote areas. However, with phone coverage and data transfer improving through the global internet access underway, this will soon become easier and less costly. Having the remote camera identify the species, using AI, and sending just the resulting metadata, but not the images themselves, could prove energy and cost-efficient enough to enable real-time monitoring. This would be especially useful for detecting poachers and highly endangered species, and monitoring the performance of sensors over the long term. Prototypes of this kind of system have been developed but are not yet widely available.

Over the last century, advances in camera trapping science have been tightly tied to advances in camera technology: better flashes, film, and lenses. However, in today's digital era, improvements come from diverse fields, creating more useful images and other sensor data, and being used at large scales never imagined by scientists even a decade ago. This has created a big data challenge, leading to a new wave of improvements around automated processing and statistical interpretation (see also Wearn & Glover-Kapfer, 2019). These advancements will ultimately result in greater and more timely contributions of camera trapping to conservation, which is critical at a time of rapid changes and species declines. Meanwhile, and aside from its scientific applications, better quality camera trap images and video will continue to trigger people's imagination and inspire a passion for wildlife across broad audiences.

5.8 Acknowledgements

We are grateful to the book editors for inviting us to write this chapter, and to Ollie Wearn and an anonymous reviewer for constructive comments on the manuscript.

References

Aguhob, J. C., Shah, J. N., Elhassan, E. E. M., et al. (2018). First record of the endangered Arabian Tahr Arabitragus jayakari (Thomas, 1894) in the Hatta Mountain Conservation Area, Dubai, United Arab Emirates. *Journal of Threatened Taxa*, 10, 12561–12565.

Ahumada, J. A., Fegraus, E., Birch, T., et al. (2019). Wildlife insights: a platform to maximize the potential of camera trap and other passive sensor wildlife data for the planet. *Environmental Conservation*, 47(1), 1–6.

Ahumada, J. A., Hurtado, J., & Lizcano, D. (2013). Monitoring the status and trends of tropical forest terrestrial vertebrate communities from camera trap data: a tool for conservation. *PLoS One*, 8, e73707.

Ahumada, J. A., O'Brien, T. G., Mugerwa, B., & Hurtado, J. (2016). Camera trapping as a monitoring tool at national and global levels. In Rovero, F., & Zimmermann, F. (eds.). *Camera Trapping for Wildlife Research*. Pelagic Publishing, Exeter, UK (pp. 196–218).

Amin, R., Wacher, T., Bowkett, A. E., Ogwoka, M., & Agwanda, B. (2018). Africa's forgotten forests: the conservation value of Kenya's northern coastal forests for large mammals. *Journal of East African Natural History*, 107, 41–61.

Anton, V., Hartley, S., & Wittmer, H. U. (2018). Evaluation of remote cameras for monitoring multiple invasive mammals in New Zealand. *New Zealand Journal of Ecology*, 42, 74–79.

Azevedo, F. C., Lemos, F. G., Rocha, D. G., Costa, A. N., & Freitas-Júnior, M. C. (2016). Novo registro do Cachorro-Vinagre Speothos venaticus em uma. *Bioscience Journal*, 32, 7.

Balbuena, D., Alonso, A., Panta, M., Garcia, A., & Gregory, T. (2019). Mitigating tropical forest fragmentation with natural and semi-artificial canopy bridges. *Diversity*, 11, 66.

Beaudrot, L., Ahumada, J., O'Brien, T. G., & Jansen, P. A. (2019). Detecting tropical wildlife declines through camera-trap monitoring: an evaluation of the tropical ecology assessment and monitoring protocol. *Oryx*, 53, 126–129.

Beaudrot, L., Ahumada, J. A., O'Brien, T., et al. (2016). Standardized assessment of biodiversity trends in tropical forest protected areas: the end is not in sight. *PLoS Biology*, 14, e1002357.

Bisi, F., Gagliardi, A., Cremonesi, G., et al. (2019). Distribution of wildlife and illegal human activities in the Lampi Marine National Park (Myanmar). *Environmental Conservation*, 46, 163–170.

Boron, V., Tzanopoulos, J., Gallo, J., et al. (2016). Jaguar densities across human-dominated landscapes in Colombia: the contribution of unprotected areas to long term conservation. *PLoS ONE*, 11, e0153973.

Bowler, M. T., Tobler, M. W., Endress, B. A., Gilmore, M. P., & Anderson, M. J. (2017). Estimating mammalian species richness and occupancy in tropical forest canopies with arboreal camera traps. *Remote Sensing in Ecology and Conservation*, 3, 146–157.

Cavada, N., Worsøe Havmøller, R., Scharff, N., & Rovero, F. (2019). A landscape-scale assessment of tropical mammals reveals the effects of habitat and anthropogenic disturbance on community occupancy. *PLoS One*, 14, e0215682.

Chen, C., Quan, R., Cao, G., et al. (2019). Effects of law enforcement and community outreach on mammal diversity in a biodiversity hotspot. *Conservation Biology*, 33, 612–622.

Clarin, B. M., Bitzilekis, E., Siemers, B. M., & Goerlitz, H. R. (2014). Personal messages reduce vandalism and theft of unattended scientific equipment. *Methods in Ecology and Evolution*, 5, 125–131.

Clevenger, T., Anthony, P., & Huijser, M. P. (2011). *Wildlife Crossing Structure Handbook: Design and Evaluation in North America. No. FHWA-CFL-TD-11-003.* United States. Federal Highway Administration. Central Federal Lands Highway Division. Available at: https://nrm.dfg.ca.gov/FileHandler.ashx?DocumentID =134712&inline

Comer, S., Speldewinde, P., Tiller, C., et al. (2018). Evaluating the efficacy of a landscape scale feral cat control program using camera traps and occupancy models. *Scientific Reports*, 8, 5335.

Da Silva, M. X., Paviolo, A., Tambosi, L. R., & Pardini, R. (2018). Effectiveness of Protected Areas for biodiversity conservation: mammal occupancy patterns in the Iguaçu National Park, Brazil. *Journal for Nature Conservation*, 41, 51–62.

De Souza, J. C., Da Silva, R. M., Gonçalves, M. P. R., Jardim, R. J. D., & Markwith, S. H. (2018). Habitat use, ranching, and human-wildlife conflict within a fragmented landscape in the Pantanal, Brazil. *Biological Conservation*, 217, 349–357.

Dirzo, R., Young, H. S., Galetti, M., Ceballos, G., Isaac, N. J., & Collen, B. (2014). Defaunation in the anthropocene. *Science*, 345, 401–406.

Dorji, S., Rajaratnam, R., & Vernes, K. (2019). Mammal richness and diversity in a Himalayan hotspot: the role of protected areas in conserving Bhutan's mammals. *Biodiversity and Conservation*, 28, 3277–3297.

Drouilly, M. & O'Riain, M. J. (2019). Wildlife winners and losers of extensive small-livestock farming: a case study in the South African Karoo. *Biodiversity and Conservation*, 28, 1493–1511.

Efford, M. G. & Dawson, D. K. (2012). Occupancy in continuous habitat. *Ecosphere*, 3, 32.

Espinosa, S., Celis, G., & Branch, L. C. (2018). When roads appear jaguars decline: increased access to an Amazonian wilderness area reduces potential for jaguar conservation. *PLoS One*, 13, e0189740.

Einoder, L. D., Southwell, D. M., Gillespie, G. R., Fisher, A., Lahoz-Monfort, J. J., & Wintle, B. A. (2018). Optimising broad-scale monitoring for trend detection: review and re-design of a long-term program in northern Australia. In Legge, S., Lindenmayer, D. B. N., Robinson, M., Scheele, B. C., Southwell, D. M., & Wintle, B. A. (eds.). *Monitoring Threatened Species and Ecological Communities*. CSIRO Publishing, Melbourne, Australia (pp. 271–280).

Farris, Z. J., Gerber, B. D., Valenta, K., et al. (2017). Threats to a rainforest carnivore community: a multi-year assessment of occupancy and co-occurrence in Madagascar. *Biological Conservation*, 210, 116–124.

Ferreguetti, Á. C., Pereira-Ribeiro, J., Prevedello, J. A., Tomás, W. M., Rocha, C. F. D., & Bergallo, H. G. (2018).

One step ahead to predict potential poaching hotspots: modeling occupancy and detectability of poachers in a neotropical rainforest. *Biological Conservation*, 227, 133–140.

Fisher, J. T., & Burton, A. C. (2018). Wildlife winners and losers in an oil sands landscape. *Frontiers in Ecology and the Environment*, 16, 323–328.

Forrester, T. D., Baker, M., Costello, R., Kays, R., Parsons, A. W., & McShea, W. J. (2017). Creating advocates for mammal conservation through citizen science. *Biological Conservation*, 208, 98–105.

Fuda, R. K., Ryan, S. J., Cohen, J. B., Hartter, J., & Frair, J. L. (2018). Assessing the impacts of oil exploration and restoration on mammals in Murchison Falls Conservation Area, Uganda. *African Journal of Ecology*, 56, 804–817.

Garriga, R. M., Marco, I., Casas-Díaz, E., et al. (2019). Factors influencing wild chimpanzee (*Pan troglodytes verus*) relative abundance in an agriculture-swamp matrix outside protected areas. *PLoS One*, 14, e0215545.

Ghoddousi, A., Soofi, M., Hamidi, A. K., et al. (2019). The decline of ungulate populations in Iranian protected areas calls for urgent action against poaching. *Oryx*, 53, 151–158.

Glen, A. S., Cockburn, S., Nichols, M., Ekanayake, J., & Warburton, B. (2013). Optimising camera traps for monitoring small mammals. *PloS One*, 8, e67940.

Goldingay, R. L., Taylor, B. D., & Parkyn, J. L. (2019). Use of tall wooden poles by four species of gliding mammal provides further proof of concept for habitat restoration. *Australian Mammalogy*, 41, 255–261.

Gompper, M. E., Kays, R. W., Ray, J. C., Lapoint, S. D., Bogan, D. A., & Cryan, J. R. (2006). A comparison of noninvasive techniques to survey carnivore communities in northeastern North America. *Wildlife Society Bulletin*, 34, 1142–1151.

Gooliaff, T., & Hodges, K. E. (2018). Measuring agreement among experts in classifying camera images of similar species. *Ecology and Evolution*, 8(22), 11009–11021.

Green, S. E., Davidson, Z., Kaaria, T., & Doncaster, C. P. (2018). Do wildlife corridors link or extend habitat? Insights from elephant use of a Kenyan wildlife corridor. *African Journal of Ecology*, 56, 860–871.

Guisan, A., & Thuiller, W. (2005). Predicting species distribution: offering more than simple habitat models. *Ecology Letters*, 8, 993–1009.

Hand, Christine E., Elizabeth Znidersic, and Amy K. Tegeler. "First documentation of Eastern Black Rails (Laterallus jamaicensis jamaicensis) breeding in South Carolina, USA in more than a century." *Waterbirds* 42.2 (2019): 237–241.

Havmøller, R. W., Tenan, S., Scharff, N., & Rovero, F. (2019). Reserve size and anthropogenic disturbance affect the density of an African leopard (*Panthera pardus*) meta-population. *PLoS One*, 14, e0209541.

He, Z., Kays, R., Zhang, Z., et al. (2016). Visual informatics tools for supporting large-scale collaborative wildlife monitoring with citizen scientists. *IEE Circuits and Systems Magazine*, 16, 73–86.

Hearn, A. J., Ross, J., Bernard, H., Bakar, S. A., Hunter, L. T. B., & Macdonald, D. W. (2016). The first estimates of marbled cat pardofelis marmorata population density from Bornean primary and selectively logged forest. *PLoS One*, 11, e0151046.

Hobbs, M. T., & Brehme, C. S. (2017). An improved camera trap for amphibians, reptiles, small mammals, and large invertebrates. *PloS One*, 12, e0185026.

Hofmeester, T. R., Cromsigt, J. P. G. M., Odden, J., Andrén, H., Kindberg, J., & Linnell, J. D. C. (2019). Framing pictures: a conceptual framework to identify and correct for biases in detection probability of camera traps enabling multi-species comparison. *Ecology and Evolution*, 9, 2320–2336.

Hossain, A. N. M., Barlow, A., Barlow, C. G., Lynam, A. J., Chakma, S., & Savini, T. (2016). Assessing the efficacy of camera trapping as a tool for increasing detection rates of wildlife crime in tropical protected areas. *Biological Conservation*, 201, 314–319.

Howe, E. J., Buckland, S. T., Després-Einspenner, M., & Kühl, H. S. (2017). Distance sampling with camera traps. *Methods in Ecology and Evolution*, 8, 1558–1565.

Howe, E. J., Buckland, S. T., Després-Einspenner, M. L., & Kühl, H. S. (2017). Distance sampling with camera traps. *Methods in Ecology and Evolution*, 8, 1558–1565.

Ivanov, G., Karamanlidis, A. A., Stojanov, A., Melovski, D., & Avukatov, V. (2016). The re-establishment of the golden jackal (*Canis aureus*) in FYR Macedonia: implications for conservation. *Mammalian Biology*, 81, 326–330.

Jamhuri, J., Samantha, L. D., Tee, S. L., et al. (2018). Selective logging causes the decline of large-sized mammals including those in unlogged patches surrounded by logged and agricultural areas. *Biological Conservation*, 227, 40–47.

Jennings, A. P., Naim, M., Advento, A. D., et al. (2015). Diversity and occupancy of small carnivores within oil palm plantations in central Sumatra, Indonesia. *Mammal Research*, 60, 181–188.

Karanth, K. U., & Nichols, J. D. (1998). Estimation of tiger densities in India using photographic captures and recaptures. *Ecology*, 79, 2852–2862.

Kays, R. (2016). *Candid Creatures: How Camera Traps Reveal the Mysteries of Nature*. Johns Hopkins University Press, Baltimore, MD.

Kays, R., Arbogast, B. S., Baker-Whatton, M., et al. (2020a). An empirical evaluation of camera trap study design: how many, how long, and when? *Methods in Ecology and Evolution*, 11(6), 700–713.

Kays, R., McShea, W. J., & Wikelski, M. (201920b). Born digital biodiversity data: millions and billions. *Diversity and Distributions*, 26(5), 644–648.

Kays, R., Parsons, A. W., Baker, M. C., et al. (2017). Does hunting or hiking affect wildlife communities in protected areas? *Journal of Applied Ecology*, 54, 242–252.

Kays, R., Tilak, S., Kranstauber, B., et al. (2011). Monitoring wild animal communities with arrays of motion sensitive camera traps. *International Journal of Research and Reviews in Wireless Sensor Networks*, 1, 19–29.

Khwaja, H., Buchan, C., Wearn, O. R., et al. (2019). Pangolins in global camera trap data: implications for ecological monitoring. *Global Ecology and Conservation*, 20, e00769.

Kinnaird, M. F., & O'Brien, T. G. (2012). Effects of private-land use, livestock management, and human tolerance on diversity, distribution, and abundance of large African mammals: livestock and large mammals in Kenya. *Conservation Biology*, 26, 1026–1039.

Kolowski, J. M., & Alonso, A. (2010). Density and activity patterns of ocelots (*Leopardus pardalis*) in northern Peru and the impact of oil exploration activities. *Biological Conservation*, 143, 917–925.

Kramer-Schadt, S., Reinfelder, V., Niedballa, J., et al. (2016). The Borneo carnivore database and the application of predictive distribution modelling. *The Raffles Bulletin of Zoology*, suppl. 33, 18–41.

Lama, S. T., Ross, J. G., Bista, D., et al. (2019). First photographic record of marbled cat Pardofelis marmorata Martin, 1837 (Mammalia, Carnivora, Felidae) in Nepal. *Nature Conservation*, 32, 19–34.

LaPoint, S., Gallery, P., Wikelski, M., & Kays, R. (2013). Animal behavior, cost-based corridor models, and real corridors. *Landscape Ecology*, 28, 1615–1630.

Lessa, I. C. M., Ferreguetti, Á. C., Kajin, M., Dickman, C. R., & Bergallo, H. G. (2017). You can't run but you can hide: the negative influence of human presence on mid-sized mammals on an Atlantic island. *Journal of Coastal Conservation*, 21, 829–836.

Li, X., Bleisch, W. V., & Jiang, X. (2018). Using large spatial scale camera trap data and hierarchical occupancy models to evaluate species richness and occupancy of rare and elusive wildlife communities in southwest China. *Diversity and Distributions*, 24, 1560–1572.

Linkie, M., Guillera-Arroita, G., Smith, J., et al. (2013). Cryptic mammals caught on camera: assessing the utility of range wide camera trap data for conserving the endangered Asian tapir. *Biological Conservation*, 162, 107–115.

Loock, D. J. E., Williams, S. T., Emslie, K. W., et al. (2018). High carnivore population density highlights the conservation value of industrialised sites. *Scientific Reports*, 8, 16575.

MacKenzie, D., Nichols, J., Royle, J, Pollock, K., Bailey, L., & Hines, J. (2017). *Occupancy Estimation and Modeling*, 2nd edn. Academic Press, San Diego, CA.

Magle, S. B., Fidino, M., Lehrer, E. W., et al. (2019). Advancing urban wildlife research through a multi-city collaboration. *Frontiers in Ecology and the Environment*, 17, 232–239.

McShea, W. J., Forrester, T., Costello, R., He, Z., & Kays, R. (2016). Volunteer-run cameras as distributed sensors for macrosystem mammal research. *Landscape Ecology*, 31, 55–66.

Meek, P. D., Ballard, G. A., Sparkes, J., Robinson, M., Nesbitt, B., & Fleming, P. J. (2019). Camera trap theft and vandalism: occurrence, cost, prevention and implications for wildlife research and management. *Remote Sensing in Ecology and Conservation*, 5, 160–168.

Meyer, N. F. V., Moreno, R., Sutherland, C., et al. (2020). Effectiveness of Panama as an intercontinental land bridge for large mammals. *Conservation Biology*, 34(1), 207–219.

Millspaugh, J. J., Kesler, D. C., Gitzen, R. A., et al. (2012). Wildlife radiotelemetry and remote monitoring. In Silvy, N. J. (ed.). *The Wildlife Techniques Manual: Volume 1 Research*. Johns Hopkins University Press, Baltimore, MD (pp. 258–283).

Moeller, A. K., Lukacs, P. M., & Horne, J. S. (2018). Three novel methods to estimate abundance of unmarked animals using remote cameras. *Ecosphere*, 9, e02331.

Moo, S. S. B., Froese, G. Z. L., & Gray, T. N. E. (2018). First structured camera-trap surveys in Karen State, Myanmar, reveal high diversity of globally threatened mammals. *Oryx*, 52, 537–543.

Murphy, A., Kelly, M. J., Karpanty, S. M., Andrianjakarivelo, V., & Farris, Z. J. (2019). Using camera traps to investigate spatial co-occurrence between exotic predators and native prey species: a case study from northeastern Madagascar. *Journal of Zoology*, 307, 264–273.

TEAM Network (2011). Terrestrial *Vertebrate Protocol Implementation Manual, v. 3.1*. Tropical Ecology, Assessment and Monitoring Network, Conservation International. Available at: https://figshare.com/articles/TEAM_TV_protocol/9730562.

Nguyen, A., Tran, V. B., Hoang, D. M., et al. (2019). Camera-trap evidence that the silver-backed chevrotain Tragulus versicolor remains in the wild in Vietnam. *Nature Ecology and Evolution*, 3, 1650–1654.

Niedballa, J., Sollmann, R., Courtiol, A., & Wilting, A. (2016). camtrapR: an R package for efficient camera trap data management. *Methods in Ecology and Evolution*, 7, 1457–1462.

Norouzzadeh, M. S., Nguyen, A., Kosmala, M., et al. (2018). Automatically identifying, counting, and describing wild animals in camera-trap images with deep learning. *Proceedings of the National Academy of Sciences*, 115, E5716–E5725.

O'Brien, T. G., Baillie, J. E. M., Krueger, L., & Cuke, M. (2010). The wildlife picture index: monitoring top trophic levels: the wildlife picture index. *Animal Conservation*, 13, 335–343.

Oberosler, V., Tenan, S., Zipkin E., & Rovero, F. (2020b). When parks work: effect of anthropogenic disturbance on occupancy of tropical forest mammals. *Ecology and Evolution*, 10(9), 3881–3894.

Oberosler, V., Tenan, S., Zipkin, E. F., & Rovero, F. (2020a). Poor management in protected areas is associated with lowered tropical mammal diversity. *Animal Conservation*, 23, 171–181.

Pacifici, K., Reich, B., Miller, D., et al. (2017). Integrating multiple data sources in species distribution modeling: a framework for data fusion. *Ecology*, 98, 840–850.

Paolino, R. M., Royle, J. A., Versiani, N. F., et al. (2018). Importance of riparian forest corridors for the ocelot in agricultural landscapes. *Journal of Mammalogy*, 99, 874–884.

Pardo, L. E., Campbell, M. J., Cove, M. V., Edwards, W., Clements, G. R., & Laurance, W. F. (2019). Land management strategies can increase oil palm plantation use by some terrestrial mammals in Colombia. *Scientific Reports*, 9, 7812.

Pardo, L. E., Roque, F. de O., Campbell, M. J., Younes, N., Edwards, W., & Laurance, W. F. (2018). Identifying critical limits in oil palm cover for the conservation of terrestrial mammals in Colombia. *Biological Conservation*, 227, 65–73.

Parsons, A. W., Forrester, T., Baker-Whatton, M. C., et al. (2018). Mammal communities are larger and more diverse in moderately developed areas. *eLife*, 7, e38012.

Parsons, A. W., Forrester, T., McShea, W. J., Baker-Whatton, M. G., Millspaugh, J. J., & Kays, R. (2017). Do occupancy or detection rates from camera traps reflect deer density? *Journal of Mammalogy*, 98, 1547–1557.

Paviolo, A., De Angelo, C., Ferraz, K. M., et al. (2016). A biodiversity hotspot losing its top predator: the challenge of jaguar conservation in the Atlantic Forest of South America. *Scientific Reports*, 6, 37147.

Porras, L. P., Vazquez, L.-B., Sarmiento-Aguilar, R., Douterlungne, D., & Valenzuela-Galván, D. (2016). Influence of human activities on some medium and large-sized mammals' richness and abundance in the Lacandon Rainforest. *Journal of Nature Conservation*, 34, 75–81.

Pozo, R. A., Coulson, T., McCulloch, G., Stronza, A., & Songhurst, A. (2019). Chilli-briquettes modify the temporal behaviour of elephants, but not their numbers. *Oryx*, 53, 100–108.

Puechagut, P. B., Politi, N., De Los Llanos, E. R., et al. (2018). Association between livestock and native mammals in a conservation priority area in the Chaco of Argentina. *Mastozoología Neotropical*, 25, 407–418.

Rich, L. N., Miller, D. A. W., Muñoz, D. J., Robinson, H. S., McNutt, J. W., & Kelly, M. J. (2019). Sampling design and analytical advances allow for simultaneous density estimation of seven sympatric carnivore species from camera trap data. *Biological Conservation*, 233, 12–20.

Rich, L. N., Miller, D. A. W., Robinson, H. S., McNutt, J. W., & Kelly, M. J. (2016). Using camera trapping and hierarchical occupancy modelling to evaluate the spatial ecology of an African mammal community. *Journal of Applied Ecology*, 53, 1225–1235.

Robinson, N. M., Scheele, B. C., Legge, S., et al. (2018). How to ensure threatened species monitoring leads to threatened species conservation. *Ecological Management and Restoration*, 19, 222–229.

Rogan, M. S., Balme, G. A., Distiller, G., et al. (2019). The influence of movement on the occupancy–density relationship at small spatial scales. *Ecosphere*, 10(8), e02807.

Ross, S., Al Jahdhami, M. H., & Al Rawahi, H. (2019). Refining conservation strategies using distribution modelling: a case study of the endangered Arabian tahr *Arabitragus jayakari*. *Oryx*, 53, 532–541.

Rostro-García, S., Kamler, J. F., Ash, E., et al. (2016). Endangered leopards: range collapse of the Indochinese leopard (*Panthera pardus delacouri*) in Southeast Asia. *Biological Conservation*, 201, 293–300.

Rovero, F., & Ahumada, J. (2017). The Tropical Ecology, Assessment and Monitoring (TEAM) Network: an early warning system for tropical rain forests. *Science of the Total Environment*, 574, 914–923.

Rovero, F., Ahumada, J., Jansen, P. A., et al. (2019). A standardized assessment of forest mammal communities reveals consistent functional composition and vulnerability across the tropics. *Ecography*, 43(1), 75–84.

Rovero, F., & Marshall, A. R. (2009). Camera trapping photographic rate as an index of density in forest ungulates. *Journal of Applied Ecology*, 46, 1011–1017.

Rovero, F., & Rathbun, G. B. (2006). A potentially new giant sengi (elephant-shrew) from the Udzungwa Mountains, Tanzania. *Journal of East Africa Natural History*, 95, 111–115.

Rovero, F., & Zimmermann, F. (2016). *Camera Trapping for Wildlife Research*. Pelagic Publishing, Exeter, UK.

Rowcliffe, J. M., & Carbone, C. (2008). Surveys using camera traps: are we looking to a brighter future? *Animal Conservation*, 11, 185–186.

Rowcliffe, J. M., Carbone, C., Kays, R., Kranstauber, B., & Jansen, P. A. (2014). Density estimation using camera trap surveys: the random encounter model. In Meek, P., & Fleming, P. (eds.). *Camera Trapping: Wildlife Management and Research*, pp. 317–323. Pelagic Publishing, Exeter, UK.

Rowcliffe, J. M., Jansen, P., Kays, R., Kranstauber, B., & Carbone, C. (2016). Speed cameras for mammals: measuring travel speed and day range using camera traps. *Remote Sensing in Ecology and Conservation*, 2, 84–94.

Rowcliffe, J. M., Kays, R., Carbone, C., & Jansen, P. A. (2013). Clarifying assumptions behind the estimation of animal density from camera trap rates. *Journal of Wildlife Management*, 77(5), 876.

Rowcliffe, M. J., Carbone, C., Jansen, P. A., Kays, R., & Kranstauber, B. (2011). Quantifying the sensitivity of camera traps using an adapted distance sampling approach. *Methods in Ecology and Evolution*, 2, 467–476.

Ruykys, L., & Carter, A. (2019). Removal and eradication of introduced species in a fenced reserve: quantifying effort, costs and results. *Ecological Management and Restoration*, 20, 239–249.

Sadhu, A., Jayam, P. P. C., Qureshi, Q., Shekhawat, R. S., Sharma, S., & Jhala, Y. V. (2017). Demography of a small, isolated tiger (*Panthera tigris tigris*) population in a semi-arid region of western India. *BMC Zoology*, 2, 16.

Sandbrook, C., Luque-Lora, R., & Adams, W. (2018). Human bycatch: conservation surveillance and the social implications of camera traps. *Conservation and Society*, 16, 493.

Schank, C. J., Cove, M. V., Kelly, M. J., et al. (2017). Using a novel model approach to assess the distribution and conservation status of the endangered Baird's tapir. *Diversity and Distributions*, 23, 1459–1471.

Schlichting, P. E., Dombrovski, V., & Beasley, J. C. (2019). Use of abandoned structures by Przewalski's wild horses and other wildlife in the Chernobyl Exclusion Zone. *Mammal Research*, https://doi.org/10.1007/s13364-019-00451-4.

Schuttler, S. G., Sears, R. S., Orendain, I., et al. (2018). Citizen science in schools: students collect valuable mammal data for science, conservation, and community engagement. *Bioscience*, https://doi.org/10.1093/biosci/biy141.

Shamoon, H., Maor, R., Saltz, D., & Dayan, T. (2018). Increased mammal nocturnality in agricultural landscapes results in fragmentation due to cascading effects. *Biological Conservation*, 226, 32–41.

Sharma, K., Bayrakcismith, R., Tumursukh, L., et al. (2014). Vigorous dynamics underlie a stable population of the endangered snow leopard Panthera Uncia in Tost Mountains, South Gobi, Mongolia. *PLoS One*, 9, e101319.

Sharma, R. K., Bhatnagar, Y. V., & Mishra, C. (2015). Does livestock benefit or harm snow leopards? *Biological Conservation*, 190, 8–13.

Shrestha, Y., & Lapeyre, R. (2018). Modern wildlife monitoring technologies: conservationists versus communities? A case study: the Terai-arc landscape, Nepal. *Conservation and Society*, 16, 91–101.

Smit, J., Pozo, R., Cusack, J., Nowak, K., & Jones, T. (2019). Using camera traps to study the age–sex structure and behaviour of crop-using elephants Loxodonta africana in Udzungwa Mountains National Park, Tanzania. *Oryx*, 53, 368–376.

Sollmann, R., Mohamed, A., Samejima, H., & Wilting, A. (2013). Risky business or simple solution—relative abundance indices from camera-trapping. *Biological Conservation*, 159, 405–412.

Spehar, S. N., Loken, B., Rayadin, Y., & Royle, J. A. (2015). Comparing spatial capture–recapture modeling and nest count methods to estimate orangutan densities in the Wehea Forest, East Kalimantan, Indonesia. *Biological Conservation* 191, 185–193.

Steenweg, R., Hebblewhite, M., Kays, R., et al. (2017). Scaling-up camera traps: monitoring the planet's biodiversity with networks of remote sensors. *Frontiers in Ecology and the Environment*, 15, 26–34.

Steenweg, R., Hebblewhite, M., Whittington, J., Lukacs, P., & McKelvey, K. (2018). Sampling scales define occupancy and underlying occupancy-abundance relationships in animals. *Ecology* 99, 172–183.

Stewart, F. E., Volpe, J. P., Eaton, B. R., Hood, G. A., Vujnovic, D., & Fisher, J. T. (2019). Protected areas alone rarely predict mammalian biodiversity across spatial scales in an Alberta working landscape. *Biological Conservation*, 240, 108252.

Šver, L., Bielen, A., Križan, J., & Gužvica, G. (2016). Camera traps on wildlife crossing structures as a tool in gray wolf (*canis lupus*) management—five-years monitoring of wolf abundance trends in Croatia. *PLoS One*, 11, e0156748.

Swanson, A., Kosmala, M., Lintott, C., Simpson, R., Smith, A., & Packer, C. (2015). Snapshot Serengeti, high-frequency annotated camera trap images of 40 mammalian species in an African savanna. *Scientific Data*, 2, 150026.

Tobler, M. W., Garcia Anleu, R., Carrillo-Percastegui, S. E., et al. (2018). Do responsibly managed logging concessions adequately protect jaguars and other large and medium-sized mammals? Two case studies from Guatemala and Peru. *Biological Conservation*, 220, 245–253.

Urbanek, R. E., Ferreira, H. J., Olfenbuttel, C., Dukes, C. G., & Albers, G. (2019). See what you've been missing: an assessment of Reconyx® PC900 hyperfire cameras. *Wildlife Society Bulletin*, 43(4), 630–638.

Velho, N., Srinivasan, U., Singh, P., & Laurance, W. F. (2016). Large mammal use of protected and community-managed lands in a biodiversity hotspot: mammal use across protection regime. *Animal Conservation*, 19,199–208.

Wang, T., Royle, A. J., Smith, J. L. D., et al. (2018). Living on the edge: opportunities for Amur tiger recovery in China. *Biological Conservation*, 217, 269–279.

Wearn, O. R., & Glover-Kapfer, P. (2017).Camera-trapping for conservation: a guide to best-practices. *WWF Conservation Technology Series*. Available at: https://www.wwf.org.uk/sites/default/files/2019-04/Camera Traps-WWF-guidelines.pdf

Wearn, O. R., & Glover-Kapfer, P. (2019). Snap happy: camera traps are an effective sampling tool when compared with alternative methods. *Royal Society Open Science*, 6, 181748.

Wearn, O. R., Rowcliffe, J. M., Carbone, C., Pfeifer, M., Bernard, H., & Ewers, R. M. (2017). Mammalian species abundance across a gradient of tropical land-use intensity: a hierarchical multi-species modelling approach. *Biological Conservation* 212, 162–171.

Welbourne, D. J., Claridge, A. W., Paull, D. J., & Lambert, A. (2016). How do passive infrared triggered camera traps operate and why does it matter? Breaking down common misconceptions. *Remote Sensing in Ecology and Conservation* 2, 77–83.

Whitfield, J. (1998). A saola poses for the camera. *Nature*, 396, 410.

Whitworth, A., Beirne, C., Pillco Huarcaya, R., et al. (2019). Human disturbance impacts on rainforest mammals are most notable in the canopy, especially for larger-bodied species. *Diversity and Distributions*, 25, 1166–1178.

Whitworth, A., Braunholtz, L. D., Huarcaya, R. P., MacLeod, R., & Beirne, C. (2016). Out on a limb: arboreal camera traps as an emerging methodology for inventorying elusive rainforest mammals. *Tropical Conservation Science*, 9, 675–698.

Yaap, B., Magrach, A., Clements, G. R., et al. (2016). Large mammal use of linear remnant forests in an industrial pulpwood plantation in Sumatra, Indonesia. *Tropical Conservation Science*, 9, 194008291668352.

Yamada, T., Yoshida, S., Hosaka, T., & Okuda, T. (2016). Logging residues conserve small mammalian diversity in a Malaysian production forest. *Biological Conservation*, 194, 100–104.

Young, J. K., Golla, J. M., Broman, D., et al. (2019). Estimating density of an elusive carnivore in urban areas: use of spatially explicit capture-recapture models for city-dwelling bobcats. *Urban Ecosystems*, 22, 507–512.

Young, S., Rode-Margono, J., & Amin, R. (2018). Software to facilitate and streamline camera trap data management: a review. *Ecology and Evolution*, 8, 9947–9957.

Yousif, H. (2019). *Deep Neural Networks for Animal Object Detection and Recognition in the Wild*. Ph.D. dissertation. University of Missouri, Columbia, MO.

Zapata-Ríos, G., & Branch, L. C. (2018). Mammalian carnivore occupancy is inversely related to presence of domestic dogs in the high Andes of Ecuador. *PLoS One*, 13, e0192346.

CHAPTER 6

Animal-borne technologies in wildlife research and conservation

Kasim Rafiq, Benjamin J. Pitcher, Kate Cornelsen,
K. Whitney Hansen, Andrew J. King, Rob G. Appleby,
Briana Abrahms, and Neil R. Jordan

6.1 Introduction

Animal-borne technologies are devices carried by animals that collect or transmit data for research. The diversity and use of animal-borne devices, and the species and contexts in which they are applied, have all rapidly increased in the last few decades. We now have a wide variety of devices that are developed and customized to allow a suite of measurements to be collected on both the animals bearing the units and the environments they inhabit (Wilmers et al., 2015). These devices are, for example, being used to explore the movements and behaviours of species that were once too difficult to study due to the costs and challenges of acquiring data through traditional means, including in topics such as species' migrations, social interactions, and energetics. Specific examples of research questions will be considered throughout the chapter.

Prior to animal-borne technologies, traditional methods to study wild animal movements and behaviours involved marking individuals (e.g. with coloured bands) during repeated sampling efforts and/or collecting data during visual observations, often following intensive manual tracking efforts or via opportunistic encounters (e.g. Schaller, 1976; Bailey, 2005). These early studies provided insights into the ecology of many free-ranging species. They led to important discoveries on animal movements, habitat preferences, foraging strategies, and other key behaviours, and they provided fundamental insights into more comprehensive studies. However, data collected in this way are often accompanied by several important limitations. For many cryptic and wide-ranging species, considerable tracking effort (using footprints and marks left in the environment) is often required to find animals and is limited to areas where such signs are easily visible. Thus, the costs, including time and resources (e.g. vehicle fuel and research hours), of acquiring these data are high, with days to weeks of searching sometimes required to find specific animals of the study species (KR, personal observation). This can limit the sample sizes that inferences are made on and limit the ability of many researchers to collect such data themselves, for example restricting such activities to researchers from certain socioeconomic backgrounds. As well as the logistical challenges of navigating certain environments, direct observations may also be biased towards environments where animals can be easily seen, such as the short grasslands of the Serengeti (Schaller, 1976) than habitats where behaviours are easily obscured, such as riparian forests. This limits our understanding of how animal behaviour changes across habitats. These challenges (and others) ultimately limit the spatial and temporal (time) scales that traditional movement and behaviour studies can be carried out. In turn, this limits our ability to understand

Kasim Rafiq et al., *Animal-borne technologies in wildlife research and conservation*. In: *Conservation Technology*.
Edited by Serge A. Wich and Alex K. Piel, Oxford University Press.
© Oxford University Press 2021. DOI: 10.1093/oso/9780198850243.003.0006

animal movements and behaviours that occur over relatively large scales, such as species' migrations.

The advent of animal-borne technologies was a key milestone in studies of animal ecology. It addressed many of the limitations of traditional tracking approaches by lowering the costs of acquiring data over greater spatial and temporal scales. This chapter will thus focus on how animal-borne technologies, and particularly tracking technologies, are currently used in wildlife conservation and ecology. We start by reviewing some of the key tracking technologies used today. These technologies range from simple transmitters that emit pulsed radio frequencies, allowing researchers to manually locate tagged wildlife, to satellite trackers that can transmit global positioning system (GPS) locations remotely to researchers anywhere in the world. Other animal-borne devices, such as accelerometers and audio recorders, that can be used in combination with tracking technologies are also discussed in this chapter. We do not provide an exhaustive list of the animal-borne technologies being used in wildlife studies, but instead we review the state-of-the-art and emerging technologies that can be used in wildlife research and conservation. We also provide applications of the tools we describe throughout the chapter, including an in-depth case study on marine species. We end by considering the primary challenges and future prospects of using animal-borne tracking technologies in wildlife research.

6.2 New technology

6.2.1 Animal-borne tracking technologies

6.2.1.1 Very high frequency (VHF) tracking

VHF systems consist of transmitters that send radio signals of specific frequencies and receiver-antennae units that detect transmitted signals (Kenward, 2000). Radio signals are often transmitted within set bands, varying depending on country, but commonly include the 148–152 MHz, 163–165 MHz, and 216–220 MHz ranges (e.g. Kolz, 1983; Federal Communications Commission, 2020). VHF transmitters are usually designed to pulse transmissions, with the primary pulse characteristics being pulse width (how long each individual transmission operates) and pulse interval (the time that elapses between transmissions). These and

other features are often adjusted to extend battery life, increase transmitter detectability, and encode individual identity (i.e. by assigning different transmission frequencies to different animals) (Taylor et al., 2017).

Modern VHF transmitters can weigh as little as 0.3 g, with size varying depending on elements such as battery and antenna type (Holohil, 2020). Transmitters with unique frequencies are typically battery-powered, attached to animals, and encased within housing to protect them from the environment, and the animals themselves, during their deployments. Depending on the study species' characteristics and environment, housing and attachment methods vary considerably, from backpacks and collars to direct on- and under-skin attachments (Figure 6.1).

Combined with directional antennae, VHF receivers can be used to detect the direction of a VHF transmitter based on received signal strength. This approach can be used to coarsely estimate the location of a tagged animal by triangulating the VHF signal. This was historically a common use of VHF and was used to gain early insights into the space-use of many wide-ranging species (e.g. lions, *Panthera leo*, Schaller, 1976). However, the advent of other tracking technologies that now remotely record locations means that VHF receivers are rarely used in this way today, except where very lightweight instruments are required and other technologies are not feasible (Kissling et al., 2014). Instead, VHF transmitters are now commonly used to find tagged individuals and collect data that require researchers to be near the animal, for example, to observe behaviour (Rafiq et al., 2020a), collect biological samples (Waugh et al., 2016), or retrieve data that are stored onboard other animal-borne devices (Wilson et al., 2013).

By allowing scientists to efficiently track animals, rather than waiting for opportunistic sightings or investing in intensive tracking efforts, relying on footprints and other signs left directly by the target animal, VHF transmitters provide a simple, reliable, and cost-effective approach to tracking wildlife (Kenward, 2000). Nevertheless, compared to other forms of tracking technologies (see next), collecting data using VHF systems is still relatively labour-intensive and expensive, as most studies manually track individuals on foot or with a vehicle across

Figure 6.1 Examples of housing and attachment methods for animal-borne devices. (a) F2HKv2 collar housing a GPS tag (GiPSy-4, www.technosmart.eu/) and inertial sensors (Daily Diary http://wildbytetechnologies.com/) fitted to a male chacma baboon (*Papio ursinus*) in South Africa to study urban foraging behaviour (image courtesy of Gaelle Fehlmann). (b) Homing pigeon (*Columba livia*) carrying a backpack with air pollution sensors to track urban pollution (http://pigeonairpatrol.com) (image courtesy of DigitasLBi). (c) Gentoo penguin (*Pygoscelis papua*) with a data-logger taped to its back (recording heading, speed, depth, temperature, and light intensity) to examine its foraging dynamics (image courtesy of Rory Wilson). (d) A radar transponder attached by adhesive to a bee (*Bombus terrestris audax*), used to document its entire foraging career (image courtesy of Lars Chittka). (e) Accelerometer (AXY-4 www.technosmart.eu/) attached with adhesive to an agama (*Stellagama stellio*) to study its behavioural states (image courtesy of Savvas Zotos). (f) A female harlequin frog (*Atelopus limosus*) carrying a miniature (0.3 g) radio transmitter (https://www.holohil.com/) for tracking survival rate following release to the wild (image courtesy of Brian Gratwicke).

challenging terrain. One solution to this problem is the use of drones for: (i) localizing the positions of VHF transmitters, (ii) streamlining the process of downloading animal-borne data, and/or (iii) collecting observational data, for example, storing video data onboard for researcher retrieval or using real-time video streaming (Dos Santos et al., 2014). Another solution is the use of VHF receiver tower arrays to automate the process of estimating transmitter locations. The Motus Wildlife Tracking System, for example, is an international network of tower arrays that is independently managed by collaborating researchers (Taylor et al., 2017). The distribution of these arrays across large scales has been particularly useful in studying the migratory movements of small flying species, including birds, bats, and insects. A detailed overview on Motus can be found in Taylor et al.(2017).

6.2.1.2 Satellite tracking

Satellite navigation systems use satellites to autonomously map the location of devices on the Earth's surface. Tracking animals using satellite systems became widespread in wildlife research in the 1980s (Benson, 2012), initially with Argos, an environmentally focused satellite data collection system, and then with the GPS.[1] The advent of satellite tracking allowed researchers to remotely collect locational

---

[1] The GPS satellite network, originally deployed by the United States Department of Defence, has more recently been joined by networks from other countries including GLONASS (Russia), BeiDou (China) and Galileo (European Union), with others planned (e.g. IRNSS, India). As it remains common to refer to devices utilizing any of these networks generically as 'GPS' devices, we will also, but it is worth noting that many satellite receiver manufacturers now support data from multiple networks.

Table 6.1 Example uses and recommended reading for animal-borne devices discussed in this chapter

Sensor type	Research context examples	Target species examples	Suggested review papers
VHF	Habitat selection, invasive species migration, resource selection	Monarch butterfly: Fisher, Adelman & Bradbury, 2020; Asian hornet: Kennedy et al., 2018	(Kenward, 2000; Mech & Barber, 2002; Habib et al., 2014)
Argos	Habitat selection, hunting/foraging, migration, resource selection	Pinnipeds: Costa et al., 2010; Green turtle: Godley et al., 2002; Humpback whale: Dulau et al., 2017	(Hebblewhite & Haydon, 2010; Tomkiewicz et al., 2010)
GPS	Habitat selection, hunting/foraging, human-wildlife conflict, migration, resource selection, social interaction	Red foxes: Bischof et al., 2019; Green turtle: Godley et al., 2002; Egyptian vulture: García-Ripollés, López-López & Urios, 2010; leopard: Rafiq et al., 2019	(Hebblewhite & Haydon, 2010; Tomkiewicz et al., 2010; Hofman et al., 2019)
RFID	Disease transmission, hunting/foraging, human-wildlife conflict, invasive species, resource selection, social interaction	Honeybee: Nunes-Silva et al., 2019; hummingbird: Bandivadekar et al., 2018; Tasmanian devil: Hamede et al., 2009;	(Bonter & Bridge, 2011)
Light-level geolocators	Long-distance migration, habitat loss, environmental contaminant risk	Seabirds: González-Solís et al., 2007; Montevecchi et al., 2012	None specific to animal-borne light-level geolocators
Accelerometer	Activity budgets, energy expenditure, hunting/foraging, resource selection	Polar bear: Pagano et al., 2020; African wild dog: Hubel et al., 2016; great white shark: Watanabe et al., 2019	(Brown et al., 2013)
Audio	Activity budgets, anthropogenic disturbance, hunting/foraging, resource selection, social interactions	African lion: Wijers et al., 2018; chipmunk: Couchoux et al., 2015; sperm whale: Fais et al., 2016; Eurasian jackdaw: Stowell et al., 2017	None specific to animal-borne audio
Video	Energy expenditure, hunting/foraging, invasive species, social interaction	Tasmanian devil: Andersen et al., 2020; falcon: Kane & Zamani, 2014; great white sharks: Watanabe et al., 2019	None specific to animal-borne video

data without triangulating from VHF signals. This innovation led to insights into a diversity of themes in ecology and conservation, including in studies of wildlife corridor mapping, animal behaviour, and human-wildlife conflict (Table 6.1).

While Argos and GPS both rely on satellites to estimate positional data, they differ in the methods used to acquire and transmit those data. This can impact each systems' relative suitability to specific applications in wildlife research, which we discuss further later. Argos satellites detect the location of platform transmitter terminals using the Doppler effect, which in this instance is the change in frequency of the platform transmitter terminal (PTT) waves as the satellite moves relative to the transmitter (Figure 6.2A). In animal-borne systems, PTTs are attached to wildlife to transmit the location of individuals to orbiting satellites. These data are retransmitted by satellites to receiving stations on the Earth and then sent to data processing centres before becoming available to researchers (Costa et al., 2010).

GPS animal-borne devices triangulate locations by measuring the time taken to receive signals from three or more satellites, whose precise locations are known (Figure 6.2B) (Tomkiewicz et al., 2010). In this way, receivers communicating with three satellites can be localized in two dimensions, and those communicating with over three satellites can be localized in three dimensions. These data are then stored on the animal-borne device for future download and/or remotely transmitted to researchers (discussed next). Given that both Argos and GPS systems require communication with satellites, both can be impacted by animal behaviour, such as burrowing or diving, and environmental features that obscure visibility of the sky.

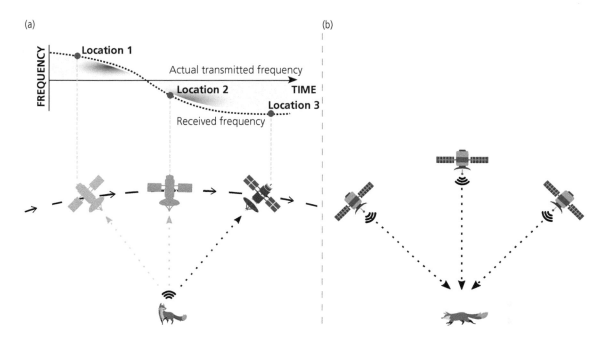

Figure 6.2 Principles of how Argos and GPS systems estimate locations. (a) An Argos satellite requires at least two successive communications with an animal-borne transmitter to calculate the tag's approximate location using the Doppler effect (changes in transmitted and received frequency due to satellite movement) (adapted from Argos, 2020). (b) Animal-borne GPS devices triangulate their location based on time and location data simultaneously transmitted to receivers by at least three GPS satellites (watercolour vectors created by freepik, https://www.freepik.com).

In practical terms, there are three key differences between the use of Argos and GPS in wildlife studies: sampling frequency, data precision, and data acquisition. Argos satellites are polar orbiting. This means that the frequency that transmitters are visible to satellites and able to transmit locations depends on the latitude of transmitters. For example, transmitters at the equator have approximately 6–7 daily opportunities to transmit locations while those at the poles have 14. In contrast, GPS satellites are uniformly distributed across 6 orbits and so communication with animal-borne GPS tags is near ubiquitous, meaning locations can be recorded at higher frequencies, with some studies collecting locational data several times per second (e.g. Wilson et al., 2013).

Due to the way that locations are calculated, locational data from Argos are also coarser, with locational errors of hundreds of metres, compared to GPS locations, which are <1 m in some cases (Tomkiewicz et al., 2010; Thomson et al., 2017).

The sampling frequency and positional error typical of Argos mean that these units are often only used when researchers study broad-scale movements, for example, in magnitudes of kilometres, where frequent sampling and high-resolution spatial data are not required, such as in continental species migration studies (e.g. Godley et al., 2002). In contrast, GPS-enabled devices are ideal for studies of finer-scale movements and behaviours, such as hunting and social interactions (Wilson et al., 2013; Elbroch & Quigley, 2017). For individuals that only move into areas with a clear view of the sky for brief periods, such as sea turtles and marine mammals, snapshot GPS receivers, where minimal *raw* GPS signal data are recorded on the unit and post-processed when the units are recovered, are increasingly being used (Tomkiewicz et al., 2010).

Data acquisition is also critical to consider when selecting devices. In Argos systems, data are collected by satellites and relayed to Earth-based centres. This makes the system well-suited for monitoring elusive animals or those that live in environments logistically difficult for researchers to negotiate. Indeed, Argos has been widely used in many tracking studies of marine and Antarctic species, such as Weddell seals (*Leptonychotes weddellii*) and humpback whales (*Megaptera novaeangliae*) (Nachtsheim et al., 2019; Riekkola et al., 2019). In contrast, with GPS, locational data are first stored on animal-borne devices. Researchers must then either physically retrieve tags, remotely download data using portable download units, or transmit data from the animal-borne device using satellite (e.g. Iridium[2]), cellular, or other networks (e.g. LoRa[3]). Each of these data acquisition methods has its advantages and disadvantages, which impact their suitability to specific research scenarios, and these have been reviewed elsewhere (for an overview, we recommend reviews by Tomkiewicz et al., 2010 and Hofman et al., 2019).

6.2.1.3 Radio frequency identification (RFID)

RFID tags represent one of the smallest forms of animal tracking tags and are well suited to tracking animal movements at or in the vicinity of known locations. While the satellite(s) in Argos and GPS systems function as spatial reference points for locational data, transponder readers play that role in RFID systems and are usually deposited at fixed locations to record when animals fitted with transponder tags come into close proximity of them. Specifically, animal-borne RFID transponder tags transmit a unique tag identification code, which is detected and stored by a transponder reader within close proximity (Chetouane, 2015). In this way, the occurrence of *encounters* between tags and readers can be used to investigate a wide range of topics, including foraging strategies and human-wildlife conflict (Bandivadekar et al., 2018; Testud et al., 2019).

Generally, RFID tags are attached (or implanted in) to animals, and readers are placed within the environment at places of interest, such as food or water sources or highway crossing structures (Dexter et al., 2016). However, some tag systems are also capable of operating as readers, so that a tag both transmits its unique identification number and records the tags it comes into close proximity with (hence they are often called 'proximity loggers'). In this way, encounters between individuals can be recorded, which is useful in various contexts, including studies of disease transmission and species interactions (Hamede et al., 2009; Drewe et al., 2012; Moyers et al., 2018).

Other information that can be stored by readers includes remaining transponder battery capacity, body temperature, and transponder signal strength, which can be used as a proxy for distance. The use of this technology to accurately record distances between readers and tags is limited, however, as it is generally only feasible to determine tags as being 'close' to readers (e.g. <50 m), and even then, results can be unreliable (see Krull et al., 2018).

There are two broad types of RFID tags—passive and active—a distinction based on their power source, which in turn affects their relative suitability for specific applications. Passive tags draw power from electromagnetic waves emitted by externally powered readers, and as a result they can only transmit information over short distances (<3 m) (Chetouane, 2015). This also provides passive tags with a longer operational life, which can sometimes exceed the lifespan of tagged individuals, making them well suited for collecting long-term data. In contrast, active tags require their own power source, which allows them to transmit information over relatively large distances, from over 100 m (Chetouane, 2015).

Active tags are typically larger than passive tags and can weigh from 200 mg, although advances in circuit designs may soon bring this down to <95 mg (Kumari & Hasan, 2020). In contrast, passive tags can weigh from ~4 mg and measure ~1 mm^2, and they have often been applied to smaller species and in situations where individuals are likely to come into close contact with readers. For example, they have been applied to amphibian and insect species to monitor their use of wildlife road-crossing structures (Testud et al., 2019). They have also been widely used in bee foraging studies (Nunes-Silva et al., 2019), with readers

---

[2] Iridium is a satellite network that provides two-way transmission for voice and data (Tomkiewicz et al., 2010)

[3] LoRa (long range) is a low-power, long-range wireless data transfer system, which is particularly popular in machine to machine and internet-of-things applications (Allan et al., 2018)

placed strategically in likely foraging patches. In contrast, active tags have often been applied to larger species, where the additional burden of bulkier RFID tags is less pronounced, and for research questions where a greater distance between tags and readers is required or more practical (Dexter et al., 2016).

In addition to their use in studies that are locational in nature, RFID systems can also be used to trigger other devices. For example, they can be programmed to trigger audio or visual playbacks upon detecting specific RFID identification codes to test animal responses to external stimuli experimentally. In this way, experimental playback manipulations can be automated and carried out remotely. This addresses the limitations of traditional playback experiments that require researchers to manually commence playbacks, such as limited sample sizes due to research effort and human impacts on behaviour (see Lendvai et al. (2015) for an example with tree swallows, *Tachycineta bicolor*).

6.2.1.4 Other animal-borne trackers

Two additional categories of animal-borne trackers worth briefly mentioning are acoustic transmitters and light-level geolocators. Aquatic acoustic telemetry is used to convey locational information on marine animal movements, since sound travels more efficiently in water than radio frequencies (Heupel et al., 2018). Like VHF, acoustic transmitters require specialized equipment, in this case hydrophones, to receive signals from acoustic tags. These signals allow tagged individuals to be manually located, using directional hydrophones, or have their positions triangulated within hydrophone arrays (Cooke et al., 2013). Like Motus in the terrestrial realm, large-scale tracking projects, such as the Australian Integrated Marine Observing System Animal Tracking Facility, track marine species at continental scales using permanent acoustic receiver arrays (Hoenner et al., 2018).

Light-level geolocators store data on light intensity levels (sunrise and sunset times) and measure current light levels (Thomas et al., 2011). By comparing both stored and current light data, a location can be calculated with a locational error of <1 km, depending on sensor orientation, direction of movement, and shading (Phillips et al., 2004).

Geolocators are lightweight (<2 g) and have low-power consumption, allowing long-distance migrations of small birds to be tracked (e.g. Bächler et al., 2010; Stanley et al., 2012). These tags are particularly well suited to seabird studies because they typically occur in flat environments with good views of the horizon. However, they are used less frequently in tracking studies of terrestrial species than the methods discussed earlier, and the emergence of smaller, low-power GPS tags mean that their future use is likely limited.

6.2.2 Non-tracking animal-borne technologies

6.2.2.1 Accelerometers

Accelerometers are devices that measure acceleration changes by using piezoelectric sensors, which are sensors that use electrical charges that are generated following mechanical stress (Brown et al., 2013). Modern accelerometers typically measure changes resulting from both gravitational acceleration and acceleration caused by movement, and do so on one to three axes simultaneously at rates of 0.5–10,000 Hz (Brown et al., 2013). These measurements can be used to reconstruct specific behaviours, gain insights into biomechanics, and serve as a proxy for energetic expenditure (Figure 6.3) (Wilmers et al., 2015).

Accelerometers have commonly been used to deduce animal behavioural budgets across different time series, including for marine mammals (Ladds et al., 2018), primates (Sha et al., 2017), birds (Hernández-Pliego et al., 2017), and more. Depending on the study species, however, calibrating accelerometers to distinguish between behaviours can be difficult. This is because such calibrations have often been accomplished by directly observing animals and taking time-stamped videos that can later be cross-referenced with accelerometery data (e.g. Shuert et al., 2018), which for many species, such as marine mammals and cryptic, wide-ranging carnivores, presents obvious logistical challenges. However, in recent years, other forms of ground-truthing using animal-borne devices have become more popular (see Sections 6.2.2.2 and 6.2.2.3) and have helped alleviate these issues (Pagano et al., 2017). By simultaneously recording accelerometery data alongside time-stamped behavioural data,

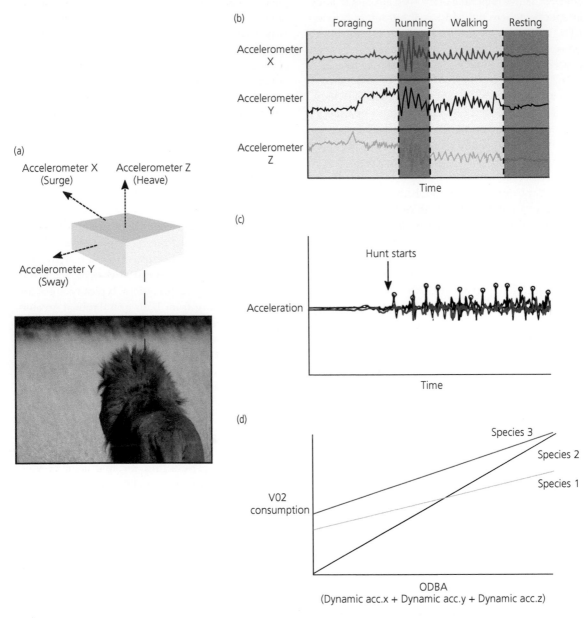

Figure 6.3 Example orientation of accelerometer axes and examples of how accelerometer data can be used. (a) Common tri-axial accelerometer orientation on a quadruped (lion image courtesy of Kasim Rafiq). (b) Raw accelerometery data can be used to differentiate between behaviours that have distinct acceleration profiles (adapted from Fehlmann et al., 2017). (c) Raw accelerometery data can be used to provide detailed insights into the biomechanics of specific movements, presented are raw accelerometery data of a partial cheetah hunt from Wilson et al. (2013). (d) The linear relationship between overall dynamic body acceleration (ODBA), which is the sum of acceleration caused by movement (dynamic acceleration) across all acceleration axes, and oxygen consumption (VO^3) means that accelerometery data can be used as a proxy for energetic expenditure. The slope of this relationship, however, differs between species and must be calibrated (Wilson et al., 2020).

acceleration profiles may be identified and associated with different behaviours, such as resting, walking, and feeding (e.g. Fehlmann et al., 2017). These data can then be used to add behavioural contexts to locational data, such as linking foraging behaviour with different habitats (Watanabe et al., 2019).

Accelerometers can also be integrated with other animal-borne devices to control sampling frequencies based on animal behaviours. Wilson et al. (2013), for example, used accelerometers to trigger the collection of high-resolution sensor data on cheetah (*Acinonyx jubatus*) runs (Figure 6.3C), useful for understanding biomechanics, during periods when acceleration passed a peak threshold, and the collection of lower-resolution data, useful for broader ecological studies, such as on resource selection and competition, during periods of low acceleration movement (e.g. Abrahms et al., 2016; Rafiq et al., 2020b). In this way, device deployment durations were maximized by collecting high-resolution data, which requires more power to collect and store, only during relevant events.

Additionally, accelerometers have frequently been applied to studies of animal energetics because of the linear relationship between bodily acceleration and metabolic rate (e.g. Pagano et al., 2020), which effectively allows acceleration to serve as a proxy for energetic expenditure (see Wilson et al., 2020). Further, accelerometers, often in combination with magnetometers,[4] can be used to refine and reconstruct fine-scale movement information and thus enhance GPS-based tracking, particularly at scales where movement might be concealed by GPS error or in environments with poor GPS signal, such as underwater (Bidder et al., 2015; Dewhirst et al., 2016). A review on the use of accelerometers in wildlife studies can be found in Brown et al. (2013).

6.2.2.2 Audio recorders

Audio recorders detect changes in pressure (sound) at the sampling frequencies of interest and convert

these to numbers that can be stored digitally. Early deployments of animal-borne audio recorders were limited by power and storage requirements, with units being either large or limited in deployment times, which prevented their use for many species (Alkon et al., 1989). Although the trade-off between unit size and longevity persists, miniaturization has driven the increased use of animal-borne audio recorders in wildlife conservation and research.

Animal-borne acoustic recorders have been used to study activity patterns in terrestrial and marine species (Couchoux et al., 2015), provide context to species' vocalizations (O'Bryan et al., 2019), and ground-truth/calibrate data collected from other devices (Wijers et al., 2018). While accelerometer data can only be used to differentiate between a small number of behaviours with distinct acceleration profiles (Brown et al., 2013), audio data from animal-borne devices may allow a greater number of behaviours to be identified. This is because context-specific sounds, such as feeding and mating, can be easier to identify from audio rather than from abstract acceleration profiles, and because behaviours cryptic to accelerometers, such as territorial vocalizations, can be captured with audio. Audio recorders attached to eastern chipmunk (*Tamias striatus*), for example, were able to capture and differentiate foraging, begging, and self-scratching behaviours (Couchoux et al., 2015).

Audio recorders can also be used to cost-effectively calibrate data from other devices, which would be more challenging and expensive to achieve through direct observations. For example, 20.5 hours of audio data generated enough data points to train accelerometers to predict lion behaviours, including feeding and drinking, with 94.3% accuracy (Wijers et al., 2018). This increased to 98.5% when predicting behaviours using audio, accelerometer, and magnetometer data together. For marine species, DTAG units collect data using a combination of devices, including hydrophones (acoustic), GPS, and accelerometers, and have also been used to provide detailed insights into the behaviour of species such as sperm whales (*Physeter macrocephalus*) (Fais et al., 2016). These devices are, for example, being used to answer questions on the use of acoustics in hunting strategies (Fais et al., 2016), migratory behaviours (Owen et al.,

[4] Magnetometers measure the strength and direction of the Earth's magnetic field from one (single-axis) to three (tri-axial) axes, in a manner analogous to accelerometers (Williams et al., 2017).

2017), and social interactions (Tyson et al., 2012). The ability to collect georeferenced acoustic data onboard animals can also allow for the exposure and responses of species to anthropogenic noises to be quantified (Mikkelsen et al., 2019). Animal audio recorders are thus being increasingly used to understand the impact of acoustic environments on animal behaviour, which can, in turn, be used to inform legislation (Wisniewska et al., 2018).

6.2.2.3 Video cameras

Video cameras are another example of technology used in animal-borne studies for decades but that have only recently begun to gain widespread traction. This has yet again been driven by the increasing accessibility of devices to researchers (both in terms of reductions in cost and complexity) and the increasing miniaturization of these technologies. Today, animal-borne cameras are being used to study a wide range of topics on both marine and terrestrial species (Pagano et al., 2017; Watanabe et al., 2019). Considered alongside tracking data, they can help assign behaviours to specific locations (Bruce et al., 2019), capture data on the surrounding environment (e.g. see *TurtleCam*), and ground-truth data from other devices (Watanabe et al., 2019).

Linking animal-borne videos to accelerometer data can be particularly useful for validating data collected from animals or environments where direct observations are rarely possible. Back-mounted cameras on free-swimming Adélie penguins (*Pygoscelis adeliae*), for example, have been used to detect unique accelerometer signatures associated with prey capture (Watanabe & Takahashi, 2013). Video data can also be particularly valuable in detecting and quantifying events that cannot be detected by other means, and so they can provide additional clarity to animal behaviours. For example, they can allow social behaviours to be captured, such as carcass feeding and aggression in Tasmanian devils (*Sarcophilus harrisii*) (Andersen et al., 2020). As another example, predator feeding can sometimes also be detected by accelerometer signature profiles and/or the clustering of GPS locations (Anderson & Lindzey, 2003; Grünewälder et al., 2012). However, hunting behaviours can be difficult to identify by these devices because

hunts are often quick, meaning that they may occur between sampled GPS locations and because not all high acceleration signatures can be linked to hunts; for example, animals may be fleeing competitors or playing. As a result, quantifying the hunting success of predators and understanding how the environment affects hunting success is a challenge. However, animal-borne video cameras can be used to provide these additional contextual data and have been used to provide insights into hunting behaviours in a variety of species in terrestrial (McGregor et al., 2015), marine (Watanabe & Takahashi, 2013), and aerial environments (Kane & Zamani, 2014).

Similar to other devices, a common challenge with video recorders is how the animal's environment and behaviour may impact the data collection. For example, dust and mud can obscure what is happening, and for nocturnal species, video can be of limited use. To date, most animal-borne video studies have relied on standard (non-infrared) camera modules, presumably due to size and power consumption limitations. However, moving forward, we expect that infrared cameras, better suited to nocturnal species, will become more widely used.

6.2.2.4 Other animal-borne technologies

In addition to the devices discussed in this chapter so far, there are a range of other devices that can be applied to wildlife research. In Oceanography, for example, animal-borne devices paired with pressure/depth sensors can be used to collect high-resolution data on ocean environments, including temperature, salinity, chlorophyll content, and fluorescence. They can turn animals into collectors of fine-scale spatial and temporal data (Harcourt et al., 2019). GPS trackers have also been combined with animal-borne radar detectors attached to seabirds to detect illegal fishing vessels in the Southern Ocean (Weimerskirch et al., 2020).

The examples provided above are not exhaustive but are given to provide an understanding of the diversity of sensors available, their potential applications, and how their use alongside tracking devices can benefit research and conservation. For further detail, a review of animal-borne devices for ocean observations can be found in Harcourt et al. (2019), and a general review of animal-borne devices can be found in Wilmers et al. (2015).

6.2.3 Data analyses: software in use

As animal-borne technologies improve, so do the software used to process and analyse their data. High-resolution data collected by animal-borne devices can violate many of the underlying assumptions of classical statistical tests and may require the use of advanced statistical approaches that consider unusual properties of the data. Accelerometer data collected at high sampling frequencies, for example, is often temporally autocorrelated. Further, researchers are often interested in using animal-borne data to model specific processes, such as habitat selection or energetic expenditure, requiring highly specialized statistical approaches. Consequently, R is a particularly popular programming language when analysing animal-borne data because it is open-source and has a range of prewritten code, contributed by other users and packaged as *functions*. These functions can be used to perform relatively complex computations, such as defining species' home ranges and quantifying interactions between individuals in a few lines of code. Many of the approaches to analyse animal-borne data can be found in *R,* and you will likely find that many of the animal-borne study papers you read will use R *(for a review, see* Joo et al., 2020).

GIS packages, including QGIS (open-source) and ArcGIS (paid), are also commonly used software by researchers and allow animal-borne data to be analysed and visualized spatially. For example, by overlaying animal-borne data with other layers of information, such as satellite acquired vegetation maps, animal movements can be visualized relative to environmental features. While this spatial visualization can also be carried out *via* statistical languages, such as *R,* the difficulty and length of time in learning statistical programming languages generally versus the intuitive interfaces of many GIS software mean that a combination of software is often used. For example, data might be initially processed and analysed in *R* before being exported for visualization in some GIS software. These software examples are non-exhaustive but are provided to give a general introduction to the most commonly used software that one might expect to use with animal-borne data.

6.3 Case study: using tracking technologies to identify key habitats for conservation

In conservation, tracking technology is widely used to identify key habitats and areas for species protection (Cushman et al., 2018). One strategy for habitat conservation is to prioritize the protection of areas with high ecological value. Areas of Ecological Significance (AES), for example, are indicators of high levels of lower-trophic biomass and biodiversity. In other words, these areas have significant overlap in features that are ecologically important and so support higher biodiversity (Hindell et al., 2020). However, identifying important habitat areas at an ecosystem scale presents a significant challenge because data must be collected at large spatial scales from many different individuals and over long periods. This challenge is further amplified in environments where the logistical challenges of finding and monitoring animals make it difficult to collect such tracking data.

The harsh marine environment of the Southern Ocean is home to a diverse range of endemic species and is critical in many global biogeochemical cycles and climate systems (Chown & Brooks, 2019). However, many marine mammal and fish species in the Southern Ocean are recovering from historic human overexploitation, and the region is still the focus of large-scale squid, toothfish, and krill harvesting, species that are key to the Southern Ocean food web (Chown & Brooks, 2019). Further, the area faces rapid changes due to anthropogenic greenhouse gas emissions (Swart et al., 2018).

The following case study shows how satellite tracking data from multiple species can be aggregated to help guide ecosystem-level conservation efforts. Specifically, Hindell et al. (2020) used spatial aggregations of predators in the Southern Ocean to identify areas important to the predator species themselves, and that can be classed as AES. To understand the impacts of human exploitation and climate change, they then assessed the extent of

AES overlap with fishing effort and modelled likely changes to the distribution of these areas under future climate scenarios.

6.3.1 Methods

To identify AES, Hindell et al. (2020) combined data from over 70 international contributors and research programmes, spanning 25 years from 1991 to 2016 (Ropert-Coudert et al., 2020). The data set included 4060 individual animals fitted with animal-borne devices, including 12 species of seabird and 5 species of marine mammal, encompassing over 2.9 million at-sea locations. Given the range of species, body sizes, and technological developments over the 25-year period, data were collected from various tracking technologies that employed one of three positioning techniques: light-level geolocators, Argos platform terminal transmitters, and GPS tags. The species selected for this data set had circumpolar distributions and represented top and mesopredators with a variety of ecological niches and life-history traits. By focusing on widely distributed species that accessed the breadth of the ocean (such as shallow and deep divers and inshore and offshore foragers) and excluding species with limited distributions, it was possible to assess AES Significance across the full expanse of the Southern Ocean.

To accurately model habitat selection for each species, tracking data were identified to the different life-history stages, for example, breeding or non-breeding, because habitat use and importance vary between these stages. Habitat importance values were generated for each species and life-history stage, and these were combined into community-level habitat importance values. To account for differences in species richness, the researchers calculated habitat importance values separately for the sub-Antarctic and Antarctic regions and then combined them. The top 10% of habitat importance values were used to identify AES.

6.3.2 Results/discussion

The AES identified by Hindell et al. (2020) in the Southern Ocean coincided with regions of known high productivity in krill and lower-trophic level species. Human disturbance from fishing and climate change was disproportionately distributed within these areas' boundaries rather than outside. Hindell et al. (2020) found that many of the existing and proposed Marine Protected Areas occurred within AES, but they also recommended that protected areas be extended to include other identified areas and incorporate predicted movements driven by climate change.

This study highlights the power of animal-borne tracking technologies to inform conservation and management actions. The combination of data sets obtained over nearly three decades of tracking enabled researchers to examine AESs at the scale of an entire ocean and reinforces the importance of data sharing when addressing conservation objectives. As the barriers to tracking technologies (e.g. cost and size) decrease, this push towards larger aggregated data sets for understanding species' movements beyond individual wildlife populations and for informing ecosystem-level conservation is something that we predict will become increasingly popular and powerful. We will explore this further in the final section of this chapter.

6.4 Limitations and constraints

In this chapter, we have only briefly alluded to some of the challenges of using animal-borne technologies in wildlife research, focusing instead on the opportunities that they provide. In the following section, we discuss four of the most common challenges and considerations researchers face in using these devices. This list is not exhaustive and there are also many species- and environment-specific challenges that can arise. However, the following overview will introduce the challenges that we feel are most commonly encountered.

6.4.1 Sample sizes

In the case of tracking studies, sample size refers to the number of individuals (or social units) from whom tracking data are collected. In studies of animal movements, depending on the study questions, sample sizes of 20 to 100 are recommended to make accurate inferences about the study population (Latham et al., 2015). However, sample sizes are often constrained by expected (e.g. project budget, rarity of species) and unexpected (e.g. equipment failure) challenges (Thomas et al., 2011; Latham et al., 2015). As a result, many studies fall well short of the recommended sample sizes (e.g. Gese et al., 2016; Hubel et al., 2018).

One solution to overcoming small sample sizes is through data collaboration and sharing, for example, through open data initiatives such as Movebank, a free online data repository for wildlife tracking data (movebank.org; Kranstauber et al., 2011). The '*Jaguar Movement Database*' is also a good example of data collaboration (Morato et al., 2018). The database compiles locational data from 117 jaguars (*Panthera onca*) and covers a large portion of the species' geographical range, offering greater opportunities for understanding the ecology of this wide-ranging apex predator than any single study alone. In many cases, however, data sharing is limited by concerns around data security, accessibility, and publication rights—the latter resulting from the academic output system (Hampton et al., 2013). Future prospects on data sharing are discussed later in the chapter.

Another potential solution to small sizes, depending on the scale of the question and the study system, is changing the data collection method. For example, commercial GPS radio collar costs for terrestrial carnivores range from USD $1,200 to $8,000[5], and for studies on resource selection, sample sizes of > 30 are recommended. This means that costs for resource selection analyses for one species can exceed $36,000 (Latham et al., 2015). A reasonable alternative in this scenario might be to use camera traps, which can estimate resource selection metrics for multiple species, have lower unit costs, and are non-invasive (Wearn & Glover-Kapfer, 2017).

6.4.2 Device weight limitations and animal welfare

The impacts of animal-borne technologies are device- and species-specific and depend on several considerations, such as device weights, dimensions,

[5] Costs from Latham et al. (2015)

attachment methods, species' locomotion, and habitat types (Kenward, 2000; Peniche et al., 2011; Bodey et al., 2018). Although some studies provide guidelines to mitigate long-term harm to animals, such as 2–5% of device to body mass thresholds (Kenward, 2000; Smircich & Kelly, 2014), some studies ignore such guidelines, and devices within this threshold can still impact fitness for some species (O'Mara et al., 2014; Rasiulis et al., 2014). A meta-analysis of 214 studies using animal-borne devices on birds, for example, found that units negatively impacted species survival, reproduction, and parental care (Bodey et al., 2018).

Handling animals to attach devices can also negatively impact species, for example, by increasing animal stress levels and injury during struggles (Dennis & Shah, 2012). For some species, such as large carnivores, the potential risks that animals pose to themselves and researchers while being handled necessitate the use of anaesthesia for the safe attachment of animal-borne devices. This can, however, also present immediate risks to the sedated animal. Adverse reactions to anaesthesia can cause mortality (Arnemo et al., 2014), and can have longer-term negative impacts (hours to weeks). For example, Cattet et al., (2008) found that black bear (*Ursus americanus*) and grizzly bear (*Ursus arctos*) movements decreased for several weeks after handling and that animal body condition deteriorated as the number of times they had been captured increased. As a result, ethics permits are typically required by research institutes and wildlife management authorities, depending on the region, and a series of guidelines are typically administered, such as the use of a registered veterinarian for anaesthesia use.

Even in cases where no deleterious effects of animal-borne devices have been demonstrated, researchers have an ethical obligation to limit device sizes, weights, and any other associated impacts, where possible, so that devices are appropriate for both the species studied and research question. Although the minimum sizes of devices have decreased significantly over time, a trade-off typically exists between device size and battery longevity (Mitchell et al., 2019). Smaller, lighter devices typically have reduced battery life and/or functionality. For many species, device weights are

solely on satellite tracking (personal observation). Further, although satellite data may be encrypted, it can also be intercepted during transfer or by hacking into servers. In this way, there is a risk that near real-time locational data can be accessed and used for nefarious means. While we have not found published examples of this, broader data on animal locations have been used to bring some species close to extinction (e.g. Stuart et al., 2006). Additionally, the possibility of such data getting into the wrong hands extends to the well-meaning sharing of information in scientific papers. While some open-accesses journals (e.g. *PLoS One*) have data-sharing exemptions for threatened species, scientists themselves should carefully consider the conservation impacts of indiscriminate open-data sharing (Lindenmayer & Scheele, 2017).

In each of the contexts described earlier, data management is key. In this regard, Lennox et al., (2020) provide a comprehensive overview and framework for data protection. There is also a need to adopt clear guidelines that protect at-risk animals and consider privacy concerns of the people that might be affected by the deployments of animal-borne devices. We expect significant discussion and advances in this regard over the coming years.

6.6 Future directions

In this chapter, we have covered the main animal-borne technologies currently in use in wildlife studies and provided examples of how these technologies are being used in research and conservation. In the final part of the chapter, we will discuss some of what we consider to be the most promising developments in animal-borne technologies, with specific examples of active organizations and developments in these areas.

6.6.1 Open-source hardware

Open-source hardware, hereafter open hardware, refers to devices whose design specifications are openly available for anyone with the appropriate resources and skillsets to create, modify, or distribute. Compared to commercially manufactured devices, open hardware provides opportunities for

conservationists to more easily customize technologies for project- and species-specific uses and to acquire devices at lower costs (Hill et al., 2019). Although open hardware is already being used within research efforts (e.g. Bridge et al., 2019; Rafiq et al., 2019), barriers including user inexperience in device manufacture and limited support from creators' limits device uptake (Hill et al., 2019). As a result, there tends to be much research duplication around open hardware devices, with separate research groups often developing devices that solve similar needs rather than developing upon existing solutions (Joppa, 2015).

Audiomoth, an acoustic logger, is undoubtedly one of the most successful examples of open hardware in conservation (Hill et al., 2018). This is largely due to its partnership with the Arribada Initiative (arribada.org), which provides development, manufacturing, and customer-facing support for researchers developing open hardware for conservation. This partnership's success is highlighted by the fact that from October 2017 to March 2019, 5242 Audiomoth devices were purchased for research on a range of species across the globe (Hill et al., 2019). While not an animal-borne device, Audiomoth is a useful example of how open hardware, when done correctly, can be used to increase research capacity, and it has recently led to the development of an open hardware animal-borne audio recorder, μMoth.[6]

In the realm of tracking technologies, promising open hardware devices currently under development include OpenCollar[7] and the Arribada Initiatives' Horizon GPS Argos tag.[8] The latter has recently gone into its first round of production and offers access to a range of onboard sensors, such as an accelerometer and pressure sensor, as well as Bluetooth and cell connectivity, all for a significantly lower cost compared to commercial units.

A common problem with open hardware is that development and support for a particular solution can wane over time, particularly if community uptake is minimal or development funding

---

[6] https://www.openacousticdevices.info/mmoth
[7] https://opencollar.io/
[8] http://arribada.org/product/arribada-horizon-artic-r2-developers-kit/

is not renewed. This is particularly pertinent for animal-borne devices given the welfare and ethical implications of attaching devices that directly or indirectly compromise animal welfare, for example, through inappropriate use or device failure. This is also a risk with commercial technologies. It could be argued that, at this stage, the higher levels of support provided by companies and their greater investments into research and development mean that their commercial devices are less likely to compromise animal welfare in research applications. However, as the conservation technology discipline matures, collaborations strengthen, and funding sources become more secure (e.g. through innovative open-source revenue models), we predict the barriers to open hardware use will continue to decrease, with organizations[9] and members of the conservation technology community following the Audiomoth model.

6.6.2 Open data in animal tracking

One of the challenges of using data collected from animal-borne devices is collecting data from enough individuals and over large enough scales to make reliable population and ecosystem-level inferences. In some cases, this challenge can be overcome with the reduced per-unit costs offered by open hardware and the collaborative purchasing of components. Another solution is through greater research collaboration and data sharing between researchers, allowing research questions and conservation management challenges to be addressed at greater spatial and temporal scales (Hampton et al., 2013). However, while this collaborative approach is needed for research on some of society's most pressing challenges, such as climate change, barriers surrounding data collection, storage, analyses, and ownership have largely limited its use in ecology (Farley et al., 2018).

Open data initiatives, such as Movebank, improve the potential for collaboration, which we feel has not reached its potential. For a large part, this is due to the competitive nature of the industries in which conservation technology occurs. Both technology development and academia reward success in a competitive market and thus stifle collaboration, which ultimately impedes the speed of progress (Joppa, 2015). Addressing this challenge requires a change in the culture of the institutions paying for the data and research staff, and the broader academic environment. At the international level, organizations in the private and public sectors are already actively supporting the shift to open data initiatives, such as Microsoft's *Planetary Computer* initiative[10] and the United Nations call for a digital ecosystem for the planet (UNEP, 2019).

As device costs fall and capabilities increase, the quantity and diversity of data generated are also rapidly increasing. This is leading to an increasing understanding of and requirement for '*big data*' approaches in ecology and conservation. Big data is a term that typically refers to large, complex data sets that '*exceed the capacity or capability of current or conventional methods and systems*' (Ward & Barker, 2013). As such, it presents several challenges in terms of storage and processing power that may be particularly profound in wildlife research broadly and with animal-borne devices in particular.

We expect that the barriers to big data will be largely overcome within the next decade and that the use of big data in ecology will follow trends seen in other areas of society and become much more commonplace. Low-cost devices are already increasing the accessibility of animal-borne devices to wider audiences and contributing to the generation of larger data sets. Icarus, for example, is an initiative between the Max Planck Institute of Animal Behaviour and state-space agencies (e.g. NASA and the ESA) and uses receivers on the International Space Station to collect locations from animal-borne devices.[11]

6.6.3 Machine learning on the edge

Machine learning is a subset of artificial intelligence and allows applications to improve automatically through experience (see Chapter 11). Machine

---

[9] Key organizations ins field include Conservation X Labs (conservationxlabs.com), the Arribada Initiative (arribada.org) and WILDLABS (wildlabs.net),

[10] https://innovation.microsoft.com/en-us/planetary-computer
[11] https://www.icarus.mpg.de/en

learning algorithms are thus a series of rules or processes for one particular use (e.g. identifying whales from images or gunshots from audio) that improve automatically as more data are fed into them. In the last decade, and particularly within the last few years, machine learning applications have increasingly made their way into wildlife research and conservation (Valletta et al., 2017). They have, for example, been used in applications to reduce the time required to pre-process and curate data (Norouzzadeh et al., 2018) and improve inferences of complex data sets (Shoemaker et al., 2018).

Traditionally, machine learning algorithms have been computationally and power-intensive, meaning that their use with animal-borne devices has largely been restricted to when the data has already been collected and can be processed on personal computers or the cloud. Edge machine learning, in contrast, is the use of machine learning algorithms directly on deployed hardware (Yazici et al., 2018). By processing data on local hardware, edge machine learning reduces costs (time and money) of transferring data from devices to end-users and lowers security risks. Increasingly, many of these algorithms are being run on smaller and smaller consumer devices, such as smartwatches and smartphones, and recent developments in low-power machine learning algorithms for miniature internet-of-things sensors offer exciting opportunities for animal-borne devices (Warden & Situnayake, 2020). For example, machine learning could be used to more efficiently switch tracking technologies between different sampling frequencies, based on behaviours of interest, to maximize data resolutions and deployment durations. Alternatively, machine learning could be used to turn animals into environmental sensors for illegal activities by, for example, detecting gunshot signals using audio recorders and satellite transmitting locations to authorities.

6.7 Closing remarks

Animal-borne technologies have revolutionized our capacity to study species' movement and behaviour across the globe, and to collect data about the environments in which they live and move. Collectively, these insights have helped to guide conservation actions. The increasingly rapid development

and miniaturization of sensors are likely to significantly influence what is likely to be a critical period in wildlife research and conservation management. As a result of this miniaturization and the improved capability of individual sensors, it is increasingly feasible to combine multiple sensors in the same animal-borne device. This expands the number of species and scenarios to which they can be applied, essentially broadening the diversity of both applied and pure problems that conservation technology can contribute to solving. In addition, the mass production of device components for industry applications is helping to drive down costs and access. Coupled with improving collaboration through open-source projects, these devices are providing extraordinary opportunities to obtain comprehensive, data-rich information about animals' lives relevant for their management and conservation. These new advances also allow a move from relatively passive uses of animal-borne data and retrospective analyses to real-time responses, such as those needed in the study and management of human-wildlife conflict.

In the past 70 years, conservation technology has advanced considerably. While animal-borne devices are not a panacea in themselves, when intelligently integrated with other disciplines, including ecology, climate science, and social science, animal-borne technologies have almost unlimited potential in wildlife research and conservation.

References

Abrahms, B., Jordan, N. R., Golabek, K. A., McNutt, J. W., Wilson, A. M., & Brashares, J. S. (2016). Lessons from integrating behaviour and resource selection: activity-specific responses of African wild dogs to roads. *Animal Conservation*, 19(3), 247–255.

Alkon, P. U., Cohen, Y., & Jordan, P. A. (1989). Towards an acoustic biotelemetry system for animal behavior studies. *The Journal of Wildlife Management*, 53, 658–662.

Allan, B. M., Nimmo, D. G., Ierodiaconou, D., VanDerWal, J., Pin, K. L., & Ritchie, E. G. (2018). Futurecasting ecological research: the rise of technoecology. *Ecosphere*, 9, e02163.

Andersen, G. E., McGregor, H. W., Johnson, C. N., & Jones, M. E. (2020). Activity and social interactions in a wide-ranging specialist scavenger, the Tasmanian devil (*Sarcophilus harrisii*), revealed by animal-borne video collars. *PLoS One*, 15, e0230216.

Anderson, C. R., & Lindzey, F. G. (2003). Estimating cougar predation rates from GPS location clusters. *The Journal of Wildlife Management*, 67, 307–316.

Argos (2020). *Argos User Manual*. Available at: https://www.argos-system.org/manual/

Arnemo, J. M., Evans, A. L., Fahlman, Å., & Caulkett, N. (2014). Field Emergencies and Complications. In West, G., Heard, D., & Caulkett, N. (eds.). *Zoo animal and wildlife immobilization and anesthesia*. John Wiley & Sons, Hoboken, NJ (pp. 139–147).

Bächler, E., Hahn, S., Schaub, M., et al. (2010). Year-round tracking of small trans-Saharan migrants using light-level geolocators. *PLoS One* 5, e9566.

Bailey, T. N. (2005). *The African Leopard: Ecology and Behavior of a Solitary Felid*. The Blackburn Press, Caldwell, NJ.

Bandivadekar, R. R., Pandit, P. S., Sollmann, R., et al. (2018). Use of RFID technology to characterize feeder visitations and contact network of hummingbirds in urban habitats. *PLoS One*, 13, e0208057.

Benson, E. (2012). One infrastructure, many global visions: the commercialization and diversification of Argos, a satellite-based environmental surveillance system. *Social Studies of Science*, 42, 843–868.

Bidder, O. R., Walker, J. S., Jones, M. W., et al. (2015). Step by step: reconstruction of terrestrial animal movement paths by dead-reckoning. *Movement Ecology*, 3, 23.

Bischof, R., Gjevestad, J. G. O., Ordiz, A., Eldegard, K., & Milleret, C. (2019). High frequency GPS bursts and path-level analysis reveal linear feature tracking by red foxes. *Scientific Reports*, 9, 1–13.

Bodey, T. W., Cleasby, I. R., Bell, F., et al. (2018). A phylogenetically controlled meta-analysis of biologging device effects on birds: deleterious effects and a call for more standardized reporting of study data. *Methods in Ecology and Evolution*, 9, 946–955.

Bonter, D. N. & Bridge, E. S. (2011). Applications of radio frequency identification (RFID) in ornithological research: a review: RFID applications in ornithology. *Journal of Field Ornithology*, 82, 1–10.

Bridge, E. S., Wilhelm, J., Pandit, M. M., et al. (2019). An Arduino-based RFID platform for animal research. *Frontiers in Ecology and Evolution*, 7, 257.

Brown, D. D., Kays, R., Wikelski, M., Wilson, R., & Klimley, A. P. (2013). Observing the unwatchable through acceleration logging of animal behavior. *Animal Biotelemetry*, 1, 20.

Bruce, S. J., Zito, S., Gates, M. C., et al. (2019). Predation and risk behaviors of free-roaming owned cats in Auckland, New Zealand via the use of animal-borne cameras. *Frontiers of Veterinary Sciences*, 6, 205.

Cattet, M., Boulanger, J., Stenhouse, G., Powell, R. A., & Reynolds-Hogland, M. J. (2008). An evaluation of long-term capture effects in ursids: implications for wildlife welfare and research. *Journal of Mammalogy*, 89, 973–990.

Chetouane, F. (2015). An overview on RFID technology instruction and application. *IFAC-PapersOnLine*, 15th IFAC Symposium on Information Control Problems in Manufacturing, 48 (pp. 382–387).

Chisholm, S., Stein, A. B., Jordan, N. R., et al. (2019). Parsimonious test of dynamic interaction. *Ecology and Evolution*, 9, 1654–1664.

Chown, S. L., & Brooks, C. M. (2019). The state and future of Antarctic environments in a global context. *Annual Review of Environment and Resources*, 44, 1–30.

Cooke, S. J., Midwood, J. D., Thiem, J. D., et al. (2013). Tracking animals in freshwater with electronic tags: past, present and future. *Animal Biotelemetry*, 1, 5.

Costa, D. P., Robinson, P. W., Arnould, J. P. Y., et al. (2010). Accuracy of ARGOS locations of pinnipeds at-sea estimated using Fastloc GPS. *PLoS One* 5, e8677.

Couchoux, C., Aubert, M., Garant, D., & Réale, D. (2015). Spying on small wildlife sounds using affordable collar-mounted miniature microphones: an innovative method to record individual daylong vocalisations in chipmunks. *Sci Rep*, 5, 10118.

Cushman, S. A., Elliot, N. B., Bauer, D., et al. (2018). Prioritizing core areas, corridors and conflict hotspots for lion conservation in southern Africa. *PLoS One*, 13, e0196213.

Dennis, T. E., & Shah, S. F. (2012). Assessing acute effects of trapping, handling, and tagging on the behavior of wildlife using GPS telemetry: a case study of the common brushtail possum. *Journal of Applied Animal Welfare Science*, 15, 189–207.

Dewhirst, O. P., Evans, H. K., Roskilly, K., Harvey, R. J., Hubel, T. Y., & Wilson, A. M. (2016). Improving the accuracy of estimates of animal path and travel distance using GPS drift-corrected dead reckoning. *Ecology and Evolution*, 6, 6210–6222.

Dexter, C. E., Appleby, R. G., Edgar, J. P., Scott, J., & Jones, D. N. (2016). Using complementary remote detection methods for retrofitted eco-passages: a case study for monitoring individual koalas in south-east Queensland. *Wildlife Research*, 43, 369.

Dos Santos, G. A. M., Barnes, Z., Lo, E., et al. (2014). Small Unmanned Aerial Vehicle System for Wildlife Radio Collar Tracking. In 2014 IEEE 11th International Conference on Mobile Ad Hoc and Sensor Systems. Presented at the 2014 IEEE 11th International Conference on Mobile Ad Hoc and Sensor Systems (MASS), Philadelphia, PA, USA, IEEE. (pp. 761–766).

Drewe, J. A., Weber, N., Carter, S. P., et al. (2012). Performance of proximity loggers in recording intra- and inter-species interactions: a laboratory and field-based validation study. *PLoS One*, 7, e39068.

Du Preez, B., Hart, T., Loveridge, A. J., & Macdonald, D. W. (2015). Impact of risk on animal behaviour and habitat transition probabilities. *Animal Behaviour*, 100, 22–37.

Dulau, V., Pinet, P., Geyer, Y., et al. (2017). Continuous movement behavior of humpback whales during the breeding season in the southwest Indian Ocean: on the road again! *Movement Ecology*, 5, 11.

Elbroch, L. M., & Quigley, H. (2017). Social interactions in a solitary carnivore. *Current Zoology*, 63, 357–362.

Fais, A., Johnson, M., Wilson, M., Aguilar Soto, N., & Madsen, P. T. (2016). Sperm whale predator-prey interactions involve chasing and buzzing, but no acoustic stunning. *Scientific Reports*, 6, 1–13.

Farley, S. S., Dawson, A., Goring, S. J., & Williams, J. W. (2018). Situating ecology as a big-data science: current advances, challenges, and solutions. *BioScience*, 68, 563–576.

Federal Communications Commission (2020). FCC online table of frequency allocations. Available at: https://transition.fcc.gov/oet/spectrum/table/fcctable.pdf

Fehlmann, G., O'Riain, M. J., Kerr-Smith, C., et al. (2017). Extreme behavioural shifts by baboons exploiting risky, resource-rich, human-modified environments. *Scientific Reports*, 7, 15057.

Fieberg, J., Matthiopoulos, J., Hebblewhite, M., Boyce, M. S., & Frair, J. L. (2010). Correlation and studies of habitat selection: problem, red herring or opportunity? *Philosophical Transactions of the Royal Society of London B: Biological Sciences*, 365, 2233–2244.

Fisher, K. E., Adelman, J. S., & Bradbury, S. P. (2020). Employing very high frequency (vhf) radio telemetry to recreate monarch butterfly flight paths. *Environmental Entomology*, 49, 312–323.

García-Ripollés, C., López-López, P., & Urios, V. (2010). First description of migration and wintering of adult Egyptian Vultures Neophron percnopterus tracked by GPS satellite telemetry. *Bird Study*, 57, 261–265.

Gese, E. M., Terletzky, P. A., & Cavalcanti, S. M. C. (2016). Identification of kill sites from GPS clusters for jaguars (*Panthera onca*) in the southern Pantanal, Brazil. *Wildlife Research*, 43, 130–139.

Godley, B. J., Richardson, S., Broderick, A. C., Coyne, M. S., Glen, F., & Hays, G. C. (2002). Long-term satellite telemetry of the movements and habitat utilisation by green turtles in the Mediterranean. *Ecography*, 25, 352–362.

González-Solís, J., Croxall, J. P., Oro, D., & Ruiz, X. (2007). Trans-equatorial migration and mixing in the wintering areas of a pelagic seabird. *Frontiers in Ecology and the Environment*, 5, 297–301.

Grünewälder, S., Broekhuis, F., Macdonald, D. W., et al. (2012). Movement activity based classification of animal behaviour with an application to data from cheetah (*Acinonyx jubatus*). *PLoS One* 7, e49120.

Habib, B., Shrotriya, S., Sivakumar, K., Sinha, P. R., & Mathur, V. B. (2014). Three decades of wildlife radio telemetry in India: a review. *Animal Biotelemetry*, 2, 4.

Hamede, R. K., Bashford, J., McCallum, H., & Jones, M. (2009). Contact networks in a wild Tasmanian devil (*Sarcophilus harrisii*) population: using social network analysis to reveal seasonal variability in social behaviour and its implications for transmission of devil facial tumour disease. *Ecology Letters*, 12, 1147–1157.

Hampton, S. E., Strasser, C. A., Tewksbury, J. J., et al. (2013). Big data and the future of ecology. *Frontiers in Ecology and the Environment*, 11, 156–162.

Harcourt, R., Sequeira, A. M. M., Zhang, X., et al. (2019). Animal-borne telemetry: an integral component of the ocean observing toolkit. *Frontiers in Marine Science*, 6, 326.

Hebblewhite, M., & Haydon, D. T. (2010). Distinguishing technology from biology: a critical review of the use of GPS telemetry data in ecology. *Philosophical Transactions of the Royal Society of London B: Biological Sciences*, 365, 2303–2312.

Hellgren, E. C. (1988). Use of breakaway cotton spacers on radio collars. *Wildlife Society Bulletin*, 16, 216–218.

Hernández-Pliego, J., Rodríguez, C., Dell'Omo, G., & Bustamante, J. (2017). Combined use of tri-axial accelerometers and GPS reveals the flexible foraging strategy of a bird in relation to weather conditions. *PLoS One*, 12(6), e0177892.

Heupel, M. R., Kessel, S. T., Matley, J. K., & Simpfendorfer, C. A. (2018). Acoustic telemetry. In Carrier, J. C., Heithaus, M. R., & Simpfendorfer, C. A. (eds.). *Shark Research: Emerging Technologies and Applications for the Field and Laboratory*. CRC Press, Boca Raton, FL (pp. 133–156).

Hill, A. P., Davies, A., Prince, P., Snaddon, J. L., Doncaster, C. P., & Rogers, A. (2019). Leveraging conservation action with open-source hardware. *Conservation Letters*, 12, e12661.

Hill, A. P., Prince, P., Covarrubias, E. P., Doncaster, C. P., Snaddon, J. L., & Rogers, A. (2018). AudioMoth: evaluation of a smart open acoustic device for monitoring biodiversity and the environment. *Methods in Ecology and Evolution*, 9, 1199–1211.

Hindell, M. A., Reisinger, R. R., Ropert-Coudert, Y., et al. (2020). Tracking of marine predators to protect Southern Ocean ecosystems. *Nature*, 580, 87–92.

Hoenner, X., Huveneers, C., Steckenreuter, A., et al. (2018). Australia's continental-scale acoustic tracking database and its automated quality control process. *Scientific Data*, 5, 1–10.

Hofman, M. P. G., Hayward, M. W., Heim, M., et al. (2019). Right on track? Performance of satellite telemetry in terrestrial wildlife research. *PLoS One*, 14, e0216223.

Holohil (2020). *Holohil Transmitters List*. Available at: https://www.holohil.com/transmitters/#tableall

Hooten, M. B., Scharf, H. R., & Morales, J. M. (2019). Running on empty: recharge dynamics from animal movement data. *Ecology Letters*, 22(2), 377–389.

Hubel, T. Y., Golabek, K. A., Rafiq, K., McNutt, J. W., & Wilson, A. M. (2018). Movement patterns and athletic performance of leopards in the Okavango Delta. *Proceedings of the Royal Society B: Biological Sciences*, 285, 20172622.

Hubel, T. Y., Myatt, J. P., Jordan, N. R., Dewhirst, O. P., McNutt, J. W., & Wilson, A. M. (2016). Energy cost and return for hunting in African wild dogs and cheetahs. *Nature Communications*, 7, 11034.

Joo, R., Boone, M. E., Clay, T. A., Patrick, S. C., Clusella-Trullas, S., & Basille, M. (2020). Navigating through the r packages for movement. *Journal of Animal Ecology*, 89, 248–267.

Joppa, L. N. (2015). Technology for nature conservation: an industry perspective. *Ambio*, 44, 522–526.

Jordan, N. R., Buse, C., Wilson, A. M., et al. (2017). Dynamics of direct inter-pack encounters in endangered African wild dogs. *Behavioral Ecology and Sociobiology*, 71, 115.

Kane, S. A., & Zamani, M. (2014). Falcons pursue prey using visual motion cues: new perspectives from animal-borne cameras. *Journal of Experimental Biology*, 217, 225–234.

Kennedy, P. J., Ford, S. M., Poidatz, J., Thiéry, D., & Osborne, J. L. (2018). Searching for nests of the invasive Asian hornet (*Vespa velutina*) using radio-telemetry. *Communications Biology*, 1, 1–8.

Kenward, R. E. (2000). *A Manual for Wildlife Radio Tagging*. Academic Press, Cambridge, MA.

Kissling, W. D., Pattemore, D. E., & Hagen, M. (2014). Challenges and prospects in the telemetry of insects. *Biological Reviews*, 89, 511–530.

Kolz, A. L. (1983). Radio frequency assignments for wildlife telemetry: a review of the regulations. *Wildlife Society Bulletin*, 11(1), 4.

Kranstauber, B., Cameron, A., Weinzerl, R., Fountain, T., Tilak, S., Wikelski, M., & Kays, R. (2011). The Movebank data model for animal tracking. *Environmental Modelling & Software* 26, 834–835.

Krull, C. R., McMillan, L. F., Fewster, R. M., et al. (2018). Testing the feasibility of wireless sensor networks and the use of radio signal strength indicator to track the movements of wild animals. *wilr* 45, 659–667.

Kumari, M., & Hasan, S. M. R. (2020). A new CMOS implementation for miniaturized active RFID insect tag and VHF insect tracking. *IEEE Journal of Radio Frequency Identification* 99, 1–1.

Ladds, M. A., Salton, M., Hocking, D. P., et al. (2018). Using acceleromgets of wild fur seals from captive surrogates. *PeerJ*, 6, e5814.

Latham, A. D. M., Latham, M. C., Anderson, D. P., Cruz, J., Herries, D., & Hebblewhite, M. (2015). The GPS craze: six questions to address before deciding to deploy GPS technology on wildlife. *New Zealand Journal of Ecology* 39, 11.

Lendvai, Á. Z., Akçay, Ç., Weiss, T., Haussmann, M. F., Moore, I. T., & Bonier, F. (2015). Low cost audiovisual playback and recording triggered by radio frequency identification using Raspberry Pi. *PeerJ*, 3, e877.

Lennox, R. J., Harcourt, R., Bennett, J. R., et al. (2020). A novel framework to protect animal data in a world of ecosurveillance. *BioScience*, 70(6), 468–476.

Lindenmayer, D., & Scheele, B. (2017). Do not publish. *Science*, 356, 800–801.

Loyd, K. A. T., Hernandez, S. M., Carroll, J. P., Abernathy, K. J., & Marshall, G. J. (2013). Quantifying free-roaming domestic cat predation using animal-borne video cameras. *Biological Conservation*, 160, 183–189.

Matthews, A., Ruykys, L., Ellis, B., et al. (2013). The success of GPS collar deployments on mammals in Australia. *Australian Mammalogy*, 35(1), 65.

McGregor, H., Legge, S., Jones, M. E., & Johnson, C. N. (2015). Feral cats are better killers in open habitats, revealed by animal-borne video. *PLoS One*, 10, e0133915.

Mech, D., & Barber, S. (2002). *A critique of wildlife radio tracking*. US National Parks Service, Washington, DC.

Meek, P. D., & Butler, D. (2014). Now we can 'see the forest and the trees too' but there are risks: camera trapping and privacy law in Australia. In camera trapping in wildlife research and management. In Meek, P., & Fleming, P. (eds.). *Camera Trapping: Wildlife Management and Research*. CSIRO Publishing, Melbourne, Australia (pp. 331–345).

Mikkelsen, L., Johnson, M., Wisniewska, D. M., et al. (2019). Long-term sound and movement recording tags to study natural behavior and reaction to ship noise of seals. *Ecology and Evolution*, 9, 2588–2601.

Mitchell, L. J., White, P. C. L., & Arnold, K. E. (2019). The trade-off between fix rate and tracking duration on estimates of home range size and habitat selection for small vertebrates. *PLoS One* 14, e0219357.

Montevecchi, W. A., Hedd, A., McFarlane Tranquilla, L., et al. (2012). Tracking seabirds to identify ecologically important and high risk marine areas in the western North Atlantic. *Biological Conservation*, 156, 62–71.

Morato, R. G., Thompson, J. J., Paviolo, A., et al. (2018). Jaguar movement database: a GPS-based movement dataset of an apex predator in the Neotropics. *Ecology*, 99, 1691–1691.

Moyers, S. C., Adelman, J. S., Farine, D. R., Thomason, C. A., & Hawley, D. M. (2018). Feeder density enhances house finch disease transmission in experimental epidemics. *Philosophical Transactions of the Royal Society B: Biological Sciences*, 373, 20170090.

Munden, R., Börger, L., Wilson, R. P., et al. (2019). Making sense of ultrahigh-resolution movement data: a new algorithm for inferring sites of interest. *Ecology and Evolution*, 9, 265–274.

Nachtsheim, D. A., Ryan, S., Schröder, M., et al. (2019). Foraging behaviour of Weddell seals (*Leptonychotes weddellii*) in connection to oceanographic conditions in the southern Weddell Sea. *Progress in Oceanography*, 173, 165–179.

Norouzzadeh, M. S., Nguyen, A., Kosmala, M., et al. (2018). Automatically identifying, counting, and describing wild animals in camera-trap images with deep learning. *Proceedings of the National Academy of Sciences*, 115, E5716–E5725.

Nunes-Silva, P., Hrncir, M., Guimarães, J. T. F., et al. (2019). Applications of RFID technology on the study of bees. *Insectes Sociaux*, 66, 15–24.

O'Bryan, L. R., Abaid, N., Nakayama, S., et al. (2019). Contact calls facilitate group contraction in free-ranging goats (*Capra aegagrus hircus*). *Frontiers of Ecology and Evolution*, 7, 73.

O'Mara, M. T., Wikelski, M., & Dechmann, D. K. N. (2014). 50 years of bat tracking: device attachment and future directions. *Methods in Ecology and Evolution*, 5, 311–319.

Owen, K., Kavanagh, A. S., Warren, J. D., Noad, M. J., Donnelly, D., Goldizen, A. W., & Dunlop, R. A. (2017). Potential energy gain by whales outside of the Antarctic: Prey preferences and consumption rates of migrating humpback whales (*Megaptera novaeangliae*). *Polar Biology*, 40(2), 277–289.

Pagano, A. M., Atwood, T. C., Durner, G. M., & Williams, T. M. (2020). The seasonal energetic landscape of an apex marine carnivore, the polar bear. *Ecology*, 101, e02959.

Pagano, A. M., Rode, K. D., Cutting, A., et al. (2017). Using tri-axial accelerometers to identify wild polar bear behaviors. *Endangered Species Research*, 32, 19–33.

Peniche, G., Vaughan-Higgins, R., Carter, I., Pocknell, A., Simpson, D., & Sainsbury, A. (2011). Long-term health effects of harness-mounted radio transmitters in red kites (*Milvus milvus*) in England. *Vet Record*, 169, 311.

Phillips, R. A., Silk, J. R. D., Croxall, J. P., Afanasyev, V., & Briggs, D. R. (2004). Accuracy of geolocation estimates for flying seabirds. *Marine Ecology Progress Series*, 266, 265–272.

Potts, J. R., Börger, L., Scantlebury, D. M., Bennett, N. C., Alagaili, A., & Wilson, R. P. (2018). Finding turning-points in ultra-high-resolution animal movement data. *Methods in Ecology and Evolution*, 9, 2091–2101.

Rafiq, K., Appleby, R. G., Edgar, J. P., et al. (2019). Open-DropOff: an open-source, low-cost drop-off unit for animal-borne devices. *Methods in Ecology and Evolution*, 10, 1517–1522.

Rafiq, K., Jordan, N. R., Meloro, C., et al. (2020a). Scent-marking strategies of a solitary carnivore: boundary and road scent marking in the leopard. *Animal Behaviour*, 161, 115–126.

Rafiq, K., Jordan, N. R., Wilson, A. M., et al. (2020b). Spatio-temporal factors impacting encounter occurrences between leopards and other large African predators. *Journal of Zoology*, 310(3), 191–200.

Rasiulis, A. L., Festa-Bianchet, M., Couturier, S., & Côté, S. D. (2014). The effect of radio-collar weight on survival of migratory caribou: collar weight effect on survival. *The Journal of Wildlife Management*, 78, 953–956.

Riekkola, L., Andrews-Goff, V., Friedlaender, A., Constantine, R., & Zerbini, A. N. (2019). Environmental drivers of humpback whale foraging behavior in the remote Southern Ocean. *Journal of Experimental Marine Biology and Ecology*, 517, 1–12.

Ropert-Coudert, Y., Van de Putte, A. P., Reisinger, R. R., et al. (2020). The retrospective analysis of Antarctic tracking data project. *Scientific Data*, 7, 94.

Schaller, G. B. (1976). *The Serengeti Lion: A Study of Predator-Prey Relations*. University of Chicago Press, Chicago, IL.

Sha, J. C. M., Kaneko, A., Suda-Hashimoto, N., et al. (2017). Estimating activity of Japanese macaques (*Macaca fuscata*) using accelerometers. *American Journal of Primatology*, 79, e22694.

Shoemaker, K. T., Heffelfinger, L. J., Jackson, N. J., Blum, M. E., Wasley, T., & Stewart, K. M. (2018). A machine-learning approach for extending classical wildlife resource selection analyses. *Ecology and Evolution*, 8, 3556–3569.

Shuert, C. R., Pomeroy, P. P., & Twiss, S. D. (2018). Assessing the utility and limitations of accelerometers and machine learning approaches in classifying behaviour during lactation in a phocid seal. *Animal Biotelemetry*, 6, 14.

Silva, R., Afán, I., Gil, J. A., & Bustamante, J. (2017). Seasonal and circadian biases in bird tracking with solar GPS-tags. *PLoS One*, 12, e0185344.

Smircich, M. G., & Kelly, J. T. (2014). Extending the 2% rule: the effects of heavy internal tags on stress physiology, swimming performance, and growth in brook trout. *Animal Biotelemetry*, 2, 16.

Stanley, C. Q., MacPherson, M., Fraser, K. C., McKinnon, E. A., & Stutchbury, B. J. M. (2012). Repeat tracking of individual songbirds reveals consistent migration timing but flexibility in route. *PLoS One*, 7, e40688.

Stowell, D., Benetos, E., & Gill, L. F. (2017). On-bird sound recordings: automatic acoustic recognition of activities and contexts. *IEEE/ACM Transactions on Audio, Speech, and Language Processing*, 25, 1193–1206.

Stuart, B. L., Rhodin, A. G. J., Grismer, L. L., & Hansel, T. (2006). Scientific description can imperil species. *Science*, 312, 1137.

Swart, N. C., Gille, S. T., Fyfe, J. C., & Gillett, N. P. (2018). Recent Southern Ocean warming and freshening driven by greenhouse gas emissions and ozone depletion. *Nature Geoscience*, 11, 836–841.

Taylor, P., Crewe, T., Mackenzie, S., et al. (2017). The Motus Wildlife Tracking System: a collaborative research network to enhance the understanding of wildlife movement. *Avian Conservation and Ecology*, 18(1), 8.

Testud, G., Vergnes, A., Cordier, P., Labarraque, D., & Miaud, C. (2019). Automatic detection of small PIT-tagged animals using wildlife crossings. *Animal Biotelemetry*, 7, 21.

Thomas, B., Holland, J. D., & Minot, E. O. (2011). Wildlife tracking technology options and cost considerations. *Wildlife Research*, 38, 653.

Thomson, J. A., Börger, L., Christianen, M. J. A., Esteban, N., Laloë, J.-O., & Hays, G. C. (2017). Implications of location accuracy and data volume for home range estimation and fine-scale movement analysis: comparing Argos and Fastloc-GPS tracking data. *Marine Biology*, 164(10), 204.

Tomkiewicz, S. M., Fuller, M. R., Kie, J. G., & Bates, K. K. (2010). Global positioning system and associated technologies in animal behaviour and ecological research. *Philosophical Transactions of the Royal Society of London B: Biological Sciences*, 365, 2163–2176.

UNEP (2019). Foresight *Brief: Building* a *Digital Ecosystem* for the *Planet*. Available at: https://wedocs.unep.org/handle/20.500.11822/30612

Valletta, J. J., Torney, C., Kings, M., Thornton, A., & Madden, J. (2017). Applications of machine learning in animal behaviour studies. *Animal Behaviour*, 124, 203–220.

Voas, J., & Kshetri, N. (2017). Human tagging. *Computer*, 50, 78–85.

Ward, J. S., & Barker, A. (2013). Undefined by data: a survey of big data definitions. *arXiv: 1309*. Available at: https://www.adambarker.org/papers/bigdata_definition.pdf

Warden, P., & Situnayake, D. (2020). *Machine Learning with TensorFlow Lite on Arduino and Ultra-Low-Power Microcontrollers*. O'Reilly, Newton, MA (p. 149).

Watanabe, Y. Y., Payne, N. L., Semmens, J. M., Fox, A., & Huveneers, C. (2019). Hunting behaviour of white sharks recorded by animal-borne accelerometers and cameras. *Marine Ecology Progress Series*, 621, 221–227.

Watanabe, Y. Y., & Takahashi, A. (2013). Linking animal-borne video to accelerometers reveals prey capture variability. *Proceedings of the National Academy of Sciences*, 110, 2199–2204.

Waugh, C., Khan, S. A., Carver, S., et al. (2016). A prototype recombinant-protein based chlamydia pecorum vaccine results in reduced chlamydial burden and less clinical disease in free-ranging koalas (*Phascolarctos cinereus*). *PLoS One*, 11(1), e0146934.

Wearn, O. R., & Glover-Kapfer, P. (2017). Camera-*Trapping For Conservation: A Guide* to *Best-Practices*. WWF Conservation Technology Series. WWF, Woking, UK.

Weimerskirch, H., Collet, J., Corbeau, A., et al. (2020). Ocean sentinel albatrosses locate illegal vessels and provide the first estimate of the extent of nondeclared fishing. *Proceedings of the National Academies of Sciences United States of America*, 117, 3006–3014.

Weise, F. J., Hauptmeier, H., Stratford, K. J., et al. (2019). Lions at the gates: trans-disciplinary design of an early warning system to improve human-lion coexistence. *Frontiers in Ecology and Evolution*, 6, 242.

Wijers, M., Trethowan, P., Markham, A., et al. (2018). Listening to lions: animal-borne acoustic sensors improve bio-logger calibration and behaviour classification performance. *Frontiers in Ecology and Evolution*, 6, 171.

Williams, H. J., Holton, M. D., Shepard, E. L. C., et al. (2017). Identification of animal movement patterns using tri-axial magnetometry. *Movement and Ecology*, 5, 6.

Wilmers, C. C., Nickel, B., Bryce, C. M., Smith, J. A., Wheat, R. E., & Yovovich, V. (2015). The golden age of bio-logging: how animal-borne sensors are advancing the frontiers of ecology. *Ecology*, 96, 1741–1753.

Wilson, A. M., Lowe, J. C., Roskilly, K., Hudson, P. E., Golabek, K. A., & McNutt, J. W. (2013). Locomotion dynamics of hunting in wild cheetahs. *Nature*, 498, 185–189.

Wilson, R. P., Börger, L., Holton, M. D., et al. (2020). Estimates for energy expenditure in free-living animals using acceleration proxies: a reappraisal. *Journal of Animal Ecology*, 89, 161–172.

Wisniewska, D. M., Johnson, M., Teilmann, J., et al. (2018). High rates of vessel noise disrupt foraging in wild harbour porpoises (*Phocoena phocoena*). *Proceedings of the Royal Society B: Biological Sciences*, 285, 20172314.

Yazici, M. T., Basurra, S., & Gaber, M. M. (2018). Edge machine learning: enabling smart internet of things applications. *big data and Cognitive Computing*, 2, 26.

CHAPTER 7

Field and laboratory analysis for non-invasive wildlife and habitat health assessment and conservation

Cheryl D. Knott, Amy M. Scott, Caitlin A. O'Connell, Tri Wahyu Susanto, and Erin E. Kane

7.1 Introduction

As human encroachment, habitat destruction, and climate change increasingly impact animal populations, conservationists need fast, reliable methods to study animal health status and physiology. Methodological advances have enabled lab-based sample preparation and some laboratory tests in field settings. The field labs we describe here are spaces involving sample preparation and preservation, with some on-site analysis, allowing researchers and conservationists to obtain unique and valuable information on animal health and physiology. Though some samples must still be analysed in traditional laboratories, field labs have continued to improve storage and transportation of samples for later analysis. Field conservationists can take advantage of technological advances by subjecting samples to cutting-edge techniques at off-site laboratories, aided by improved sample preservation *in situ*. Laboratory analyses of field-collected samples can examine habitat health, for example, by determining floristic species composition and food nutritional content. Field labs also improve inclusion and training of in-country students and researchers, facilitate capacity building, and develop local expertise.

Here we discuss the application of some of these techniques, including nutritional, faecal, and urine studies, hormonal analysis, parasitology, and genetics. Our particular focus is on conservation-relevant,

non-invasive analysis. We describe equipment that can be brought into the field to conduct these analyses and their range of applications, including sample preparation and on-site lab work. We show how these types of analyses have been applied to conservation, focusing on our area of expertise, wild primates. We provide a case study, based on our application of these methods, to research and conserve wild Bornean orangutans (*Pongo pygmaeus wurmbii*) in Gunung Palung National Park, Indonesia. We end by discussing some current limitations and constraints on field analysis, recommending approaches to overcome these problems, and predicting areas of future innovation.

7.1.1 What questions can be asked using field laboratory methods?

Understanding an animal's habitat requirements entails examining diets, the nutritional content of different foods, nutritional intake across habitats, and habitat quality in terms of temporal and spatial availability of macro- and micronutrients (Birnie-Gauvin et al., 2017). This enables prioritization of habitat for conservation, development of nutritionally relevant corridors through fragmented areas, and meeting the dietary needs of captive animals in *ex situ* conservation programmes (Robbins et al., 2004; Rode et al., 2006; Rodgers, 2017; Madliger et al., 2018).

Studying endangered species' physiological and disease status provides critical information on the effect of anthropogenic disturbances on wildlife health, distribution, and long-term viability. Establishing baseline health values allows field scientists to monitor populations for fluctuations that might reflect changing conservation status. Determining the social and ecological sources and impacts of stress is critical to monitoring animal health. Although stress responses are adaptive in the short-term, chronic stress activation can lead to long-term health deficits, which may contribute significantly to adult mortality (Sapolsky, 1992). Fluctuations in hormones like glucocorticoids, and measurement of urinary proteins such as C-peptide, reflect energy mobilization and provide important health status biomarkers. Tracking oestrus cycling to detect pregnancy and monitoring reproductive status in both females and males allow informed management decisions about the health and viability of endangered populations.

Other health indicators include parasites, broadly defined. When severe, parasites can impose significant immunological and metabolic costs and increase the risk of death and species extinction (Stoner, 1996; Smith et al., 2006). Intestinal parasites can be assessed non-invasively and provide a potential window into the health of threatened wild animal populations. In recent years, evidence in wild primates has emerged to support the presumption that intestinal parasites themselves are physiologically costly, rather than a by-product of other immune stressors (Friant et al., 2016; Akinyi et al., 2019; Müller-Klein et al., 2019). It is important to first establish species-typical infection patterns, including interindividual variation within a population. Because many wild animals are otherwise healthy despite parasite infection (Ashford et al., 1990; Stuart et al., 1993; Nunn & Altizer, 2006; O'Connell, 2018), establishing how other health measures covary with the presence of particular parasites provides insight into the significance of parasitic intestinal infection.

Genetic analysis is important to identify individuals, document species and subspecies distribution, monitor genetic diversity within and between populations, and track wildlife crime (Baker & Palumbi, 1994; Dalton & Kotze, 2011; Gonçalves et al., 2015; Mondol et al., 2015). Identifying the subspecies of illegally captured or poached animals can narrow down the location where they were captured. Genetic variation within a population is also an important measure of population health and viability. Documenting and preserving genetic variation and diversity are critical aspects of conservation (Frankham, 1995) because high genetic diversity is important for the sustainable conservation of a population and a species (Frankham, 1995; Lacy, 1997).

7.1.2 Traditional methods and limitations

When systematic wildlife studies first began over 50 years ago, these questions were unanswerable. Laboratory methods for analysing hormones, nutrition, and genetics did not exist, so field samples were not collected for these purposes. Research questions were answered through observation or animal capture. For example, questions about reproductive cycles and paternity were addressed through visual observations of mating, and where possible in some species, oestrous swelling cycles (Tutin, 1979; Wallis, 1992; Wallis, 1995b). This provided limited (and often erroneous) information because both fecundity and paternity had to be inferred from behaviour (Knott, 2001; Knott, 2005a). Comparison of genetically determined paternity against behaviourally and hormonally determined paternity show that the latter does not produce accurate results (de Ruiter et al., 1994). The development of laboratory analysis for steroid hormones and other biomarkers made it possible to examine reproductive status in captive settings, but the inherent fragility of these molecules initially made collecting and preserving samples from wild organisms functionally impossible (Behringer & Deschner, 2017).

Our understanding of diet was historically limited to species lists of plants consumed (Rothman et al., 2012). Caloric and nutrient intake could only be qualitatively assessed by reporting the percentage of different food categories eaten, but pooling food items by category obscures significant variation (Knott, 2005a; Lambert & Rothman, 2015). Time spent feeding was a common measure of food intake, but this is a particularly poor estimate of calorie intake as animals may feed longer on items that have low nutritional value and feeding rates vary between food items (Knott, 1998). Lucas and colleagues (2001) developed a field-kit for physico-chemical characterization of wild foods, including foods' physical properties (e.g. colour, mechanics), nutritional chemistry (e.g. phenols, amino acids), and spatial ecology (e.g. dispersion). Adopting these field kits helped propel knowledge, especially of foods' mechanical properties, but had limitations for examining other nutrients.

Faecal parasite samples are typically collected and preserved in fixative solution in the field for later analysis in a traditional lab setting. Parasites are concentrated from faecal samples using flotation and/or sedimentation, followed by microscope-based identification of parasite taxa using morphological characteristics. Examining intestinal parasites using microscopy alone cannot identify parasites to the species level in many cases, thus limiting reporting to general taxonomic classifications (Stuart & Strier, 1995; Gillespie, 2006) and preventing researchers from fully appreciating the extent of parasite richness within and between animal populations.

7.2 Development of new technology for sample preservation and analysis

With the advent of new and improved laboratory procedures to measure plant nutritional composition and assess animal physiological status through urine and faeces, new methods were developed to collect and preserve these samples in the field for laboratory analysis (Whitten et al., 1998; Conklin-Brittain et al., 1999; Rothman et al., 2012). Thus, field labs were established to process and preserve samples for later analysis or, where possible, conduct analyses in the field. In setting up field labs for wildlife conservation, researchers have adapted existing technology into a field setting and taken advantage of new and more portable devices for data collection and analysis. Many of these innovations are due to a readiness to try something for the first time and to think 'outside the box' about what is possible in the field.

Here we review three types of technological advances facilitating analysis of field-collected samples. The first are new methods of sample preservation, including drying or adding chemicals, for *ex situ* analysis. The second is the application of devices or methods, often relatively low tech, to directly analyse samples in the field. The third is the use of new portable and battery-operated devices that have allowed researchers to conduct some high-tech analyses in remote locations. As laboratory equipment becomes smaller and more compact, and electricity is more readily available in some remote field sites, transporting equipment to a field setting has become more feasible. Table 7.1 lists some of the pieces of equipment that have recently been introduced, or are being tested, in field lab settings.

7.2.1 Nutritional analyses

With more widespread laboratory methods to analyse plant nutrient composition, field researchers have developed methods to dry and preserve plants for later analysis. Initial processing involves separating samples into component parts (such as seed, pulp, and husk), wet weighing, and then drying (Conklin-Brittain et al., 1998; Knott, 1998; Rothman et al., 2012). Kerosene drying ovens are typically used, although care must be taken to ensure that the drying oven maintains a constant temperature of 40–55°C to keep the sample from burning or moulding. Commercially available electric food dehydrators provide an alternative drying method. These maintain a constant, easily controlled temperature, preventing damage to the sample (Rothman et al., 2012), but require at least 24 hours of electricity, draw a considerable amount of power, and hold limited samples. Though solar dehydrators are available, they tend to take too long to reach the desired temperature and samples are prone to mould or insect infestation. After drying, samples are reweighed and stored in sealed plastic bags with silica gel. Samples can be analysed in an *ex situ* nutritional chemistry lab for the percentage of ash, lipid, available protein, crude protein, carbohydrates, starch, neutral detergent fibre, acid detergent fibre, lignin, cutin, radial diffusion tannins, condensed tannins, and minerals (Knott, 1998; Conklin-Brittain et al., 2006; Rothman et al., 2012). Wet-chemistry analysis involves considerable time and is labour intensive. However, advances in

Table 7.1 Selected examples of equipment available for use in field laboratories

Equipment	Types of Analysis	Application	Dimensions/Weight	Cost (USD)
Infrared Spectroscopy	Identification of unknown molecules, quantification of nutritional content	Rapid analysis of nutritional content from plant samples	28 × 44 × 45 cm	$10,000–$80,000
Bento Lab System	DNA amplification and visualization	DNA extraction, preparation for downstream analysis, Sex-typing, checking for PCR product, etc.	33.3 × 21.4 × 8.1 cm, weighs 3.5 kg	$1600–1950
miniPCR	DNA amplification	Preparation for downstream analysis	5.1 × 12.7 × 10.2 cm, weighs 0.45 kg	$650–800
MiniOne PCR system	DNA amplification	Preparation for downstream analysis	27.9 × 27.9 × 33.0 cm, weighs 2.72 kg	$799
Franklin Real-Time PCR	DNA quantification	Pathogen detection and quantification, environmental DNA analysis, etc.	1.5 kg	$6000
Liberty 16 Real-Time PCR	DNA quantification	Pathogen detection and quantification, environmental DNA analysis, etc.	3.0 kg	contact manufacturer
blueGel Electrophoresis	DNA visualization	Sex-typing, checking for PCR product, etc.	23 × 10 × 7 cm, weighs 0.35 kg	$350
MiniOne Electrophoresis system	DNA visualization	Sex-typing, checking for PCR product, etc.	33 × 25.4 × 10.1 cm, weighs 1.36 kg	$279
MinION nanopore sequencer	DNA sequencing	DNA barcoding for species or subspecies identification	10 × 3.2 × 2 cm, weighs 90 g	$1000
Labomed CxL Trinocular Microscope	Observation and identification of parasites	Parasite species richness, virulence (egg counts), and prevalence	38.1 × 22.9 × 25.4 cm, weighs 4.1 kg	$800
Motic Moticam BTU8 microscope camera	Photographs and videos of faecal parasites	Documenting parasites observed on-site, identification of parasites off-site	21.5 × 12.8 × 0.09 cm	$640

conservation technology can speed analysis of field preserved samples through new laboratory methods and equipment.

One such method is to subject field-dried samples to near and mid-infrared spectroscopy (NIRS, MIRS). These techniques irradiate a substance with infrared light to record peaks in reflectance. Peaks are associated with molecular bonds present in different macro- and micronutrients; the spectra can thus be used to calculate nutrient composition (McKelvy et al., 1996; Coates, 2000; Pasquini, 2003; Dunham et al., 2016). These methods are widely applied in the agro-food industry (Teixeira dos Santos et al., 2013), and wildlife biologists have used this technology to more quickly and efficiently

analyse the nutrient content of foods eaten by wild primates (Foley et al., 1998; Felton et al., 2009; Rothman et al., 2009; Dunham et al., 2016). Infrared nutritional analyses rely on predictive equations based on models created from wet-chemistry analysis of a subset of plant data. These models must be calibrated for individual habitats of interest (Rothman et al., 2012) and rely on complete wet-chemistry analysis of hundreds of samples. After this calibration, it is possible to rapidly process several hundred samples per day. Spectrometers require instrument-specific software, but a chemometric program such as Pirouette (Infometrix Inc., Woodville, WA) facilitates statistical analysis and model creation (Dunham et al., 2016).

To date, nutritional analyses using IR spectroscopy for conservation applications are laboratory-based. However, a new phase of on-site nutritional analysis may be possible with recent advances in the development of smaller, portable benchtop NIRS machines and handheld devices (e.g. Agilent Technology's 4300 Handheld FTIR). As projects build large data sets of lab-based wet-chemistry nutritional analyses for important food items, researchers can bring infrared spectroscopic techniques into the field. This could revolutionize our understanding of animal nutrition and habitat variability as such devices could be used on fresh samples, and hundreds of samples could be processed per day (Vance et al., 2016).

Figure 7.1 Example of field lab setup for hormonal testing and sample preparation. Cheryl Knott pipetting wild orangutan urine on filter paper, with urinary chemstrips and hormonal test kits shown. Photograph © Tri Wahyu Susanto, Gunung Palung Orangutan Project.

7.2.2 Hormonal analyses

Reproductive hormones were first measured in human blood samples, using radioactively labelled antigens in a radioimmunoassay (Yalow & Berson, 1960). Eventually, methods were developed to measure hormones in urine using radioimmunoassay, and then enzyme immunoassays (which do not use radioactively-labelled antigens). These techniques were applied to samples from captive animals (Shideler et al., 1983; Monfort et al., 1986; Munro & Lasley, 1988; Kirkpatrick et al., 1990) and later applied to field-collected samples (Robbins & Czekala, 1997; Czekala & Sicotte, 2000). However, the application of these techniques in the wild was limited because samples needed to be kept frozen, which was not possible at most remote locations. Thus, one new method developed for hormone preservation was drying urine on filter paper for later laboratory analysis (Campbell, 1994; Shideler et al., 1995). This methodology was subsequently developed for use with wildlife in the field (Knott, 1997; Knott et al., 2007) (Figure 7.1). With the availability of electricity at some remote field sites and the development of small, portable freezers, urinary assays using frozen samples have become more common (Higham, 2016). Because freezing samples is not possible at many locations, techniques using new sample collection media, such as cotton swabs, are still being developed and tested (Danish et al., 2015).

Methods to analyse faecal hormones were developed concurrently with urinary analyses. Faeces is often both easier to collect than urine and can be preserved by drying, although gut retention time based on diet and body size can be confounding variables (Wasser et al., 1991; Shideler et al., 1994; Savage et al., 1997; Whitten et al., 1998; Foerster & Monfort, 2010). Solid-phase extraction (SPE) cartridges, small polypropylene cartridges with an absorbent material, provide another method for sample collection and preservation (Stavisky et al., 1995; Beehner & Whitten, 2004; Pappano et al., 2010; Santymire & Armstrong, 2010; Edwards et al., 2014). Promisingly, using SPE cartridges for field-based hormone extractions can yield results comparable to lab extraction techniques (Edwards et al., 2014). Faecal samples are collected, weighed, suspended in 90% methanol, hand-shaken for 5 minutes, and then filtered through moistened filter paper (Edwards et al., 2014). After samples are loaded into SPE cartridges using a solvent optimized for the hormone of interest, they can be stored at room temperature and laboratory analysis can proceed as though extraction also occurred in the lab (Edwards et al., 2014). Because some studies using SPE cartridges to store extracts have had poor recovery success upon re-extraction, the technique and loading solvents must be validated for each taxa and hormone metabolite of interest (Santymire & Armstrong, 2010).

There are currently no methods to quantitatively measure hormones directly in field labs. However, another potential use of infrared spectroscopy is its application to analyse urinary hormones. In a study by Kinoshita et al. (2016), urine was collected from captive and wild orangutans and frozen. Frozen samples were then applied to filter paper, analysed for creatinine and oestrogen using NIRS and ELISA, and the values compared. Values measured by ELISA and predicted by NIRS were highly correlated (Kinoshita et al., 2016). NIRS has also been used to track oestrogen in captive giant pandas (Kinoshita et al., 2010; Kinoshita et al., 2012) and parturition in captive snow leopards (Vance et al., 2016), suggesting that IR spectroscopy may be a promising technique for rapid non-invasive monitoring of reproductive status in wild settings, pending further validation.

Another area of innovation is the development of quick assay test kits. For example, human over-the-counter pregnancy test kits have been successfully used to detect pregnancy in wild orangutans (Knott et al., 2019a). There are also test kits using human saliva, urine, or blood from finger pricks to assess ovulation, testosterone, oestrogen, and follicle-stimulating hormone. It may be possible to use these tests for rapid hormonal assessment in non-humans. However, tests would need to be non-invasive and validated for these other taxa. An important caveat for all endocrine research is that comparing or combining endocrine values obtained using different types of samples, different preservation techniques, and different analytical methods is problematic and, if done, needs to show that the methods are highly correlated.

7.2.3 Urine and faecal health measures

Urinary dipsticks, or chemstrips, are commonly used in human medical laboratories and provided an early field technique to assess primate health status. They have been used successfully to examine health status in wild primates, particularly apes (Knott, 1998; Kaur & Huffman, 2004; Krief et al., 2005; Harrison et al., 2009; Leendertz et al., 2010; Erb et al., 2018), and some monkeys (MacIntosh et al., 2012). Dipstick tests are done immediately after urine collection in the field, measuring specific gravity and pH, and detecting the presence of leukocytes, nitrite, ketones, protein, glucose, urobilinogen, bilirubin, and blood in the urine. Depending on the taxa of interest, they may not be a sensitive enough measure to identify periods of low-to-moderate immunological or metabolic stress (Leendertz et al., 2010; Behringer et al., 2017; Cancelliere et al., 2018). However, chemstrips have been successfully validated and used to detect ketosis, periods of negative energy balance when fat is metabolized, in wild orangutans (Knott, 1998; Naumenko et al., 2019). Although specific gravity can also be measured using dipsticks, this measure is not very precise (de Buys Roessing et al., 2001). A better field solution is to measure urine samples' specific gravity using a handheld digital specific gravity refractometer, such as those manufactured by Atago. Urinary specific gravity, which is not affected by muscle mass, can be used to assess urinary water content and thus hydration (Emery Thompson et al., 2012a).

Another technique to assess health status in the field is measuring faecal temperature as a non-invasive way to determine internal body temperature. This technique was first developed by Jensen et al. (2009) for use with wild chimpanzees. Recently, the technique has been further validated in another chimpanzee population (Negrey et al., 2020) and in orangutans (Harwell et al., 2019). This method shows promise as a way to reveal signs of infection indicated by body temperatures elevated above baseline values. A handheld digital data-logging thermometer is inserted into the centre of the faecal deposit immediately after defecation. The peak temperature recorded, or repeated measures fit to a sigmoidal curve, is used to determine temperature. Time since defecation, time of measurement, animal height, ambient temperature, weight of faecal deposit, age/sex, and associated behavioural variables may all impact faecal temperature in addition to illness and should thus be considered (Jensen et al., 2009; Harwell et al., 2019; Negrey et al., 2020). Faecal temperature, together with laboratory-based analyses, can yield important health information and guide management decisions related to population health (Lorich, 2020).

Though some potentially useful biomarkers of immune function, such as circulating cytokine levels, can be most accurately assessed in plasma (Hoffman et al., 2011), there are continuing efforts to identify non-invasive indicators of immune system activation. Recent work has demonstrated that urinary biomarkers including neopterin, urokinase plasminogen activator receptor (suPAR), and cytokine metabolite reactive proteins also facilitate non-invasive assessment of health status by indicating immunological activation, particularly when examined with other health indicators including glucocorticoid concentrations (Downs & Steward, 2014; Heistermann & Higham, 2015; Higham et al., 2015, Higham et al., 2020).

7.2.4 Field microscopy for parasites

Having a microscope on-site for parasite analysis can eliminate the need for fixative solutions that may distort some parasites (Nurcahyo et al., 2017) and analysis of fresh samples may better indicate true parasitic infections (O'Connell, 2018). While microscopic analysis alone usually cannot identify parasites to the species level, on-site lab analysis provides a rapid assessment of interindividual variation in infection by different parasite taxa (Stuart & Strier, 1995). Given climatic risks to expensive equipment under field conditions, relatively inexpensive microscopes with high-quality optics are now available and can be brought into the field (Figure 7.2). These technological changes make analysis of fresh samples in the field possible. At our site, we have used a Labomed CxL trinocular microscope, a battery-powered LED light microscope, for over 5 years (O'Connell, 2018). A Moticam BTU8, an Android tablet with its own software attached to a microscopy imaging camera (5 MP), allows for easy and quick photography, measurement, editing, and sharing of microscope images of parasites. This setup is fairly inexpensive and lasts well under humid field conditions by storing in a sealed container with silica gel.

Molecular techniques can also more precisely classify parasites from wild animal populations (Nurcahyo et al., 2017), but these genetic analyses typically take place off-site in traditional lab

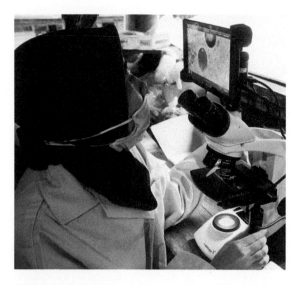

Figure 7.2 Ishma Fatiha Karimah using a Labomed CxL trinocular compound light microscope with a Moticam BTU8 microscope camera system attached for field lab parasitology. The tablet screen of the BTU8 allows for clear real-time viewing of sample slides, which also facilitates on-site training in parasitological techniques. Photograph © Tim Laman.

settings. New portable technologies such as the miniPCR™ and MinION™, along with advances in DNA barcoding capabilities (Section 7.2.5, next), may soon allow species-level identification of parasites in the field.

7.2.5 Genetic analysis

The innovation of PCR (polymerase chain reaction) in the 1980s revolutionized genetic methods by enabling rapid amplification of DNA, permitting genetic analysis of non-invasive samples with only small amounts of DNA (Woodruff, 2004). In the late 1990s, field primatologists began using genetics to study mating systems, social structure, dispersal patterns, relatedness, phylogeny, and biogeography (Di Fiore et al., 2011). Because DNA degrades rapidly, leading to failed experiments and erroneous results if not preserved, proper preservation is paramount. Hair, faeces, and buccal cells from food wadges are possible non-invasive DNA sources that can be preserved in 90% ethanol, RNAlater, or dried using a two-step alcohol-silica method (Nsubuga et al., 2004; Di Fiore et al., 2011). These preserve genetic material in samples at ambient temperature for months or years until they can be transported for analysis. In addition, portable PCR thermocyclers, electrophoresis systems, and DNA sequencers have come onto the market in the past decade, enabling some genetic analysis in field labs.

PCR thermocyclers amplify extracted DNA for downstream analyses, including capillary electrophoresis for individual identity and relatedness or sequencing of nucleotides for species identification and genome mapping. The miniPCR™ thermocycler (MiniPCR), which can be powered by a wall outlet, portable battery pack, or solar-powered battery, amplifies DNA effectively (Guevara et al., 2017; Pomerantz et al., 2018; González-González et al., 2019). The MiniOne PCR system is another portable field-appropriate DNA amplifier, with the additional benefit of keeping samples at 4°C (Maestri et al., 2019).

Real-time PCR (RT-PCR) thermocyclers, such as the Liberty 16 Real-Time PCR, monitor the quantity of DNA product as it is amplified. RT-PCR applications include quantifying PCR product, testing for the presence of pathogens, and measuring environmental DNA. The Franklin™ Real-Time PCR Thermocycler (Biomeme) is battery powered and has been validated, but has higher rates of false negatives in detecting environmental DNA than lab-based methods (Sepulveda et al., 2018). Gel electrophoresis can be used to visualize DNA fragment lengths and to check that conventional PCR has successfully amplified DNA for downstream analysis. Each of the companies producing portable PCR thermocyclers also produce electrophoresis systems. The portable blueGel™ (MiniPCR) electrophoresis system has been tested in the field and can be powered by a wall outlet, portable battery packs, or solar-powered battery (Guevara et al., 2017). The portable MiniOne electrophoresis system has also been used to verify PCR product in the field (Maestri et al., 2019).

Another new device is the MinION™ nanopore sequencer (Oxford Nanopore Technologies), a portable, rapid, and relatively inexpensive sequencer powered by a USB 3.0 cable through a laptop. The MinION™ uses nanopore technology to determine the order of nucleotide bases in DNA or RNA. While the minION™ does have a high error rate (Mikheyev & Tin, 2014; Laver et al., 2015), consensus sequencing can greatly reduce this (Krehenwinkel et al., 2019a). Consensus sequencing generates hundreds of base pair sequences and compares them to determine which base call is most likely correct. Developments in software and downstream analysis (pipelines) are rapidly improving the accuracy of consensus sequencing using the MinION™ nanopore sequencer (Menegon et al., 2017; Maestri et al., 2019; Seah et al., 2020). Still, high error rates limit application for minION™. It is currently not accurate enough for small variant calling such as STR (short tandem repeats) or SNP (single nucleotide polymorphisms) calling (Magi et al., 2017). These may be possible in the future (Cornelis et al. 2017; Cornelis et al., 2018; Asogawa et al., 2019; Cornelis et al., 2019; Tytgat et al., 2020).

The MinION™ can be used, however, for DNA barcoding (Magi et al., 2017), which utilizes species-specific (or subspecies-specific) portions of DNA to identify a sample's origin. In order to test a sample, the sequence of the species-specific region of DNA must already be known. DNA barcoding for documenting species and tracking wildlife crime can be accomplished in the field with a portable PCR thermocycler and the MinION™ nanopore sequencer. DNA metabarcoding and next-gen sequencing also show promise for examining diets when detailed observations of feeding and foraging are not possible, especially if animals are not well-habituated, or to identify species of invertebrates eaten (Valentini et al., 2008; Pomapnon et al., 2012; Quéméré et al., 2013). However, the precision with which dietary composition can be examined is currently limited by incomplete reference databases for plant and invertebrate species of interest (Lyke et al., 2019).

Genetic analysis of faecal samples can also identify species composing gut microbiota which is useful for assessing impact of habitat disturbance or fragmentation on gut flora and, potentially, digestive efficiency and energy intake (Amato et al., 2013; Amato et al., 2014). Several grams of faeces in 95% ethanol solution can btemperatures for later genetic analysis to describe microbial taxonomy and function (Song et al., 2016; Mallott et al., 2019).

7.3 Applications to conservation

Over the last 30 years, tremendous strides have been made in understanding animal health and physiology through the analysis of field-collected samples, both in the field and in the laboratory. We review some of these applications, focusing primarily on our experience with non-invasive techniques applied to the study of wild primates, but also drawing more broadly where possible. We highlight specific examples of how data gathered from analysis of field samples have been applied to solve conservation problems.

7.3.1 Nutritional analysis

All organisms require adequate macro- and micronutrients to grow, develop, and successfully reproduce. Inadequate caloric intake can suppress reproduction, and if prolonged, can lead to starvation and death. Identifying whether habitats contain the foods needed to meet animal nutrient requirements is fundamental to conservation (Raubenheimer et al., 2012). Combining observational fieldwork with laboratory analyses of dietary nutritional content has yielded important conservation outcomes. For example, identifying the nutritional consequences of wildlife feeding on wild or cultivated foods in anthropogenic landscapes has proven important to understanding ranging behaviour and human–wildlife conflict in vervet monkeys (*Chlorocebus aethiops*) (Cancelliere et al., 2018), black bears (*Ursus americanus*) (Baldwin & Bender, 2009), grizzly bears (*Ursos arctos*) (Coogan & Raubenheimer, 2016), foxes (*Vulpes macrotis mutic*) (Newsom et al., 2010), Samoan fruit bats (*Pteropus samoensis*) (Nelson et al., 2008), and koalas (*Phascolarctos cinereus*) (Moore et al., 2005).

In the past 30 years, ecologists have recognized that anthropogenic changes influence food availability and food quality, and pose new challenges for digestive physiology (Birnie-Gauvin et al., 2017). New approaches to analysing and interpreting nutritional ecology, including applying the geometric framework, have helped address specific questions, such as the role of individual nutrients in wildlife foraging decisions (Raubenheimer, 2011). An illustrative example comes from critically endangered mountain gorillas (*Gorilla beringei beringei*), whose diet and behaviour have been the focus of long-term field and laboratory research analyses (Rothman et al. 2006; Rothman et al., 2011). In light of crop-raiding by gorillas, which increased human-wildlife conflict around Volcanoes National Park, researchers quantified gorillas' sodium intake and discovered that cultivated foods allow gorillas to more easily meet their sodium needs than wild foods (Grueter et al., 2018). Researchers planted sodium-rich cultivated foods at forest edges surrounded by a nutritionally poor buffer zone in the hopes of mitigating crop-raiding, although the gorillas' response to this is not yet known (Grueter et al., 2018).

7.3.2 Energetic and social stress

Hormone analyses have allowed field biologists to examine how environmental stressors may impact animal physiology, health, and survival. Evidence from primates and many other organisms suggests that acute social stressors, such as predation and aggression, and importantly, metabolic stress, are associated with elevations in faecal and urinary metabolites of glucocorticoids (Cavigelli, 1999; Pride, 2005; Engh et al., 2006; Arlet & Isbell, 2009; Wittig et al., 2015). These are now widely used measures of metabolic stress in wild animals, and are particularly valuable because they can be non-invasively measured from urine, saliva, faeces, feathers, and hair (Behringer et al., 2017). Hormone excretion and metabolism vary dramatically by species; thus methods and assays must be validated for the hormone of interest for each study species (Touma & Palme, 2005; Higham, 2016).

C-peptide, a by-product of insulin conversion, has been used as a proxy of insulin secretion to monitor net differences between caloric intake and energy expenditure (Emery Thompson & Knott, 2008). While C-peptides are present in serum, they are secreted into urine at a constant rate (Emery Thompson, 2016). Thus, analysis of urinary C-peptide concentrations provides a non-invasive measure of energy balance, as demonstrated in wild orangutans (*Pongo pygmaeus*) (Emery Thompson & Knott, 2008), chimpanzees (*Pan troglodytes*) (Emery Thompson et al., 2012b), macaques (*Macaca mulatta, M. fascicularis*) (Girard-Buttoz et al., 2011), mountain gorillas (*Gorilla beringei*) (Grueter et al., 2014), and bonobos (*Pan paniscus*) (Emery Thompson et al., 2012b; Grueter et al., 2014; Surbeck et al., 2015). Analytes are preserved in the field by drying, freezing, or storing on filter paper for later laboratory analysis (Hunt & Wasser, 2003; Emery Thompson & Knott, 2008).

7.3.3 Reproductive hormones

Documenting reproductive parameters, including cycle length, gestation, and endocrine indicators of reproductive status, is fundamental to understand a species social and reproductive strategies, life history, and the relationship between reproduction and the environment. From an applied perspective, a clear understanding of reproductive function is fundamental to effective conservation and management of wild and captive populations (Wildt & Wemmer, 1999; Wikelski & Cooke, 2006). For example, research on the reproductive physiology of female primates in degraded habitats has demonstrated the importance of ecological and behavioural flexibility for maintaining reproductive function (Milich et al., 2014). Captive breeding efforts such as gamete banking, artificial insemination, and release of captive-born individuals into the wild have been of great importance to species including golden lion tamarins (*Leontopithecus rosalia*) (Kleiman et al., 1986), black-footed ferrets (*Mustela nigripes*) (Miller et al., 1994), and takhis (*Equus ferus przewalksii*) (Monfort et al., 1991).

While female reproductive success is ultimately measured by the successful production of offspring, this is a very long-term process. Thus, over the short term, variation in female reproductive health can be assessed by measuring variation in production of the two major female sex steroids, oestradiol and progesterone, excreted into urine and faeces in their metabolized forms. Oestradiol levels reflect the maturation of the developing oocyte, signalling the probability of ovulation and

conception, while progesterone levels in the latter half of the cycle influence the growth of the endometrial lining, affecting the ability of fertilized zygotes to implant (Ellison, 1990). Similarly, testosterone is correlated with male mating effort and to the secondary sexual characteristics that enhance mating success (Wingfield et al., 1990; Muller & Wrangham, 2004). Because mating effort is a significant energetic cost for males, high testosterone is only expected when males are in prime health (Folstad & Karter, 1992). Likewise, high testosterone has been suggested to have damaging effects on long-term health via its effects on compromising the immune system (Muehlenbein & Bribiescas, 2005). However, some studies show a positive relationship between testosterone and immune function, likely because healthy males may be more able to bear the cost of high testosterone while still maintaining immune function (Emery Thompson & Georgiev, 2014). As earlier, these analytes are preserved in the field for later laboratory analysis (Knott, 1997; Hunt & Wasser, 2003; Knott, 2005b).

7.3.4 Health and immune function

Urine test strips are widely used to assess metabolic status in non-human primates, especially great apes, using the presence of ketones to indicate negative energy balance (Naumenko et al., 2019). Though not all test strip measurements have been validated for non-human use, the ability to rapidly and non-invasively assess general health characteristics is a useful starting point to evaluate population health. For instance, the Taï Chimpanzee Project includes urine test strips as part of a long-term population monitoring scheme (Leendertz et al., 2010) and uses these results to drive other monitoring efforts and the development of non-invasive, laboratory-based, techniques (Behringer et al., 2017; Behringer et al., 2019).

In concert with laboratory analysis of urinary creatinine, specific gravity has been used to non-invasively assess muscle mass in wild chimpanzees (Emery Thompson et al., 2012a; Samuni et al., 2020). These techniques facilitate population-level health monitoring, as well as identification of particularly at-risk individuals who might require veterinary care or be at risk during disease outbreaks (Wu et al., 2018). Their utilization can guide decisions about investing time and resources to develop or adapt laboratory methods for evaluating immune responses (Boughton et al., 2011; Downs & Steward, 2014).

7.3.5 Parasite prevalence and species richness

Conservation scientists can use data on within and between-population variation in parasitic infection to determine whether factors such as host density, climate, and level of anthropogenic disturbance alter intestinal parasitic infection profiles and the implications for the long-term health of a population. While most research on the health effects of intestinal parasites has been conducted on captive and domesticated animals, there is mounting evidence that these parasites adversely impact wild primate fitness (Friant et al., 2016; Akinyi et al., 2019; Müller-Klein et al., 2019). Patterns of parasitic infection vary with degree of habitat disturbance in primates (Gillespie, 2006; Nunn & Altizer, 2006; Gillespie et al., 2008; Labes et al., 2010), ungulates (Sun et al., 2018), and murid rodents (Wells et al., 2007; Chaisiri et al., 2010). Thus, longitudinal, non-invasive monitoring of intestinal parasites from faecal samples enables identification of shifts from baseline patterns over time, serving as an indicator of population health (Gillespie, 2006).

Conservation programmes that involve reintroductions can also benefit from field parasitology. Introducing wildlife to new suitable habitats or to areas where they were extirpated is a conservation strategy used for many endangered animal taxa (Griffith et al., 1989), but requires knowledge of species and population-typical parasites to inform reintroduction plans (Viggers et al., 1993). Reintroduced European bison (*Bison bonasus*) had significantly more intense parasitic infections than their wild counterparts (Kołodziej-Sobocińska et al., 2018), and similar findings have been reported for orangutans (Warren et al., 2001; Nurcahyo et al., 2017). Immunological naivete and previous deworming may contribute to pathogenic infections in reintroduced animals (Viggers et al., 1993). Reintroductions can impact the disease dynamics of

the reintroduced animals themselves as well as the broader ecosystem. For example, when grey wolves recolonized areas in Central Europe, changes in parasite prevalence in their ungulate prey were documented (Lesniak et al., 2018). These analyses were all conducted in traditional labs, but access to field-expedient high-quality microscopy equipment and the developing portable molecular techniques that have been described earlier (Section 7.2) makes these assessments in the field feasible.

7.3.6 Genetic applications to conservation

The development of genetic analysis tools has important implications for conservation. Rapid assessment of genetic diversity from environmental DNA can demonstrate biodiversity loss in anthropogenically degraded habitats (Krehenwinkel et al., 2019b). For cryptic species and species that can only be identified through genetic differences, DNA barcoding and metagenomics can document geographic distribution and areas of concern for conservation. Identifying individual animals and genetic relatedness within and between populations are important tools for wildlife conservationists. When using faecal samples as the DNA source, microsatellite genotyping is the most cost-effective and reliable method for individual identification and genetic relatedness measurement. Microsatellite genotyping can be accomplished with DNA sequencing or capillary electrophoresis. Most researchers use capillary electrophoresis because it is much less expensive while providing accurate results. Because portable capillary electrophoresis instruments do not exist, and portable sequencing instruments are not yet accurate enough for microsatellite genotyping (Asogawa et al., 2019), most of these analyses are done off-site using field preserved samples.

The use of portable field equipment for genetic analysis is only beginning to become feasible because the lower quality of DNA that is obtained from non-invasive samples, combined with the high error rates of the MinION™ nanopore sequencer, limits the types of analyses that can be reliably and cost-effectively performed with non-invasive samples in a portable field laboratory. We know of only two published studies (Guevara et al., 2017; Watsa et al., 2019) that analysed non-invasively collected genetic samples in a field lab. Guevara et al. (2017) analysed faecal samples to sex-type sifaka lemurs (*Propithecus coquereli*) and Watsa et al. (2019) analysed primate faecal samples (howler monkey, *Alouatta seniculus*; spider monkey, *Ateles chamek*; emperor tamarins, *Saguinus imperator*; saddleback tamarins, *Leontocebus weddelli*; and night monkeys, *Aotus nigrifrons*) to examine the diversity of eukaryotic gut microbiota. Seah et al. (2020) demonstrated that non-invasive samples (faeces and hair from a deceased snow leopard, *Panthera uncia*, and faeces and feathers from a deceased cinnamon teal, *Anas cyanoptera*) can be used for DNA barcoding. However, these samples were stored at −80°C immediately after collection, aiding DNA preservation. But this preservation method is not possible in most field settings.

Most published studies utilizing portable genetic equipment in the field (primarily miniPCR™ thermocycler & MinION™ nanopore sequencer) use invasive sampling of tissue or blood from a captured and released animal, or entire individuals in the case of arthropods (Guevara et al., 2017; Menegon et al., 2017; Pomerantz et al., 2018; Blanco et al., 2019; Krehenwinkel et al., 2019a; Srivathsan et al., 2019). Invasive blood and tissue samples contain greater quantities of high-quality DNA compared with non-invasive faecal or salivary samples, and are therefore easier to analyse. DNA barcoding in the field facilitates tracking wildlife crime: identifying the species or subspecies of a recovered animal or animal part can narrow down the location from which the individual was illegally taken. Additionally, when animal parts or products are recovered, identifying the species or subspecies can prove that poaching has occurred (e.g. whale blubber, Baker & Palumbi, 1994). Remote labs from the montane rainforests of Tanzania (Menegon et al., 2017) to the Peruvian Amazon rainforest (Krehenwinkel et al., 2019a) have used DNA barcoding to identify arthropods, reptiles, and mammals, and determined species distribution of reptiles (Pomerantz et al., 2018) and mouse lemurs (*Microcebus danfossi*) (Blanco

et al., 2019). The miniPCR™ thermocycler can also amplify whole genomes in the field which can then be transported back to a traditional laboratory for analysis (Guevara et al., 2017). Many field applications for genetic analysis and other techniques are currently in development, holding promise for transforming the speed at which real-time information on wildlife physiology and health status is obtained.

7.4 Case study of orangutans in Gunung Palung National Park, Indonesia

7.4.1 Conservation and health threats faced by wild orangutans

The world's wild orangutan populations are estimated to have declined by more than half over the last 70 years (Wich et al., 2016; Santika et al., 2017; Voigt et al., 2018). This decline is driven by seven main threats: human population pressure, forest loss due to logging, conversion of forest to agricultural land, forest fires, hunting, the illegal pet trade, and weak law enforcement (Indonesian Ministry of Forestry, 2009; Spehar et al., 2018). Of remnant wild orangutan populations, 75% are located outside of protected areas (Meijaard et al., 2012) in forests that are rapidly logged and converted to agriculture, most notably plantations for oil palm and paper and pulp. These threats, and the rate of habitat loss, mean that Bornean orangutans are categorized as Critically Endangered on the IUCN Red List and listed on Appendix I of CITES (Ancrenaz et al., 2016).

Due to their semi-solitary social system, most orangutans have relatively few close interactions with other conspecifics. Using social network analysis, Carne et al. (2014) found that orangutans had significantly fewer social contacts and a correspondingly lower risk of disease transmission than did chimpanzees. Additionally, their arboreal lifestyle may put them in less contact with disease vectors compared with the African apes. This lower risk of disease transfer means that orangutans typically need to allocate less energetic resources to immune system maintenance. This is evidenced by the much lower white blood cell counts found in orangutans compared to the African apes, apparently indicative of a much less active immune system (Nunn et al., 2000). However, these lower baseline levels of immune activity may then compromise orangutans' ability to mount an immune response when exposed to increased disease risk from unnaturally high levels of contact with other wild orangutans, humans, and increased ground use due to habitat degradation. This may explain the extremely high infant mortality rates at some orangutan rehabilitation sites (Dellatore et al., 2009) in contrast to the wild, where infant mortality is rarely observed and occurs at a much lower rate than reported for the African apes. Indeed, with the longest interbirth interval of any mammal (Galdikas & Wood, 1990; van Noordwijk et al., 2018), orangutan life history is shaped by very high infant survival rates (Knott & Harwell, 2020). Data from rehabilitants thus suggests that with increased conspecific and human contact from habitat loss, infant and adult mortality could dramatically increase.

7.4.2 Gunung Palung study population

Our study population inhabits the area encompassing the Cabang Panti Research Site in Gunung Palung National Park (GPNP), West Kalimantan, Indonesia, Borneo (1°13′S, 110°7′E). GPNP and the surrounding landscape represent one of the most important blocks of orangutan habitat in the world, and a United Nations Great Ape Survival Project conservation priority area. The park contains at least 2500 individuals, with an estimated 2500 in the surrounding landscape (Johnson et al., 2005). It is one of the last viable populations of *Pongo pygmaeus wurmbii*, and one of the only remaining intact lowland alluvial rainforests in Borneo. The park serves as a water catchment area, providing clean water and buffering against climate change in the region. The deep peat forests serve as a carbon sink, mitigating flooding and tidal salinity for coastal farmlands. Despite its national park status, there are many threats to biodiversity in GPNP, including illegal logging, hunting, and encroachment from oil palm plantations.

The Gunung Palung Orangutan Project has been collecting data on orangutan behaviour, hormones, and health since 1992. As orangutans become increasingly endangered, their physiological and disease status provides critical information on how anthropogenic processes like habitat loss affect the health and long-term viability of endangered populations. We have been involved in direct conservation activities with local communities surrounding the park since 1999, and our current efforts include applied conservation research, such as drone-facilitated nest surveys. A hallmark of our project has been the development and testing of non-invasive measures to assess wild orangutans' health and disease status. Evaluating wild orangutans' health includes measuring daily caloric and nutrient intake, assessing nutritional status, determining habitat, sex, and age-specific parasite loads from faecal analysis, evaluating disease status from urinalysis, and measuring faecal temperature. Our non-invasive, field-based lab methods enable rapid, conservation-relevant research on this critically endangered species in an increasingly threatened habitat.

7.4.3 Composition and habitat distribution of orangutan foods

Orangutans inhabit tropical rainforests of Southeast Asia, an environment characterized by extreme spatial and temporal variation in fruit production (Ashton et al., 1988; Leighton, 1993; Hamilton & Galdikas, 1994). This is largely driven by the phenomenon of mast fruiting, which occurs unpredictably every 2–10 years. During masts, up to 80% of the rainforest trees fruit in synchrony, driven by trees in the Dipterocarpaceae family that occupies much of the forest biomass (Curran & Leighton, 2000). These masts are followed by extended periods of low fruit availability, during which orangutans rely on bark, leaves, and low-calorie fruits to sustain themselves through these low-fruit periods (Knott, 1998). Habitat type and plant species distribution also vary across the landscape with altitude, soil conditions, temperature, and rainfall; consequently, orangutans' food varies temporally and spatially (Richards, 1996). We assess food availability using two sets of phenological data. One data set measures fruit, flower, and young leaf availability from 60 plots and belts distributed across the Cabang Panti study site (Marshall et al., 2014). The other monitors over 1000 mature trees and lianas eaten by orangutans, distributed along 17 transects, across all habitats.

During daily orangutan follows, we collect samples of all foods eaten for identification and processing. We identify the genus and the species or morphotype of the sample based on an extensive set of botanical records at the field site. Samples are dried and processed as described earlier (Section 7.2.1) and then analysed for nutrient and caloric content in nutritional chemistry labs in Indonesia or the US. Data on the nutrient and caloric composition of the fruit eaten by orangutans, combined with phenological records, are used to determine the biomass of fruit available monthly for this species in a given area. We use the measure 'kilocalories of orangutan fruit available/hectare', which combines fruiting phenology, fruit crop size,

grams per fruit, kcals/fruit, and plant densities (Knott, 2005a). This measure takes into consideration the quality (kcals and nutrients), quantity (actual crop size), and distribution (density) of fruits. Habitat productivity for a given species can then be combined with other measures of population viability. For example, orangutan population density is determined through line-transect surveys of nests made by orangutans. Analyses consider population and habitat specific values of nests built/day and nest degradation time (van Schaik et al., 1995; Johnson et al., 2005). The number of nests found on a transect can be compared to kcals of orangutan fruit available/transect (Figure 7.3), showing the close relationship between habitat productivity and population density.

7.4.4 Health assessments from urine and faeces

Chemstrip urinalysis has revealed that during periods of low caloric intake, orangutans break down their fat deposits to use as energy, as evidenced by the presence of ketones in urine (Knott, 1998). There is a significant, though weak, negative correlation between fruit availability (kilocalories per hectare)

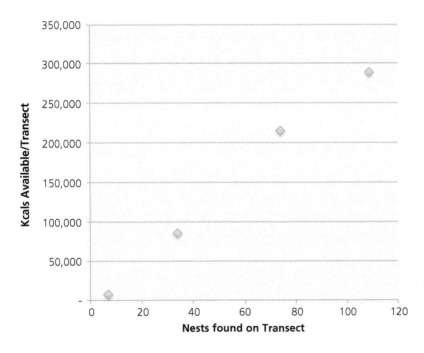

Figure 7.3 Relationship between nests found on transect and kilocalories of orangutan food available per transect from Gunung Palung National Park.

and mean monthly concentration of ketone bodies (R= −0.11, p = 0.04, N = 191), demonstrating that as fruit availability decreases, orangutans catabolize their fat stores.

One technique we have brought to the assessment of stress, and the interpretation of the glucocorticoid response, is the measurement of C-peptide in urine (Emery Thompson & Knott, 2008). C-peptide reflects insulin production and is maintained at relatively high levels in individuals with high energy balance, whose energy intake exceeds their daily caloric expenditure (Havel, 2001). Thus, measurement of cortisol and C-peptide allows us to determine the general stress response level of orangutans and, in combination with other hormonal, behavioural, and ecological variables, determine the likely sources and consequences of that stress. With more human destruction of rainforest habitat, we expect orangutan populations to become increasingly socially and energetically stressed, variables that can critically impact orangutan health. Figure 7.4 shows the close relationship between Kcals consumed and C-peptide in Gunung Palung orangutans (Knott et al., 2009).

During periods where there is a deficiency in protein intake, animals may enter a state of negative nitrogen balance as muscles are broken down to supply energy for bodily function (Martinez del Rio & Wolf, 2005). This is reflected in urine by low levels of urea, percentage nitrogen, and enriched levels of the isotope $\partial^{15}N$. We have tested this in our population, finding that urea varied according to food availability and urinary nitrogen, and that some orangutans showed low levels of $\partial^{15}N$, indicating a protein deficit (Vogel et al., 2012a; Vogel et al., 2012b). Thus, anthropogenic changes that result in extended periods of low food availability may increase orangutan vulnerability to extinction as they break down muscle tissue to support somatic maintenance (Vogel et al., 2012b).

In human-transformed landscapes, orangutans greatly increase their rate of terrestrial travel, consequently increasing their contact with humans (Campbell-Smith et al., 2012; Ancrenaz et al., 2014). Increased ground travel is also associated with increased contact with faecal material from conspecifics, humans, and other species that may transmit disease vectors and parasites (Labes et al., 2010). Furthermore, habitat compression is expected to

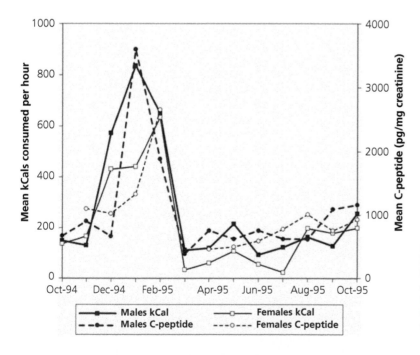

Figure 7.4 Monthly relationship between mean kilocalories consumed per hour and mean C-peptide values for orangutans in Gunung Palung National Park, West Kalimantan, Indonesia (reprinted with permission from Knott et al., 2009).

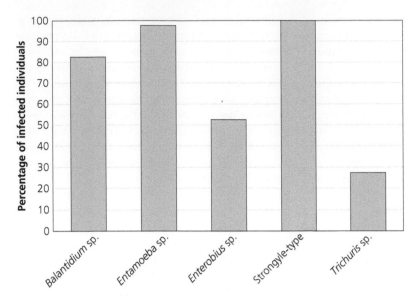

Figure 7.5 Parasite prevalence for the Gunung Palung orangutan study population from 2013 to 2014. The percent of individuals infected with each type of parasite can be assessed over time to compare infection patterns in conjunction with ecological factors such as climate, food availability, or changes in anthropogenic disturbance.

lead to increased social pressure and thus increased opportunities for disease transfer. This is demonstrated by increased parasite prevalence among orangutans in rescue centres housed in groups compared to those housed individually (Labes et al., 2010). With increasing habitat loss, it is essential to compare parasite infection patterns against baselines to identify trends and consequences. We thus monitor gastrointestinal parasite infection in our wild orangutan population to detect changes in prevalence and species richness over time. We have found that all individuals in the population are infected with parasites, but there is some age–sex class variation in species composition (O'Connell, 2018) (Figure 7.5). We continue to monitor intestinal parasites in conjunction with shifts in sociality during different periods and over the lifespan. We are currently exploring whether parasite infection patterns change over time in relation to food availability and nutritional intake to determine the impact of a changing environment on parasite infection risk.

7.4.5 Genetics

We use genetic methods to determine orangutan home range size and whether individuals that frequent our study area also range into anthropogenically disturbed areas. Although many orangutans in our study area are visually identifiable,

we also encounter individuals who are not habituated or recognizable. Because visual identification of unhabituated orangutans is difficult, but important for understanding orangutan habitat use, one aim of our genetic studies is individual identification via genotyping from non-invasive faecal samples. We take advantage of advances in field preservation methods for DNA in faeces, storing samples in RNAlater (Beja-Pereira et al., 2009) and using the two-step ethanol-silica method (Nsubuga et al., 2004) and then analyse these samples at an off-site laboratory. We compare the genotypes of orangutans seen in degraded habitats and village forests to genotypes of orangutans in the primary study area to determine if they are the same orangutans and measure relatedness. This increases our understanding of orangutan range size and habitat use. If genetic identification shows that an unhabituated orangutan is observed both within the field site and in anthropogenically disturbed areas, we better understand home range size and use.

7.5 Limitation and constraints of new technology

Field-based sample preparation and on- and off-site laboratory analyses provide data that can positively impact conservation efforts. As it becomes possible to analyse more samples directly in the field, it will dramatically reduce the time from sample collection to analysis and eliminate the need for permits to transport samples, which are often time-consuming and challenging to obtain. However, several potential obstacles and hurdles need to be overcome and considered.

A common constraint on the analysis of field samples is the remoteness of study sites. This precludes the use of freezers, or equipment needing electricity, at sites where electricity is unavailable or unreliable. Internet access to troubleshoot problems, contact colleagues for advice, and run software may also not be available. In many countries, it may be difficult to find purified water, reagents, and lab equipment. Importing equipment from abroad may also not be possible or may incur significant import fees. For this reason, equipment purchased in the study country may be many times more expensive compared to the country of manufacture. Logistically, transporting heavy equipment may be challenging, especially at sites that are accessed by foot or small watercraft.

Although field-based lab alternatives may be available, they can be prohibitively expensive. Specialized equipment and techniques require training, and instructions and documentation may not be available in local languages. Some methods, like infrared spectroscopy, require running significant numbers of reference samples in a laboratory first before more portable techniques can be used. This technique is therefore limited by the reference data set and reliant on extensive wet-chemistry analysis before field techniques are feasible.

7.5.1 Sample collection and handling

Working with biological samples involves important considerations about the collection, processing, and storage of those samples. Principal investigators should work with their institutions to ensure that field protocols have biosafety approval to safeguard human and ecosystem health. Although practical considerations may impose some limitations, many standard laboratory procedures can be followed in the field with minimal inconvenience. Proper personal protective equipment (PPE) should

be used. Gloves should always be worn when collecting or handling faeces or urine from wild animals. Masks should be worn when processing faecal or urine samples that involve centrifuging or other processes that might aerosolize the sample, such as parasite analysis. Protective clothing, including long sleeves, long pants/trousers, and closed-toe shoes, should be worn when handling biological samples. If possible, a lab coat should be added when handling chemicals. Although much of this work takes places in the tropics, where such clothing may be uncomfortable, it is important to follow these standard laboratory safety precautions.

For sample collection in the field, we recommend preparing a 'field kit' that includes gloves, a mask, sample collection vials, pipettes, and plastic bags. A small dry bag can be used to contain these items and limit contamination of other field supplies. If faecal sample processing is done at a field camp, then samples can be picked up using a glove, stored inside the inverted glove, and placed within a separate plastic bag. A permanent marker should be used to label all samples in the field. Contaminated pipettes and plastic sheets used for urine collection should be kept in a separate outer plastic bag. Plastic sheets used for urine collection can be washed in water and re-used. A thermos with ice can be used to store samples temporarily before returning to the field lab.

7.5.2 Field lab setup

When setting up a field lab, it is important to have a non-porous surface to work on that can be cleaned with bleach or 70% alcohol to decontaminate. Thick glass over wooden tables works well and is fairly easy to obtain. We also process samples on aluminium foil as this is non-porous and rigid enough to move samples on. To reduce the risk of contamination of genetic samples, precautions such as an ultraviolet (UV) light for DNA decontamination and dedicated bench space are necessary. Consider the need for adequate lighting for carrying out lab work in the evening. A movable task lamp, solar lantern, or battery-powered LEDs can solve this problem.

Freezers or refrigerators for sample storage at a field camp can be purchased either in-country or imported from abroad. Companies that specialize in equipment for camping or use on boats are good sources for such items. One important consideration is matching the operating voltage with the power source. Although computer power supplies, and some other electronics, now have universal power supplies that can work with a range of voltage from 100 to 240, this is not the case for all equipment. Thus, equipment designed for the 110-volt current, standard in North America, will require a voltage adaptor to run on 220 volt current, standard in many tropical countries. Another consideration is that small portable generators often put out inconsistent voltage levels and can damage equipment, so a voltage stabilizer is useful to increase electronics' lifespan. Running equipment continuously may be difficult if using an unreliable power source, such as a generator. Where 24-hour power is not available, hybrid systems can be used to power essential equipment. For example, we utilize a portable freezer that can be powered from an AC or DC current. It is run on a generator in the evenings, and during the day switched to being powered by two truck batteries that are charged from the generator at night. Renewable energy solutions such as solar and hydro-power may allow more consistent use of electricity and should be used if possible. A micro-hydro plant was recently installed at our site, enabling us to power freezers when water levels are sufficient.

Portable equipment may not be as reliable as standard laboratory equipment. For example, the accuracy of the MinION™ nanopore sequencer is a limitation for its application (see Section 7.2.5). Methods of downstream analysis are increasing the accuracy of the MinION™ nanopore sequencer, but the higher cost of analysis compared to certain traditional methods makes it impractical for some applications (e.g. capillary electrophoresis for STR and SNP genotyping). Additionally, in order to reduce the size and weight of portable thermocyclers and sequencers, the number of samples that can be run simultaneously is significantly reduced (e.g. portable PCR thermocyclers run 8–32 samples compared to 96 samples in the lab). Given unreliable electricity for refrigeration, all reagents must be shelf-stable. The two3 system (Biomeme) specializes in shelf-stable reagents, but has limited applications. If all reagents used for DNA extraction are not shelf-stable, then additional equipment to keep reagents cold will be necessary. Many extraction kits also require a bench top centrifuge. Contamination can also occur in a traditional laboratory setting, but maintaining a clean DNA-free work zone is especially difficult in the field (Watsa et al., 2019).

7.5.3 Use of chemicals and disposal of biological samples and contaminated supplies

The use of chemicals involves inherent risks, and chemical fume hoods are not available in the field. While field protocols must be approved by the principal investigator (PI)'s home university and local governing bodies, there are usually no safety inspections on-site at field laboratories. Thus, PIs and local research managers must prioritize lab safety management and teach and enforce safety protocols to ensure the protection of those collecting and analysing samples in remote settings. It may be possible to obtain training videos and other materials from one's institution to be used in the field. Translation of these materials into the local language may be necessary if protocols are being carried out by non-English speakers.

Many countries where wildlife is studied lack adequate facilities to handle biohazard waste. Thus, such contaminants may end up being dumped in rivers or left exposed, leading to environmental contamination and possible contact by animals or humans. All genetic analysis will create biohazard waste that must be safely removed from the field and disposed of properly. Disposal of unused biological samples as well as gloves, pipettes, vials, and other expendable lab supplies, can be a challenge at remote field locations. Separate containers for biological waste and contaminated lab supplies should be designated. Biological samples, such as faeces collected in the field, can be returned to the field. We recommend burying such samples. Contaminated plastics and glassware are more of a challenge. Such used items should be disposed of in a designated biohazard container. Red biohazard bags can be brought to the field site for this purpose. Investigators should consider whether it is possible to dispose of such items by sending them to an appropriate location in-country. Because of the lack of facilities to handle biohazard waste, the best option is often to either bury or incinerate these items if it is safe to do so. However, we do not recommend incineration because of the potential for toxic fumes being released. Samples should not be just deposited on the ground or disposed of in rivers or other bodies of water because of the potential for contamination of these areas and the likelihood of contact with other wildlife. Because of the challenges posed by disposal, such supplies should be used judiciously to limit items that need disposal. For short-term research trips, investigators should consider removing contaminated supplies to be properly disposed of elsewhere. Minimize the use of toxic reagents, in some cases by turning to the alternatives we have suggested. For example, storing faecal samples in formalin for later analysis in the field creates toxic waste that must be disposed of (Gillespie, 2006), while performing direct smear and faecal flotation on the same day as sample collection eliminates this problem. Sheather's sugar solution works well for flotation and can be prepared and disposed of easily in the field.

7.5.4 Long-term partnerships

Successful application of many of these field-based laboratory methods is contingent on long-term investment in a project. Some techniques—for example, non-invasive monitoring of female reproductive status—require sampling the same individual across several months or years. Many remote fieldwork challenges can be met by establishing close relationships with local universities, industries, and stakeholders. Local hospitals, clinics, and university labs can be consulted about where supplies and reagents may be purchased in-country. These institutions may also provide guidance as to where biohazards can be disposed of properly. Additionally, local partners are crucial to obtaining governmental permits and to adhering to in-country regulations that researchers must follow.

7.6 Social impact and benefits of field labs

Field-based laboratories can facilitate positive social relationships between conservation projects and local communities. Portable labs have much lower start-up costs than large, traditional labs, making these conservation tools more accessible to stakeholders and scientists in developing countries with high biodiversity (Blanco et al., 2019; Watsa et al., 2019). Thus, establishing field laboratories provides important capacity-building for local students, park and wildlife personnel, and NGOs who can be trained in these techniques on the ground (Watsa et al., 2019). Initial sample preparation and on-site analysis facilitate more rapid sample analysis, shortens the time between data collection and analysis, and ultimately makes communicating conservation-relevant results with stakeholders quicker and easier. It also enables rapid responses to, and interpretations of, new conservation challenges through physiological data.

Field laboratories help build local capacity, improve training for host-country researchers, and embed conservation projects in communities. Such involvement pays many dividends, including increased awareness of endangered wildlife by local communities and improved relationships with local governments. It also provides opportunities for collaboration with local universities, including training of faculty, staff, and students, and expanding sample analysis to include projects and questions relevant to local researchers and communities. Should circumstances make it challenging for non-local researchers to reach the field site, having worked to train local researchers and establish good relationships with local communities can allow projects to continue without the presence of non-local staff.

7.7 Future directions

We have explored three ways that researchers and conservationists have used field laboratories to further conservation objectives. These include developing ways to preserve samples for later off-site analysis, using low-tech on-site analyses, and taking advantage of smaller portable analytical devices for more high-tech field analysis. Before high-tech equipment is brought into the field, field samples are usually analysed using those devices at off-site laboratories. We thus anticipate that continuing to develop new methods of on-site sample preservation in field labs will be essential for applying technological advances to new or existing samples. In fact, the IUCN's guidelines for the collection of biological samples advises that samples should be 'bio-banked' for future analysis. An example is the isotopic analysis of urine and faeces. We have applied these analyses to our studies of wild orangutans, using existing samples, even though they were not initially collected with this application in mind (Knott et al., 2019b). Thus, as new laboratory procedures are developed, field conservationists should continue to find ways to preserve samples so they can take advantage of emerging technologies.

Field lab innovations are driven primarily by scientists who think creatively about applying existing technological solutions to a field setting. An early example was developing ways to dry samples for later preservation. Another newer example is the use of handheld digital thermometers to measure faecal temperature. As field scientists, we have learned that with proper storage and care, electronic scales, microscopes, and other types of equipment can be utilized in humid or dusty field conditions. Becoming familiar with work beyond one's own area of expertise and examining techniques currently used in clinical or commercial settings can provide the impetus for developing novel field applications.

We anticipate that future technological developments will continue to make high-tech laboratory equipment more portable, less expensive, less energy and chemically intensive, and more reliable for field use. For example, the ability to use NIRS technology in the field for nutrition already exists, but most field sites do not yet have the wet-chemistry analyses already completed that are necessary to utilize this technology. However, we anticipate that in the next few years, more projects will be able to implement this in the field. Field application of

NIRS for hormonal analysis is less immediate and will require more testing and development, with rigorous laboratory validation.

The use of field genetics laboratories will continue to expand rapidly as advances in technology and downstream analysis improve these methods' accuracy. Much of this will be determined by commercial or clinical applications driving investment in innovation. For example, bio-tech companies are developing new shelf-stable reagents and kits specifically for portable genetics labs. Thermocyclers and the MinION™ nanopore sequencer hold great promise for quick documentation of biodiversity, right at the source of sample collection. These portable machines have only been commercially available for a few years, so they are just beginning to be tested. As downstream analysis techniques continue to improve the accuracy of base pair calling by the MinION™ nanopore sequencer, its utility will increase.

Continued development of these methods and applications will both allow more field workers to apply technology, as it becomes easier, faster, and cheaper to use and provide the opportunity for us to ask new, currently unimaginable questions. Conservationists can collaborate with scientists in other fields like computer or chemical engineering to innovate new technologies that will further address these issues and make meaningful conservation science results faster to achieve, improving our ability to address the urgent needs of wildlife in a time of environmental crisis. Expanding the capabilities of field-based laboratories will help facilitate capacity building with local researchers and students and ensure a biodiverse, equitable future for wildlife conservation.

References

Akinyi, M. Y., Jansen, D., Habig, B., Gesquiere, L. R., Alberts, S. C., & Archie, E. A. (2019). Costs and drivers of helminth parasite infection in wild female baboons. *Journal of Animal Ecology*, 88, 1029–1043.

Amato, K. R., Leigh, S. R., Kent, A., et al. (2014). The role of gut microbes in satisfying the nutritional demands of adult and juvenile wild, black howler monkeys (*Alouatta pigra*). *American Journal of Physical Anthropology*, 155, 652–664.

Amato, K. R., Yeoman, C. J., Kent, A., et al. (2013). Habitat degradation impacts black howler monkey (*Alouatta pigra*) gastrointestinal microbiomes. *The ISME Journal*, 7, 1344–1353.

Ancrenaz, M., Gumal, M., Marshall, A. J., Meijaard, E., Wich, S. A., & Husson, S. (2016). *Pongo pygmaeus*, Bornean orangutan. *The IUCN Red List of Threatened Species* 2016, e.T17975A123809220.

Ancrenaz, M., Sollmann, R., Meijaard, E., et al. (2014). Coming down from the trees: is terrestrial activity in Bornean orangutans natural or disturbance driven?. *Scientific Reports*, 4, 4024.

Arlet, M., & Isbell, L. (2009). Variation in behavioural and hormonal responses of adult male gray-cheeked mangabeys (*Lophocebus albigena*) to crowned eagles (*Stephanoaetus coronatus*) in Kibale National Park, Uganda. *Behavioral Ecology and Sociobiology*, 63, 491–499.

Ashford, R., Reid, G., & Butynski, T. (1990). The intestinal faunas of man and mountain gorillas in a shared habitat. *Annals of Tropical Medicine & Parasitology*, 84, 337–340.

Ashton, P. S., Givnish, T. J., & Appanah, S. (1988). Staggered flowering in the Dipterocarpaceae: new insights into floral induction and the evolution of mast fruiting. *American Naturalist*, 132, 44–66.

Asogawa, M., Ohno, A., Nakagawa, S., et al. (2019). Human short tandem repeat identification using a nanopore-based DNA sequencer: a pilot study. *Journal of Human Genetics*, 65, 21–24.

Baker, C., & Palumbi, S. (1994). Which whales are hunted? A molecular genetic approach to monitoring whaling. *Science*, 265, 1538–1540.

Baldwin, R., & Bender, L. (2009). Foods and nutritional components of diets of black bear in Rocky Mounatin National Park, Colorado. *Canadian Journal of Zoology*, 87, 1000–1008.

Beehner, J. C., & Whitten, P. L. (2004). Modifications of a field method for fecal steroid analysis in baboons. *Physiology and Behavior*, 82, 269–277.

Behringer, V., & Deschner, T. (2017). Non-invasive monitoring of physiological markers in primates. *Hormones and Behavior*, 91, 3–18.

Behringer, V., Stevens, J., Leendertz, F., Hohmann, G., & Deschner, T. (2017). Validation of a method for the assessment of urinary neopterin levels to monitor health status in non-human-primate species. *Frontiers in Physiology*, 8, 51.

Behringer, V., Stevens, J., Witting, R., et al. (2019). Elevated neopterin levels in wild, healthy chimpanzees indicated constant investment in unspecific immune system. *BMC Zoology*, 4, 2.

Beja-Pereira, A., Oliveira, R., Alves, P. C., & Schwartz, M. K. (2009). Advancing ecological understandings through technological transformations in noninvasive genetics. *Molecular Ecology Resources*, 9, 1279–1301.

Birnie-Gauvin, K., Peiman, K., Raubenheimer, D., & Cooke, S. (2017). Nutritional physiology and ecology of wildlife in a changing world. *Conservation Physiology*, 5, cox30.

Blanco, M., Greene, L., Williams, R., Yoder, A., & Larsen, P. (2019). Next-generation in situ conservation and capacity building in Madagascar using a mobile genetics lab. BioRxiv, 650614.

Boughton, R. K., Joop, G., & Armitage, S. A. O. (2011). Outdoor immunology: methodological considerations for ecologists. *Functional Ecology*, 25, 81–100.

Campbell, K. L. (1994). Blood, urine, saliva and dipsticks: experiences in Africa, New Guinea, and Boston. In Campbell, K. L. & Wood, J. W. (eds.). *Human Reproductive Ecology: Interactions of Environment, Fertility, and Behavior*. New York Academy of Sciences, New York, NY (pp. 312–330).

Campbell-Smith, G., Sembiring, R., & Linkie, M. (2012). Evaluating the effectiveness of human-orangutan conflict mitigation strategies in Sumatra. *Journal of Applied Ecology*, 49, 367–375.

Cancelliere, E., Chapman, C., Twinomugisha, D., & Rothman, J. (2018). The nutritional value of feeding on crops: diets of vervet monkeys in a humanized landscape. *African Journal of Ecology*, 56, 160–167.

Carne, C., Semple, S., Morrogh-Bernard, H., Zuberbühler, K., & Lehmann, J. (2014). The risk of disease to great apes: simulating disease spread in orang-utan (*Pongo pygmaeus wurmbii*) and chimpanzee (*Pan troglodytes schweinfurthii*) association networks. *PLoS One*, 9, e95039.

Cavigelli, S. A. (1999). Behavioural patterns associated with faecal cortisol levels in free-ranging female ring-taled lemurs, Lemur catta. *Animal Behaviour*, 57, 935–944.

Chaisiri, K., Chaeychomsri, W., Siruntawineti, J., Bordes, F., Herbreteau, V., & Morand, S. (2010). Human-dominated habitats and helminth parasitism in Southeast Asian murids. *Parasitology Research*, 107, 931–937.

Coates, D. (2000). Faecal NIRS: what does it offer today's grazier?. *Tropical Grasslands*, 34, 230–239.

Conklin-Brittain, N. L., Dierenfeld, E. S., Wrangham, R. W., Norconk, M., & Silver, S. C. (1999). Chemical protein analysis: a comparison of kjeldahl crude protein and total ninhydrin protein from wild, tropical vegetation. *Journal of Chemical Ecology*, 25, 2601–2622.

Conklin-Brittain, N. L., Knott, C. D., & Wrangham, R. W. (2006). Energy intake by wild chimpanzees and orangutans: methodological considerations and a preliminary comparison. In Hohmann, G., Robbins, M.,

& Boesch, C. (eds.). *Feeding Ecology in Apes and Other Primates*. Cambridge University Press, Cambridge, UK (pp. 445–471).

Conklin-Brittain, N. L., Wrangham, R., & Hunt, K. D. (1998). Dietary response of chimpanzees and cercopithecines to seasonal variation in fruit abundance: II. Macronutrients. *International Journal of Primatology*, 19, 971–998.

Coogan, S., & Raubenheimer, D. (2016). Might macronutrient requiements influence grizzly bear-human conflict? Insights from nutritional geometry. *Ecosphere*, 7, 1–15.

Cornelis, S., Gansemans, Y., Deleye, L., Deforce, D., & Van Nieuwerburgh, F. (2017). Forensic SNP genotyping using nanopore MinION sequencing. *Scientific Reports*, 7, 41759.

Cornelis, S., Gansemans, Y., Vander Plaetsen, A. S., et al. (2019). Forensic tri-allelic SNP genotyping using nanopore sequencing. *Forensic Science International: Genetics*, 38, 204–210.

Cornelis, S., Willems, S., Van Neste, C., et al. (2018). Forensic STR profiling using Oxford Nanopore Technologies' MinION sequencer. BioRxiv, 433151. Available at: https://www.biorxiv.org/content/10.1101/433151v1.full.pdf

Curran, L. M., & Leighton, M. (2000). Vertebrate responses to spatio-temporal variation in seed production by mast-fruiting *Bornean Dipterocarpaceae*. *Ecological Monographs*, 70, 101–128.

Czekala, N., & Sicotte, P. (2000). Reproductive monitoring of free-ranging female mountain gorillas by urinary hormone analysis. *American Journal of Primatology*, 51, 209–215.

Dalton, D., & Kotze, A. (2011). DNA barcoding as a tool for species identification in three forensic wildlife cases in South Africa. *Forensic Science International*, 207, e51–e54.

Danish, L. M., Heistermann, M., Agil, M., & Engelhardt, A. (2015). Validation of a novel collection device for non-invasive urine sampling from free-ranging animals. *PLoS One*, 10, e0142051.

de Buys Roessing, A. S., Drukker, A., & Guignard, J.-P. (2001). Dipstick measurements of urine specific gravity are unreliable. *Archives of the Diseases of Childhood*, 85, 155–157.

de Ruiter, J. R., van Hooff, J. A. R. A. M., & Scheffrahn, W. (1994). Social and genetic aspects of paternity in wild long-tailed macaques (*Macaca fascicularis*). *Behaviour*, 129, 203–224.

Dellatore, D. F., Waitt, C. D., & Foitová, I. (2009). Two cases of mother-infant cannibalism in orangutans. *Primates*, 50, 277–281.

Di Fiore, A., Lawler, R. D., & Gagneux, P. (2011). Molecular primatology. In Campbell, C. J., Fuentes, A., MacKinnon, K. C., Bearder, S., & Stumpf, R. M. (eds.). *Primates in Perspective*. Oxford University Press, New York, NY (pp. 390–416).

Downs, C. J., & Steward, K. M. (2014). A primer in ecoimmunology and immunology for wildlife research and management. *California Fish and Game*, 100, 371–395.

Dunham, N., Kane, E., & Rodriguez-Saona, L. (2016). Quantifying soluble carbohydrates in tropical leaves using a portable mid-infrared sensor: implications for primate feeding ecology. *American Journal of Primatology*, 78, 701–706.

Edwards, K. L., McArthur, H. M., Liddicoat, T., & Walker, S. L. (2014). A practical field extraction method for non-invasive monitoring of hormone activity in the black rhinoceros. *Conservation Physiology*, 2, cot037.

Ellison, P. T. (1990). Human ovarian function and reproductive ecology: new hypotheses. *American Anthropologist*, 92, 952–993.

Emery Thompson, M. (2016). Energetics of feeding, social behavior, and life history in non-human primates. *Hormones and Behavior*, 91, 84–96.

Emery Thompson, M., & Georgiev, A. V. (2014). The high price of success: costs of mating effort in male primates. *International Journal of Primatology*, 35, 609–627.

Emery Thompson, M., & Knott, C. (2008). Urinary C-peptide of insulin as a non-invasive marker of energy balance in wild orangutans. *Hormones and Behavior*, 53, 526–535.

Emery Thompson, M., Muller, M. N., & Wrangham, R. W. (2012a). Technical note: variation in muscle mass in wild chimpanzees: application of a modified urinary creatinine method. *Amerian Journal of Physical Anthropology*, 149, 622–627.

Emery Thompson, M., Muller, M. N., & Wrangham, R. W. (2012b). The energetics of lactation and the return to fecundity in wild chimpanzees. *Behavioral Ecology*, 23, 1234–1241.

Engh, A. L., Beehner, J. C., Bergman, T. J., et al. (2006). Female hierarchy instability, male immigration and infanticide increase glucocorticoid levels in female chacma baboons. *Animal Behaviour*, 71, 1227–1237.

Erb, W. M., Barrow, E. J., Hofner, A. N., Utami-Atmoko, S. S., & Vogel, E. R. (2018). Wildfire smoke impacts activity and energetics of wild Bornean orangutans. *Scientific Reports*, 8, 7606.

Felton, A. M., Felton, A., Raubenheimer, D., et al. (2009). Protein content of diets dictates the daily energy intake of a free-ranging primate. *Behavioral Ecology*, 20, 685–690.

Foerster, S., & Monfort, S. (2010). Fecal glucocorticoids as indicators of metabolic stress in female Sykes' monkeys (*Cercopithecus mitis albogularis*). *Hormones and Behavior*, 58, 685–697.

Foley, W., McIlwee, A., & Lawler, I. (1998). Ecological applications of near infrared reflectance spectroscopy: a tool for rapid, cost-effective prediction of the composition of plant and animal tissues and aspects of animal performance. *Oecologia*, 116, 293–305.

Folstad, I., & Karter, A. (1992). Parasites, bright males, and the immunocompetence handicap. *The American Naturalist*, 139, 603–622.

Frankham, R. (1995). Conservation genetics. *Annual Review of Genetics*, 29, 305–327.

Friant, S., Ziegler, T. E., & Goldberg, T. L. (2016). Primate reinfection with gastrointestinal parasites: behavioural and physiological predictors of parasite acquisition. *Animal Behaviour*, 117, 105–113.

Galdikas, B. M. F., & Wood, J. W. (1990). Birth spacing patterns in humans and apes. *American Journal of Physical Anthropology*, 83, 185–191.

Gillespie, T. R. (2006). Non-invasive assessment of gastrointestinal parasite infections in free ranging primates. *International Journal of Primatology*, 27, 1129–1143.

Gillespie, T. R., Nunn, C. L., & Leendertz, F. H. (2008). Intergrative approaches to the study of primates infections disease: implications for biodiversity conservation and global health. *Yearbook of Physical Anthropology*, 51, 53–69.

Girard-Buttoz, C., Higham, J. P., Heistermann, M., Wedegärtner, S., Maestripieri, D., & Engelhardt, A. (2011). Urinary c-peptide measurement as a marker of nutritional status in macaques. *PLoS One*, 6, e18042.

Gonçalves, P. F., Oliveira-Marques, A. R., Matsumoto, T. E., & Miyaki, C. Y. (2015). DNA barcoding identifies illegal parrot trade. *Journal of Heredity*, 106, 560–564.

González-González, E., Mendoza-Ramos, J., Pedroza, S., et al. (2019). Validation of use of the miniPCR thermocycler for Ebola and Zika virus detection. *PLoS One*, 14, e0215642.

Griffith, B., Scott, J., Carpenter, J., & Reed, C. (1989). Reintroduction as a species conservation tool: status and strategy. *Science*, 245, 477–480.

Grueter, C., Deschner, T., Behringer, V., Fawcett, K., & Robbins, M. (2014). Socioecological correlates of energy balance using urinary C-peptide measurements in wild female mountain gorillas. *Physiology and Behavior*, 127, 13–19.

Grueter, C., Wright, E., Abavandimwe, D., et al. (2018). Going to extremes for sodium acquisition: use of community land and high-altitude areas by mountain gorillas *Gorilla bereingei* in Rwanda. *Biotropical*, 50, 826–834.

Guevara, E., Frankel, D., Ranaivonasy, J., et al. (2017). A simple, economical protocol for DNA extraction and amplification where there is no lab. *Conservation Genetics Resources*, 10, 119–125.

Hamilton, R. A., & Galdikas, B. M. F. (1994). A preliminary study of food selection by the orangutan in relation to plant quality. *Primates*, 35, 255–263.

Harrison, M. E., Vogel, E. R., Morrogh-Bernard, H., & van Noordwijk, M. A. (2009). Methods for calculating activity budgets compared: a case study using orangutans. *American Journal of Primatology*, 71, 353–358.

Harwell, F. S., Gotama, R., Scott, K. S., Philp, B., & Knott, C. D. (2019). Body temperature estimates for Bornean orangutans (*Pongo pygmaeus wurmbii*) from internal fecal temperature measurements. *American Journal of Physical Anthropology*, 168, 99.

Havel, P. J. (2001). Peripheral signals conveying metabolic information to the brain: short-term and long-term regulation of food intake and energy homeostasis. *Experimental Biology and Medicine*, 226, 963–977.

Heistermann, M., & Higham, J. P. (2015). Urinary neopterin, a noninvasive marker of mammalian cellular immune activation, is highly stable under field conditions. *Scientific Reports*, 5, 16308.

Higham, J. (2016). Field endocrinology of nonhuman primates: past, present, and future. *Hormones and Behavior*, 84, 145–155.

Higham, J. P., Kraus, C., Stahl-Hennig, C., Engelhardt, A., Fuchs, D., & Heistermann, M. (2015). Evaluating noninvasive markers of nonhuman primate immune activation and inflammation. *American Journal of Physical Anthropology*, 158, 673–684.

Higham, J. P., Stahl-Hennig, C., & Heistermann, M. (2020). Urinary suPAR: a non-invasive biomarker of infection and tissue inflammation for use in studies of large free-ranging mammals. *Royal Society Open Science*, 7, 191825.

Hoffman, C. L., Higham, J. P., Heistermann, M., Prendergast, B., Coe, C., & Maestripieri, D. (2011). Immune function and HPA axis activity in free-ranging rhesus macaques. *Physiology of Behavior*, 104, 507–514.

Hunt, K. E., & Wasser, S. K. (2003). Effect of long-term preservation methods on fecal glucocorticoid concentrations of grizzly bear and African elephant. *Physiological and Biochemical Zoology*, 76, 918–928.

Indonesian Ministry of Forestry (2009). *Orangutan Indonesia Conservation Strategies and Action Plan, 2007–2017: Strategi Dan Rencana Aksi Konservasi Orangutan Indonesia, 2007–2017*. Ministry of Forestry, Jakarta, Indonesia.

Jensen, S. A., Mundry, R., Nunn, C. L., Boesch, C., & Leendertz, F. H. (2009). Non-invasive body temperature measurement of wild chimpanzees using fecal temperature decline. *Journal of Wildlife Diseases*, 45, 542–546.

Johnson, A., Knott, C., Pamungkas, B., Pasaribu, M., & Marshall, A. J. (2005). A survey of the orangutan (*Pongo pygmaeus wurmbii*) population in and around Gunung Palung National Park, West Kalimantan, Indonesia based on nest counts. *Biological Conservation*, 121, 495–507.

Kaur, T., & Huffman, M. (2004). Descriptive urological record of chimpanzees (*Pan troglodytes*) in the wild and limitations associated with using multi-reagent dipstick test strips. *Journal of Medical Primatology*, 33, 187–196.

Kinoshita, K., Kuze, N., Kobayashi, T., et al. (2016). Detection of urinary estrogen conjugates and creatinine using near infrared spectroscopy in Bornean orangutans (*Pongo pygmaeus*). *Primates*, 57, 51–59.

Kinoshita, K., Miyazaki, M., Morita, H., et al. (2012). Spectral pattern of urinary water as a biomarker of estrus in the giant panda. *Scientific Reports*, 2, 856.

Kinoshita, K., Morita, H., Miyazaki, M., et al. (2010). Near infrared spectroscopy of urine proves useful for estimating ovulation in giant panda (*Ailuropoda melanoleuca*). *Analytical Methods*, 2, 1671.

Kirkpatrick, J. F., Lasley, B. L., & Shideler, S. E. (1990). Urinary steroid evaluations to monitor ovarian function in exotic ungulates: VII. Urinary progesterone metabolites in the Equidae assessed by immonoassay. *Zoo Biology*, 9, 341–348.

Kleiman, D., Beck, B., Dietz, J., Dietz, L., Ballou, J., & Coimbra-Filho, A. (1986). Conservation program for the golden lion tamarin: captive research and management, ecological studies, educational strategies, and reintroduction. In Benirshke, K. (ed.). *Primates*. Springer-Verlag, New York, NY (pp. 959–979).

Knott, C. D. (1997). Field collection and preservation of urine in orangutans and chimpanzees. *Tropical Biodiversity*, 3, 95–102.

Knott, C. D. (1998). Changes in orangutan caloric intake, energy balance, and ketones in response to fluctuating fruit availability. *International Journal of Primatology*, 19, 1061–1079.

Knott, C. D. (2001). Female reproductive ecology of the apes: implications for human evolution. In Ellison, P. T. (ed.). *Reproductive Ecology and Human Evolution*. Aldine de Gruyter, New York, NY (pp. 429–463).

Knott, C. D. (2005a). Energetic responses to food availability in the great apes: implications for Hominin Evolution. In Brockman, D. K., & van Schaik, C. P. (eds.). *Primate Seasonality: Implications for Human Evolution*. Cambridge University Press, Cambridge, UK (pp. 351–378).

Knott, C. D. (2005b). Radioimmunoassay of estrone conjugates from urine dried on filter paper. *American Journal of Primatology*, 67, 121–135.

Knott, C. D., Crowley, B., Kane, E. E., Brown, M., & Susanto, T. W. (2019b). Fecal isotopes as indicators of weaning and diet in wild Bornean orangutans. *American Journal of Physical Anthropology*, 168, 128–129.

Knott, C. D., Emery Thompson, M., & Stumpf, R. M. (2007). Sexual coercion and mating strategies of wild Bornean orangutans. *American Journal of Physical Anthropology*, Supplement 44, 145.

Knott, C. D., Emery Thompson, M., & Wich, S. A. (2009). The ecology of reproduction in wild orangutans. In Wich, S. A., Utami, S. S., Mitra Setia, T., & van Schaik, C. (eds.). *Orangutans: Geographic Variation in Behavioral Ecology and Conservation*. Oxford University Press, Oxford, UK (pp. 171–188).

Knott, C. D., & Harwell, F. S. (2020). Ecological risk and the evolution of great ape life histories. In Hopper, L., & Ross, S. (eds.). *Chimpanzees in Context*. University of Chicago Press, Chicago, IL (pp. 3–35).

Knott, C. D., Scott, A. M., O'Connell, C. A., et al. (2019a). Possible male infanticide in wild orangutans and a re-evaluation of infanticide risk. *Scientific Reports*, 9, 7806.

Kołodziej-Sobocińska, M., Demiaszkiewicz, A., Pyziel, A., & Kowalczyk, R. (2018). Increased parasitic load in captive-released European bison (*Bison bonasus*) has important implications for reintroduction programs. *Ecohealth*, 15, 467–471.

Krehenwinkel, H., Pomerantz, A., Henderson, J., et al. (2019a). Nanopore sequencing of long ribosomal DNA amplicons enables portable and simple biodiversity assessments with high phylogenetic resolution across broad taxonomic scale. *GigaScience*, 8, giz006.

Krehenwinkel, H., Pomerantz, A., & Prost, S. (2019b). Genetic biomonitoring and biodiversity assessment using portable sequencing technologies: current uses and future directions. *Genes*, 10, 858.

Krief, S., Huffman, M. A., Sevenet, T., et al. (2005). Noninvasive monitoring of the health of *Pan troglodytes schweinfurthii* in the Kibale National Park, Uganda. *International Journal of Primatology*, 26, 467–490.

Labes, E., Hegglin, D., Grimm, F., et al. (2010). Intestinal parasites of endangered orangutans (*Pongo pygmaeus*) in Central and East Kalimantan, Borneo, Indonesia. *Parasitology*, 137, 123.

Lacy, R. C. (1997). Importance of genetic variation to the viability of mammalian populations. *Journal of Mammalogy*, 78, 320–335.

Lambert, J. E., & Rothman, J. M. (2015). Fallback foods, optimal diets, and nutritional targets: primate responses to varying food availability and quality. *Annual Review of Anthropology*, 44, 493–512.

Laver, T., Harrison, J., Neill, P., et al. (2015). Biomolecular detection and quantification assessing the performance of the Oxford Nanopore Technologies MinION. *Biomolecular Detection and Quantification*, 3, 1–8.

Leendertz, S. A. J., Metzger, S., Skjerve, E., Deschner, T., Boesch, C., & Riedel, J. (2010). A longitudinal study of urinary dipstick parameters in wild chimpanzees (*Pan troglodytes verus*) in Côte d'Ivoire. *American Journal of Primatology*, 72, 689–698.

Leighton, M. (1993). Modeling dietary selectivity by Bornean orangutans: evidence for integration of multiple criteria in fruit selection. *International Journal of Primatology*, 14, 257–313.

Lesniak, I., Heckmann, I., Franz, M., et al. (2018). Recolonizing gray wolves increase parasite infection risk in their prey. *Ecology and Evolution*, 8, 2160–2170.

Lorich, T. (2020). Health *Monitoring in Great Apes: The Use of Neopterin as a Non-invasive Marker for Monitoring Diseases in Wild Chimpanzees (*Pan troglodytes verus*).* Unpublished thesis, Berlin Free University.

Lucas, P. W., Beta, T., & Darvell, B. W. (2001). Field kit to characterize physical, chemical and spatial aspects of potential primate foods. *Folia Primatologica*, 72, 11–25.

Lyke, M., Di Fiore, A., Fierer, N., Madden, A., & Lambert, J. (2019). Metagenomic analyses reveal previously unrecognized variation in the diets of sympatric Old World monkey species. *PLoS One*, 14, e0218245.

MacIntosh, A. J. J., Huffman, M. A., Nishiwaki, K., & Miyabe-Nishiwaki, T. (2012). Urinological screening of a wild group of Japanese macaques (*Macaca fuscata yakui*): investigating trends in nutrition and health. *International Journal of Primatology*, 33, 460–478.

Madliger, C., Love, O., Hultine, K., & Cox, S. (2018). The conservation physiology toolbox: status and opportunities. *Conservation Physiology*, 6, coy029.

Maestri, S., Cosentino, E., Paterno, M., et al. (2019). A rapid and accurate minION-based workflow for tracking species biodiversity in the field. *Genes*, 10, 468.

Magi, A., Semeraro, R., Mingrino, A., Giusti, B., & Aurizio, R. (2017). Nanopore sequencing data analysis: state of the art, applications and challenges. *Briefings in Bioinformatics*, 19, 1256–1272.

Mallott, E. K., Malhi, R. S., & Amato, K. R. (2019). Assessing the comparability of different DNA extraction and amplification methods in gut microbial community profiling. *Access Microbiology*, 1, e000060.

Marshall, A. J., Beaudrot, L., & Wittmer, H. U. (2014). Responses of primates and other frugivorous vertebrates to plant resource variability over space and time at Gunung Palung National Park. *International Journal of Primatology*, 35, 1178–1201.

Martinez del Rio, C., & Wolf, B. O. (2005). Mass-balance models for animal isotopic ecology. In Starck, J. M., & Wang, T. (eds.). *Physiological and Ecological Adaptations to Feeding in Vertebrates*. Science Publishers, Enfield, NH (pp. 141–174).

McKelvy, M., Britt, T., & Davis, B. (1996). Infrared spectroscopy. *Analytical Chemistry*, 68, 93–160.

Meijaard, E., Wich, S., Ancrenaz, M., & Marshall, A. J. (2012). Not by science alone: why orangutan conservationists must think outside the box. *Annals of the New York Academy of Sciences*, 1249, 29–44.

Menegon, M., Cantaloni, C., Rodriguez-Prieto, A., et al. (2017). On site DNA barcoding by nanopore sequencing. *PLoS One*, 12, e0184741.

Mikheyev, A., & Tin, M. (2014). A first look at the Oxford Nanopore MinION sequencer. *Molecular Ecology Resources*, 14, 1097–1102.

Milich, K., Stumpf, R., Chambers, J., & Chapman, C. (2014). Female red colobus monkeys maintain their densities through flexible feeding strategies in logged forests in Kibale National Park, Uganda. *American Journal of Physical Anthropology*, 154, 53–60.

Miller, B., Biggins, D., Hanebury, L., & Vargas, A. (1994). Reintroduction of the black-footed ferret (*Mustela nigripes*). In Olney, P., Mace, G., & Feistner, A. (eds.). *Creative Conservation: Interactive Management of Wild and Captive Animals*. Chapman and Hall, London, UK (pp. 455–464).

Mondol, S., Sridhar, V., Yadav, P., Gubbi, S., & Ramakrishnan, U. (2015). Tracing the geographic origin of traded leopard body parts in the Indian subcontinent with DNA-based assignment tests. *Conservation Biology*, 29, 556–564.

Monfort, S., Arthur, N., & Wildt, D. (1991). Monitoring ovarian function and pregnancy by evaluating excretion of urinary oestrogen conjugates in semi-free-ranging Przewalski's horses (*Equus przewalksii*). *Reproduction*, 91, 155–164.

Monfort, S. L., Jayaraman, S., Shideler, S. E., Lasley, B. L., & Hendrickx, A. G. (1986). Monitoring ovulation and implantation in the cynomolgus macaque (*Macaca fascicularis*) through evaluation of urinary estrone conjugates and progesterone metabolites: a technique for the routine evaluation of reproductive parameters. *Journal of Medical Primatology*, 15, 17–26.

Moore, B., Foley, W., Wallis, I., Cowling, A., & Handasyde, K. (2005). A simple understanding of complex chemistry explains feeding preferences of koalas. *Biology Letters*, 1, 64–67.

Muehlenbein, M., & Bribiescas, R. G. (2005). Testosterone-mediated immune functions and male life histories. *American Journal of Human Biology*, 17, 527–558.

Muller, M. N., & Wrangham, R. W. (2004). Dominance, aggression and testosterone in wild chimpanzees: a test of the 'challenge hypothesis'. *Animal Behaviour*, 67, 113–123.

Müller-Klein, N., Heistermann, M., Strube, C., et al. (2019). Physiological and social consequences of gastrointestinal nematode infection in a nonhuman primate. *Behavioral Ecology*, 30, 322–335.

Munro, C. J., & Lasley, B. L. (1988). Non-radiometric methods for immunoassay of steroid hormones. In Albertson, B. D., & Haseltine, F. P. (eds.). *Non-Radiometric Assays: Technology and Application in Polypeptide and Steroid Hormone Detection*. Alan R. Liss, Inc., New York, NY (pp. 289–329).

Naumenko, D., Watford, M., Atmoko, S., Erb, W., & Vogel, E. (2019). Evaluating ketosis in primate field studies: validation of urine test strips in wild Bornean orangutans (*Pongo pygmaeus wurmbii*). *Folia Primatologica*, 91, 159–168.

Negrey, J., Sandel, A. A., & Langergraber, K. E. (2020). Dominance rank and the presence of sexually receptive females predict feces-measured body temperature in male chimpanzees. *Behavioral Ecology and Sociobiology*, 74, 5.

Nelson, S., Miller, M., Heske, E., & Fahey Jr., G. (2008). Nutritional consequences of a change in diet from native to agricultural fruits for the Samoan fruit bat. *Ecography*, 23, 393–401.

Newsom, S., Ralls, K., van Horn, J. C., Fogel, M., & Cypher, B. (2010). Stable isotopes evaluate exploitation of anthropogenic foods by the endangered San Joaquin kit fox (*Vulpes macrotis mutica*). *American Journal of Mammalogy*, 91, 1313–1321.

Nsubuga, A. M., Robbins, M. M., Roeder, A. D., Morin, P. A., Boesch, C., & Vigilant, L. (2004). Factors affecting the amount of genomic DNA extracted from ape faeces and the identification of an improved sample storage method. *Molecular Ecology*, 13, 2089–2094.

Nunn, C., & Altizer, S. M. (2006). *Infectious Diseases in Primates: Behavior, Ecology and Evolution*. Oxford University Press, Oxford, UK.

Nunn, C., Gittleman, J., & Antonovics, J. (2000). Promiscuity and the primate immune system. *Science*, 290, 1168.

Nurcahyo, W., Konstanzová, V., & Foitová, I. (2017). Parasites of orangutans (primates: Ponginae): an overview. *American Journal of Primatology*, 79, e22650.

O'Connell, C. A. (2018). *The Costs and Benefits of Sociality Explored in Wild Bornean Orangutans (Pongo pygmaeus wurmbii) Anthropology*. Boston University, Boston, MA.

Pappano, D. J., Roberts, E. K., & Beehner, J. C. (2010). Testing extraction and storage parameters for a fecal hormone method. *American Journal of Primatology*, 72, 943–941.

Pasquini, C. (2003). Near infrared spectroscopy: fundamentals, practical aspects and analytical applications. *Journal of Brazilian Chemical Society*, 14, 198–219.

Pomapnon, F., Deagle, B., Symondson, W., Brown, D., Jarman, S., & Taberlet, P. (2012). Who is eating what: diet assessment using next generation sequencing. *Molecular Ecology*, 21, 1931–1950.

Pomerantz, A., Peñafiel, N., Arteaga, A., et al. (2018). Real-time DNA barcoding in a rainforest using nanopore sequencing: opportunities for rapid biodiversity assessments and local capacity building. *GigaScience*, 7, giy033.

Pride, R. (2005). High faecal glucocorticoid levels predictor mortality in ring-tailed lemurs (*Lemur catta*). *Biology Letters*, 1, 60–53.

Quéméré, E., Hibert, F., Miquel, C., et al. (2013). A DNA metabarcoding study of a primate dietary diversity and plasticity across its entire fragmented range. *PLoS One*, 8, 8, e58971.

Raubenheimer, D. (2011). Towards a quantitative nutritional ecology: the right-angled mixture triangle. *Ecological Monographs*, 81, 407–427.

Raubenheimer, D., Simpson, S., & Tait, A. (2012). Match and mismatch: conservation physiology, nutritional ecology, and the timescales of biological adaptation. *Philosophical transactions of the Royal Society B*, 367, 1629–1646.

Richards, P. W. (1996). *The Tropical Rain Forest*, 2nd edn. Cambridge University Press, Cambridge, UK.

Robbins, C., Schwartz, C., & Felicetti, L. (2004). Nutritional ecology of ursids: a review of newer methods and management implications. *Ursus*, 15, 161–171.

Robbins, M. M., & Czekala, N. M. (1997). A preliminary investigation of urinary testosterone and cortisol levels in wild male mountain gorillas. *American Journal of Primatology*, 43, 51–64.

Rode, K., Chiyo, P., Chapman, C., & McDowell, L. (2006). Nutritional ecology of elephants in Kibale National Park, Uganda, and its relationship with crop-raiding behaviour. *Journal of Tropical Ecology*, 22, 441–449.

Rodgers, E. (2017). Foraging for fast food: the changing diets of wildlife. *Conservation Physiology*, 5, cox046.

Rothman, J., Chapman, C., & Van Soest, P. (2012). Methods in primate nutritional ecology: a user's guide. *International Journal of Primatology*, 33, 542–566.

Rothman, J. M., Chapman, C. A., Hansen, J. L., Cherney, D. J. R., & Pell, A. N. (2009). Rapid assessment of the nutritional value of foods eaten by mountain gorillas: applying near-infrared reflectance spectroscopy to primatology. *International Journal of Primatology*, 30, 729–742.

Rothman, J. M., Raubenheimer, D., & Chapman, C. A. (2011). Nutritional geometry: gorillas prioritize non-protein energy while consuming surplus protein. *Biology Letters*, 7, 847–849.

Rothman, J. M., Van Soest, P. J., & Pell, A. N. (2006). Decaying wood is a sodium source for mountain gorillas. *Biology Letters*, 2, 321–324.

Samuni, L., Tkaczynski, P., Deschner, T., Löhrrich, T., Wittig, R. M., & Crockford, C. (2020). Maternal effects on offspring growth indicate post-weaning juvenile dependence in chimpanzees (*Pan troglodytes verus*). *Frontiers in Zoology*, 17, 1.

Santika, T., Meijaard, E., Budiharta, S., et al. (2017). Community forest management in Indonesia: avoided deforestation in the context of anthropogenic and climate complexities, *Global Environmental Change*, 46, 60–71.

Santymire, R. M., & Armstrong, D. M. (2010). Development of a field-friendly technique for fecal steroid extraction and storage using the African wild dog (*Lycaon pictus*). *Zoo Biology*, 29, 289–302.

Sapolsky, R. M. (1992). Cortisol concentrations and the social significance of rank instability among wild baboons. *Psychoneuroendocrinology*, 17, 701–709.

Savage, A., Shideler, S. E., Soto, L. H., et al. (1997). Reproductive events of wild cotton-top tamarins (*Saguinus oedipus*) in Colombia. *American Journal of Primatology*, 43, 329–337.

Seah, A., Lim, M. C. W., McAloose, D., Prost, S., & Seimon, T. (2020). MiniION-based DNA barcoding of preserved and non-invasively collected wildlife samples. *Genes*, 11, 445.

Sepulveda, A., Hutchins, P., Massengill, R., & Dunker, K. (2018). Tradeoffs of a portable, field-based environmental DNA platform for detecting invasive northern pike (*Esox lucius*) in Alaska. *Management of Biological Invasions*, 9, 253–258.

Shideler, S. E., Czekala, N. M., Benirschke, K., & Lasley, B. L. (1983). Urinary estrogens during pregnancy of the ruffed lemur (*Lemur variegatus*). *Biology of Reproduction*, 28, 963–969.

Shideler, S. E., Munro, C. J., Johl, H. K., Taylor, H. W., & Lasley, B. L. (1995). Urine and fecal sample collection on filter paper for ovarian hormone evaluations. *American Journal of Primatology*, 37, 305–316.

Shideler, S. E., Savage, A., Ortuno, A., Moorman, E. A., & Lasley, B. L. (1994). Monitoring female reproductive function by measurement of fecal estrogen and progesterone metabolites in the white-faced saki (*Pithecia pithecia*). *American Journal of Primatology*, 32, 95–108.

Smith, K. F., Sax, D. F., & Lafferty, K. D. (2006). Evidence for the role of infectious disease in species extinction and endangerment. *Conservation Biology*, 20, 1349–1357.

Song, S. J., Amir, A., Metcalf, J. L., et al. (2016). Preservation methods differ in fecal microbiome stability, affecting suitability for field studies, *mSystems*, 1, 3.

Spehar, S. N., Sheil, D., Harrison, T., et al. (2018). Orangutans venture out of the rainforest and into the Anthropocene. *Science Advances*, 46, e1701422.

Srivathsan, A., Hartop, E., Puniamoorthy, J., et al. (2019). Rapid, large-scale species discovery in hyperdiverse taxa using 1D MinION sequencing. *BMC Biology*, 17, 1–20.

Stavisky, R. C., Russell, E., Stallings, J., Smith, E. O., Worthman, C. M., & Whitten, P. L. (1995). Fecal steroid analysis of ovarian cycles in free-ranging baboons. *American Journal of Primatology*, 36, 285–297.

Stoner, K. E. (1996). Prevalence and intensity of intestinal parasites in mantled howling monkeys (*Alouatta palliata*) in Northeastern Costa Rica: implications for conservation biology. *Conservation Biology*, 10, 539–546.

Stuart, M., Strier, K., & Pierberg, S. (1993). A coprological survey of parasites of wild muriquis, *Brachyteles arachnoides*, and brown howling monkeys, Alouatta fusca. *Journal of Helminthological Society of Washington*, 60, 111–115.

Stuart, M. D., & Strier, K. B. (1995). Primates and parasites: a case for a multidisciplinary approach. *International Journal of Primatology*, 16, 577–593.

Sun, P., Wronski, T., Bariyanga, J., & Apio, A. (2018). Gastro-intestinal parasite infections of Ankole cattle in an unhealthy landscape: an assessment of ecological predictors, *Veterinary Parasitology*, 252, 107–116.

Surbeck, M., Deschner, T., Behringer, V., & Hohmann, G. (2015). Urinary C-peptide levels in male bonobos (*Pan paniscus*) are related to party size and rank but not to mate competition. *Hormones and Behavior*, 71, 22–30.

Teixeira dos Santos, C., Lopo, M., Pascoa, R., & Lopes, J. (2013). A review on the applications of portable near-infrared spectrometers in the agro-food industry. *Applied Spectroscopy*, 67, 1215–1233.

Touma, C., & Palme, R. (2005). Measuring fecal glucocorticoid metabolites in mammals and birds: the importance of validation. *Annals of the New York Academy of Science*, 1046, 54–74.

Tutin, C. E. G. (1979). Mating patterns and reproductive strategies in a community of wild chimpanzees (*Pan troglodytes schweinfurthii*). *Behavioral Ecology and Sociobiology*, 6, 29–38.

Tytgat, O., Gansemans, Y., Weymaere, J., Rubben, K., Deforce, D., & Van Nieuwerburgh, F. (2020). Nanopore sequencing of a forensic STR multiplex reveals loci suitable for single-contributor STR profiling. *Genes*, 11, 381.

Valentini, A., Pompanon, F., & Taberlet, P. (2008). DNA barcoding for ecologists. *Trends in Ecology and Evolution*, 24, 110–117.

van Noordwijk, M. A., Utami-Atmoko, S. S., Knott, C. D., et al. (2018). The slow ape: high infant survival and long inter-birth intervals in orangutans. *Journal of Human Evolution*, 125, 38–49.

van Schaik, C. P., Priatna, A., & Priatna, D. (1995). Population estimates and habitat preferences of orangutans based on line transects of nests. In Nadler, R. D., Galdikas, B., Sheeran, F. M., & Rosen, N. (eds.). *The Neglected Ape*. Plenum Press, New York, NY and London, UK (pp. 129–147).

Vance, C. K., Tolleson, D. R., Kinoshita, K., Rodriguez, J., & Foley, W. J. (2016). Near infrared spectroscopy in wildlife and biodiversity. *Journal of Near Infrared Spectroscopy*, 24, 1–25.

Viggers, K., Lindenmayer, D., & Spratt, D. (1993). The importance of disease in reintroduction programmes. *Wildlife Research*, 205, 687–698.

Vogel, E. R., Crowley, B. E., Knott, C. D., Blakely, M. D., Larsen, M. D., & Dominy, N. J. (2012a). A noninvasive method for estimating nitrogen balance in free-ranging primates. *International Journal of Primatology*, 33, 567–587.

Vogel, E. R., Knott, C. D., Crowley, B. E., Blakely, M. D., Larsen, M. D., & Dominy, N. J. (2012b). Bornean orangutans at the brink of protein bankruptcy. *Biology Letters*, 8.

Voigt, M., Wich, S., Ancrenaz, M., et al. (2018). Global demand for natural resources eliminated more than 100,000 Bornean orangutans. *Current Biology*, 28, 761–769.

Wallis, J. (1992). Chimpanzee genital swelling and its role in the pattern of sociosexual behavior. *American Journal of Primatology*, 28, 101–113.

Wallis, J. (1995b). Seasonal influence on reproduction in chimpanzees of Gombe National Park. *International Journal of Primatology*, 16, 435–451.

Warren, K. S., Nijman, I. J., Lenstra, J. A., Swan R. A., Heriyanto, & Den Boer, M. H. (2001). Speciation and intrasubspecific variation of Bornean orangutans, *Pongo pygmaeus pygmaeus*. *Molecular Biology and Evolution*, 18, 472–480.

Wasser, S. K., Monfort, S. L., & Wildt, D. E. (1991). Rapid extraction of faecal steroids for measuring reproductive cyclicity and early pregnancy in free-ranging yellow baboons (*Papio cynocephalus cynocephalus*). *Reproduction and Fertility*, 92, 415–423.

Watsa, M., Erkenswick, G. A., Pomerantz, A., & Prost, S. (2019). Genomics in the jungle: using portable sequencing as a teaching tool in field courses. *BioRxiv*, 581728.

Wells, K., Smales, L., Kalko, E., & Pfeiffer, M. (2007). Impact of rain-forest logging on helminth assemblages in small mammals (muridae, tupaiidae) from Borneo. *Journal of Tropical Ecology*, 23, 35–43.

Whitten, P. L., Brockman, D. K., & Stavisky, R. C. (1998). Recent advances in noninvasive techniques to monitor hormone-behavior interactions. *Yearbook of Physical Anthropology*, 27, 1–24.

Wich, S. A., Singleton, I., Nowak, M. G., et al. (2016). Land-cover changes predict steep declines for the Sumatran orangutan (*Pongo abelii*). *Science Advances*, 2, e1500789.

Wikelski, M., & Cooke, S. (2006). Conservation physiology. *Trends in Ecology and Evolution*, 21, 38–46.

Wildt, D., & Wemmer, C. (1999). Sex and wildlife: the role of reproductive science in conservation. *Biodiversity and Conservation*, 8, 965–976.

Wingfield, J. C., Hegner, R. E., Dufty, A. M., & Ball, G. F. (1990). The 'challenge hypothesis': theoretical implications for patterns of testosterone secretion, mating systems, and breeding strategies. *The American Naturalist*, 136, 829–846.

Wittig, R., Crockford, C., Weltring, A., Deschner, T., & Zuberbühler, K. (2015). Single aggressive interactions increase urinary glucocorticoid levels in wild male chimpanzees. *PLoS One*, 10, e0118695.

Woodruff, D. S. (2004). Noninvasive genotyping and field studies of free-ranging nonhuman primates. In Chapais, B., & Berman, C. M. (eds.). *Kinship and Behavior in Primates*. Oxford University Press, New York, NY (pp. 46–67).

Wu, D. F., Behringer, V., Wittig, R. W., Leendertz, F. H., & Deschner, T. (2018). Urinary neopterin levels increase and predict survival during a respiratory outbreak in wild chimpanzees (Taï National Park, Côte d'Ivoire). *Scientific Reports*, 8, 13346.

Yalow, R., & Berson, S. A. (1960). Immunoassay of endogenous plasma insulin in man. *The Journal of Clinical Investigation*, 39, 1157–1175.

CHAPTER 8

Environmental DNA for conservation

Antoinette J. Piaggio

8.1 Introduction

Biodiversity must be documented before it can be conserved. However, it may be difficult to document species with few individuals (Thompson, 2013; Goldberg et al., 2016), thus it requires a multitude of tools to detect species that occur in low numbers or are elusive (see the various chapters in this volume). One tool that has become useful for conservation efforts utilizes environmental DNA, which is DNA shed into the environment by organisms (eDNA; Taberlet et al., 2018). Typically this involves taking environmental samples such as soil, water, air, or using biological surrogates for sampling biodiversity (e.g. leeches, sponges, carrion flies, etc.; Schnell et al., 2012; Calvignac-Spencer et al., 2013; Lynggaard et al., 2019; Mariani et al., 2019) and using laboratory approaches to concentrate, isolate, and test for target DNA through polymerase chain reaction (PCR) amplification (Taberlet et al., 2018). The utilization of eDNA for species detection is part of a larger field of non-invasive DNA sampling, which more broadly includes collecting DNA passively from wildlife, through collection of faeces, saliva, feathers, hair, or other methods of sampling shed DNA. Environmental DNA has been used to document presence/absence of a target species (Ficetola et al., 2008a, 2008b; Hunter et al., 2017) or to quantify relative abundance for biodiversity from varied environments such as the arctic (e.g. Leduc et al., 2019; Von Duyke et al., 2019), marine (e.g. Port et al., 2016; Jo et al., 2017; Stoeckle et al., 2018), freshwater (e.g. Lacoursière-Roussel et al., 2016; Doi et al., 2017), and tropical (e.g. Schnell et al., 2012;

Gogarten et al., 2020) ecosystems. The application of this technology includes the detection of invasive species, pathogens (including DNA and RNA), species of conservation concern, and biodiversity (Acevedo-Whitehouse et al., 2010; Rees et al., 2014; Sakai et al., 2019).

When detection of a rare or cryptic species is the goal, often a suite of tools, such as acoustic sensors (Chapter 4), camera traps (e.g. Chapter 5; Mallet & Pelletier, 2014; Burton et al., 2015; Caravaggi et al., 2017; Wearn & Glover-Kapfer, 2019), traps such as light (Conrad et al., 2007; e.g. McLeod & Costello, 2017) or Peterson/Tomahawk or corral (e.g. Garden et al., 2007; Costello et al., 2017), trawling (Jones, 1992; Heino et al., 2011), continuous plankton recorders (e.g. Richardson et al., 2006), bait stations (e.g. Shardlow & Hyatt, 2013; Costello et al., 2017), faecal DNA mark-recapture (e.g. Waits & Paetkau, 2005; Schwartz et al., 2007), and other methods are reliably used for detection. Each method has trade-offs, including species wariness to new instruments in their environment and difficulty detecting evidence due to weather or vegetation, among other issues. Further, for species that exist at low population sizes, failure to confirm presence does not necessarily suggest its absence for many reasons, including a low DNA signal in the environment (Ellison et al., 2006; Thompson, 2013). The ability to isolate DNA from a target species without the need to capture animals provides another tool that can help biodiversity surveys (Darling, 2019). In particular, eDNA can be inexpensive to implement in the field and can be accomplished through the collection of samples easily obtained from the environment or the use of a single species to detect

Antoinette J. Piaggio, *Environmental DNA for conservation*. In: *Conservation Technology*.
Edited by Serge A. Wich and Alex K. Piel, Oxford University Press.
© Oxford University Press 2021. DOI: 10.1093/oso/9780198850243.003.0008

others (e.g. blood or carrion feeders). However, like other tools, there are trade-offs with this method, which are described in detail later in this chapter (Cristescu & Hebert, 2018).

Largely, eDNA has been used to detect the presence/absence of a target species in the environment. Presence of DNA and, in some cases, relative frequency across sites is assessed through PCR platforms that quantify the amplification of DNA but do not provide an estimate of the number of individuals present, such as real-time PCR, specifically quantitative PCR (qPCR), or droplet digital PCR (ddPCR) (Goldberg et al., 2016; Deiner et al., 2017b; Baker et al., 2018; Klymus et al., 2020). However, genomic approaches, such as next-generation sequencing (NGS) that allow for simultaneously sequencing multiple species from a single environmental sample has been used to estimate relative abundance (Valentini et al., 2009; Bohmann et al., 2014; Valentini et al., 2016; Porter & Hajibabaei, 2018). Multiple studies comparing eDNA to other tools for sampling biodiversity demonstrate that eDNA can do as well as others (Valentini et al., 2016; Gogarten et al., 2020), be an important complement to other tools, or even perform better at detection, assessing diversity, etc. than other tools (Valentini et al., 2016; Table 1, Deiner et al., 2017a; Boussarie et al., 2018). For conservationists, it is important to understand the practical applications of eDNA, what questions it can answer, its uses and limitations, and how one might implement this tool.

8.2 New technology

8.2.1 Data collection—study design

Sampling environmental DNA has rapidly become an important tool in the arsenal of non-invasive tools for detecting biodiversity and for use in conservation efforts (e.g. Thomsen & Willerslev, 2015; Cristescu & Hebert, 2018; Taberlet et al., 2018; Mize et al., 2019; Zinger et al., 2019). Endemic and elusive species such as the Rocky Mountain tailed frog (*Ascaphus montanus*) and Idaho giant salamander (*Dicamptodon atterimus*, Pilliod et al., 2013), invasive species such as bullfrogs (*Lithobates catesbeianus*, Ficetola et al., 2008a, 2008b), and overall

biodiversity of species of concern within freshwater ecosystems (Thomsen et al., 2012a) were early targets for detection using eDNA in freshwater ecosystems. In marine ecosystems, the first application of environmental DNA involved parallel sequencing of multiple species, which detected 15 fish species and even some bird species (Thomsen et al., 2012b). Compilation of data from across the now extensive use of eDNA in conservation efforts has led to the development of standard protocols for detecting eDNA in various aquatic systems (see Table S1 in Goldberg et al., 2016; Stoeckle et al., 2018; Harper et al., 2019; Jeunen et al., 2019).

Protocols for collecting eDNA samples are influenced by the research question (presence/absence, relative abundance, species composition, etc.) and the target species, among other issues (Goldberg et al., 2016; Wilcox et al., 2018; Zinger et al., 2019). Because field sites introduce unique biotic challenges (e.g. pH of the water, the water body type and size, season) to detection probabilities, pilot studies are critical (Barnes & Turner, 2016; Goldberg et al., 2016; Davis et al., 2018). Before one can determine how many samples are needed within a single water body or across a target region with variable biotic conditions, we need to know the probability of detecting a species under those conditions. Further, the processing of eDNA has several steps (DNA capture, isolation, and amplification) where detection may vary. Using multistage occupancy analyses with data from a pilot study it is possible to identify the detection probability at different water body types, under different biotic conditions, and across laboratory processes (Figure 8.1: Goldberg et al., 2016; Hunter et al., 2017; Davis et al., 2018; Doi et al., 2019; Erickson et al., 2019; Sepulveda et al., 2019).

Another complex issue for eDNA studies is calculating costs, then comparing these costs to other pertinent tools, and finding the right balance between costs and information gained. One study has explored these issues and provided a phone application that can be easily used and tailored to individual needs (Davis et al., 2020). In some cases, eDNA can be prohibitively costly and, in others, there is a balance between effort, cost, and information gained with this monitoring method. As the field progresses, so may the tools that make it easier for all levels of practitioners to assess

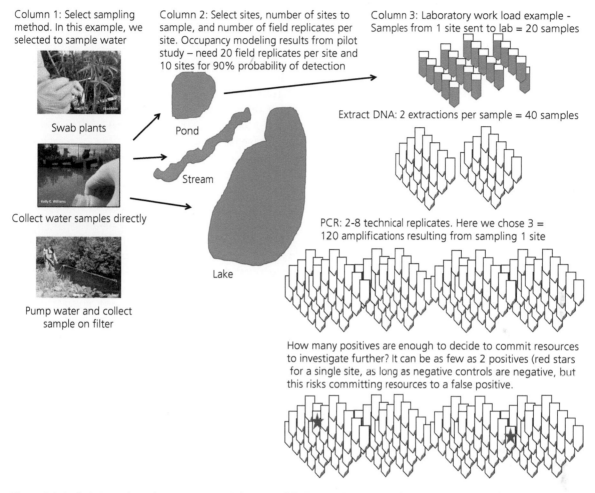

Column 1: Select sampling method. In this example, we selected to sample water

Swab plants

Collect water samples directly

Pump water and collect sample on filter

Column 2: Select sites, number of sites to sample, and number of field replicates per site. Occupancy modeling results from pilot study – need 20 field replicates per site and 10 sites for 90% probability of detection

Pond

Stream

Lake

Column 3: Laboratory work load example - Samples from 1 site sent to lab = 20 samples

Extract DNA: 2 extractions per sample = 40 samples

PCR: 2-8 technical replicates. Here we chose 3 = 120 amplifications resulting from sampling 1 site

How many positives are enough to decide to commit resources to investigate further? It can be as few as 2 positives (red stars for a single site, as long as negative controls are negative, but this risks committing resources to a false positive.

Figure 8.1 Study design and sampling strategy are critical to successfully detecting low numbers of target species using environmental DNA (eDNA). It is critical to understand the influences of each step of the process, from eDNA capture, isolation, and amplification to biotic influences (pH, water diffusion and dispersion, size and type of water body, time of year, etc.). Pilot studies (see example in this figure) conducted in the target geographic region, along with statistical assessments such as occupancy modelling (Davis et al., 2018) are necessary to determine the appropriate sample size per water body and the number of replicates required in the lab to maximize detection probability. It is also prudent to decide the parameters of what defines a positive that is relevant to management actions and tolerance for false positives or false negatives.

sample size needs, sampling design, costs, and required effort for individual needs, as has been done for faecal mark-recapture studies (Lonsinger et al., 2015). Overall, consultation with a lab experienced in eDNA studies is ideal, and there are useful resources for those wanting to begin the process of using eDNA[1].

8.2.2 Data collection—targeting single species

Many eDNA studies use DNA primers targeted to a particular species of interest. The development of an assay for a novel target species requires laboratory validation. Further, understanding detection limits and DNA persistence through the use of captive target taxa is critical before applying the assay to

---

[1]
1 https://ednaresources.science/
2 https://www.fs.fed.us/research/genomics-center/edna/
3 https://github.com/boopsboops/seadna-protocols/tree/master/docs
4 http://ednasociety.org/eDNA_manual_Eng_v2_1_3b.pdf

complex and unpredictable field conditions. Captive studies involving target taxa allow testing of eDNA assays on known quantities of shed DNA or for known periods with other biotic parameters held constant (e.g, Williams et al. 2018). Assay validation should also include the testing of primers for species-specificity, meaning they do not cross-amplify in non-target species that may be present in the target sampling area. Further, replicated dilution series of positive controls to assess limit of detection, limit of blank, and limit of quantitation is necessary for assay validation and requisite understanding of potential for false-positives and false-negatives (Goldberg et al., 2016; Klymus et al., 2020). There are standardized methods for eDNA assay validation methods and reporting of critical parameters (Klymus et al., 2020). The current body of eDNA validated assays has revealed that eDNA is ephemeral and heterogeneously distributed in the environment (Furlan et al., 2016), that sensitivity of laboratory molecular methods is affected by environmental inhibitors (e.g. turbidity, pH, water chemistry, etc.), and that there is detection variability of low quantity/quality DNA from the environment (Barnes & Turner, 2016; Goldberg et al., 2016). Thus, each assay validation provides further refined data on the utility and applicability of eDNA as a tool in varied environments and across species.

8.2.3 Data collection—targeting multiple species

Biodiversity documentation and monitoring through eDNA is best informed by the ability to identify all DNA in a single sample. The method of species-specific targeting in environmental samples does not allow for biodiversity assessment but rather provides presence/absence data or, in some cases, relative quantification of population size of a single species (Goldberg et al., 2016). NGS allows DNA fragments of a mixed-species sample to be sequenced simultaneously and thus identification of a suite of species (Figure 8.2; Thomsen et al., 2012b; Taberlet et al., 2018). The potential of NGS application to eDNA is that the number of copies of sequences represents the presence of DNA cells from individuals and can be translated to the

abundance of each species and then extrapolated to the sampling area.

The use of NGS in environmental sampling is typically to amplify a single gene across all DNA in a mixed-species sample (Figure 8.2), termed metabarcoding (Taberlet et al., 2018). The adoption of metabarcoding for biodiversity assessment of eDNA has been detailed elsewhere (Deiner et al., 2017a; Cristescu & Hebert, 2018), but an overview to allow for an applied understanding will be provided here. Multiple-species DNA detection approaches promise to be just as powerful or more powerful as traditional field tools (reviewed 2011–2019 in Shu et al., 2020) in assessing biodiversity (Valentini et al., 2016; Deiner et al., 2017a; Jo et al., 2017). It is not yet possible to quantify the number of individuals identified in a sample or ensure complete detection of every species present in a sampled habitat (Bush et al., 2019), especially rare species (Deiner et al., 2017a), but the field is rapidly moving towards addressing these and other current limitations (Valentini et al., 2016; Deiner et al., 2017a, 2017b; Taberlet et al., 2018; Knudsen et al., 2019; Shelton et al., 2019). Developing an assay to target multiple species can be quite similar to an assay for a single species (Shu et al., 2020). Methods of DNA capture and isolation may vary depending on environmental sample type and targeted community being assessed (Jeunen et al., 2019). Some primers have been validated and tested for use in eDNA metabarcoding studies (Taberlet et al., 2018), yet each can introduce biases; for example, primer biases may allow one species to be over-represented in the results, among other issues (Deiner et al., 2017a, 2017b; Taberlet et al., 2018; Zinger et al., 2019). Thus, methods developed to detect biodiversity from eDNA samples must be tested using mock communities with known amounts of DNA from various species, including rare species, to make sure they can be detected in low numbers using the chosen primers (Taberlet et al., 2018; Doi et al., 2019). In some cases, species of concern may be better monitored through species-specific approaches or a combination of approaches (Wood et al., 2019).

There are currently two primary challenges in applying metabarcoding for eDNA: (1) biases caused by laboratory processes make it challenging to ascertain quantifiable biodiversity indices; (2)

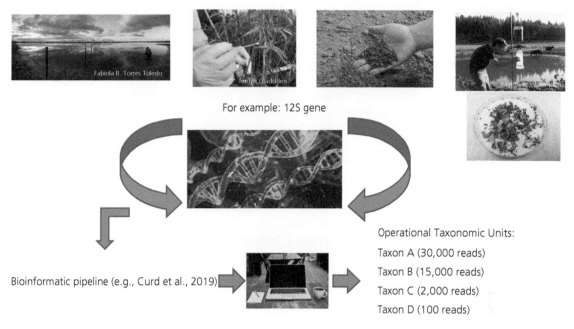

For example: 12S gene

Bioinformatic pipeline (e.g., Curd et al., 2019)

Operational Taxonomic Units:

Taxon A (30,000 reads)

Taxon B (15,000 reads)

Taxon C (2,000 reads)

Taxon D (100 reads)

Figure 8.2 Samples are taken from various environmental substrates are collected and next-generation sequencing (NGS) is used to amplify a single gene fragment or whole genomic data from all DNA in a sample. Bioinformatic pipelines such as Anacapa Toolkit (Curd et al., 2019) are used to identify Operational Taxonomic Units (OTU: Blaxter et al., 2005), groups of sequences with the chosen level of similarity and identified through comparison to publicly available DNA databases. The number of reads assigned to each OTU is the number of sequences assigned to that taxonomic designation. Additional credits: Soil in hands image courtesy of Jing/Pixabay; DNA image courtesy of geralt/Pixabay; Laptop/coffee image courtesy of Engin_Akyurt/Pixabay.

data produced are challenging to manage and analyse (see 8.2.2 Data analyses). In response to the first issue, some labs are working to find novel ways to deal with biases and use these data to estimate diversity and abundance (e.g. Shelton et al., 2019). Researchers are also developing tools to manage and analyse the data properly (e.g. Kandlikar et al., 2018). Thus, the application of NGS and approaches such as metabarcoding or even metagenomics (sequencing whole genomes rather than a single gene from a mixed-species sample) will continue to revolutionize the information available from eDNA and the utility to biodiversity assessments and thus conservation (Ruppert et al., 2019).

8.2.4 Data collection—DNA capture

Sample collection decisions are the same whether one is using an eDNA assay for single-species detection or multiple-species surveys through metabarcoding. Sample collection methodology for

collecting environmental samples advances regularly and depends on the template being targeted (e.g. water and the water type, soil, air, biological means). However, understanding the range of collection options is useful as different field conditions, target species, questions, or project resources influence the usefulness of any particular method.

Filtering water for eDNA capture is the most common approach. In freshwater systems, backpack pumps (Figure 8.1) with filtering systems are commonly used for eDNA collection (Laramie et al., 2015; Goldberg et al., 2016). Backpack pumps draw 2–10 litres of water through a filter to remove particulates and capture and concentrate DNA. Once DNA is captured on a filter, it is necessary to preserve it by applying a buffer (e.g. Longmire's buffer; Renshaw et al., 2015) or immediate freezing and maintenance of a cold chain from field to lab. In an effort to standardize collection methods, reduce field effort and field contamination in eDNA sampling, the ANDe™ (now called eDNA

Sampler Backpack) was developed (Thomas et al., 2018) for freshwater systems. The instrument has online tutorials,[2] is relatively affordable, and regularly used by field practitioners (Thomas et al., 2020). Another simple and straightforward method for smaller water bodies in freshwater systems is for field personnel to scoop water into a plastic bottle, add a buffer, and maintain at ambient temperature (this can maintain integrity for up to 56 days; Williams et al., 2016). This approach does not require maintaining a cold chain in the field or using a backpack pump. Once the water is received at the laboratory, eDNA can be concentrated and captured through filtration, or other methods such as centrifugation, chemical precipitation, or electrical charge attraction with magnetic beads, which are positive and thus attract negatively charged DNA molecules (Renshaw et al., 2015; Williams et al., 2016). The decisions between filtering in the field or lab or using other lab-based methods for eDNA capture are largely based on the time field crews have to devote to eDNA collection and procedures necessary for an optimized eDNA assay where various methods have been tested for the specific target species and environmental source (e.g., Williams et al., 2016; Williams et al., 2017).

Marine eDNA collection methods include sampling from shore with buckets (The eDNA Society, 2019) or on board ships with various pumping apparatus (Hansen et al., 2007; Costello et al., 2017; Westfall et al., 2017; The eDNA Society, 2019). More recently, robotic samplers, including the use of autonomous vehicles for sampling marine systems at extreme depths, have also been successfully tested (Yamahara et al., 2019). There are efforts to compare various methods for DNA capture and isolation to determine best practices for obtaining the highest marine species diversity as well as reliable presence/absence surveys (Stoeckle et al., 2018; Jeunen et al., 2019; Shu et al., 2020). The distribution of DNA in an ecosystem depends on the abundance of the species, movement of the substrate (e.g. diffusion and dispersion of molecules of traces of DNA through the environment) (Furlan et al.,

2016), and biotic factors that degrade detectable DNA (Barnes & Turner, 2016). The application of mechanistic models to examine fate and transport of eDNA in marine systems has shown promise for revealing from where and when eDNA was shed from the source species (Andruszkiewicz et al., 2019). Thus, as in freshwater systems, it is critical to conduct pilot studies to better understand the appropriate method for eDNA capture for the target marine ecosystem.

For collecting eDNA from aquatic systems, both marine and freshwater, there is not a standard for collecting at any particular depth (Shu et al., 2020). However, the collection of particles from substrate should be avoided as the DNA may bind colloidally to sediment particles and be retained for long periods (Barnes et al., 2014). Thus, sampling from shorelines can help avoid stirring sediment particles into the water column.

Filtering water requires identifying the balance between the volume of water one can pump through a filter, which is influenced by the degree of water turbidity and the desire to collect as many DNA particles or cells as possible. Most researchers have found that 5μm filters can handle 3–7 L of water throughput before clogging occurs and provide the optimal balance of higher throughput and DNA capture over smaller and larger pore size filters (Goldberg et al., 2017; Thomas et al., 2018; Jeunen et al., 2019). Although there is not a standard for the amount of water that needs to be collected (Shu et al., 2020), one thing that has become clear is that pooling water samples is not useful as it causes dilution and lowered probability of detection (Davis et al., 2018; Shu et al., 2020).

Environmental DNA can also be isolated from collected soil samples where the biodiversity of microorganisms (e.g. bacteria and fungi), soil invertebrate, animals, and plants can be assessed (Andersen et al., 2012; Cristescu and Hebert, 2018). Beyond collecting samples from the environment (e.g. water and soil), various substrates may have DNA traces that reveal a particular species' presence, such as browsed twigs for ungulate species (Nichols et al., 2012; Nichols & Spong, 2014; Nichols et al., 2015). To obtain such DNA simply requires swabbing the surface and isolating the

[2] https://www.smith-root.com/support/tutorials

DNA from the swab. An emerging use of eDNA is the exploitation of biological systems to subsample biodiversity, specifically targeting species that feed on blood (e.g. leeches, flies, Schnell et al., 2012; Gogarten et al., 2020), carrion (Calvignac-Spencer et al., 2013), or filter water for food (e.g. sponges, Mariani et al., 2019). Accessing these DNA sources provides a sample of biodiversity from the ecosystem where the insects and sponges live using standard DNA extraction methods. The ability to use organisms that harvest DNA from many species will continue to provide novel approaches to surveying biodiversity.

8.2.5 Data collection-DNA isolation, purification, and amplification

There is also a need to determine the ideal approach for eDNA isolation and purification (extraction from sample matrix after concentration and capture) as one approach does not fit all study species or ecosystems (Deiner et al., 2015; Jeunen et al., 2019). There are many commercially available kits designed to extract DNA (e.g. Qiagen DNeasy Kit™ among others). Pilot studies often test multiple kits or chemical methods (e.g. phenol/chloroform extractions) to see which one works best for their study system (Deiner et al., 2015; Williams et al., 2017; Shu et al., 2020). There has been no consistent pattern of best practices for DNA isolation and purification that has emerged, thus making it clear that testing for each study remains the best option (Goldberg et al., 2016; Kumar et al., 2020).

Largely, conventional PCR or qPCR is the standard DNA amplification platform (Goldberg et al., 2016; Klymus et al., 2020) for single-species detection using eDNA (for technical details, Taberlet et al., 2018), but ddPCR is becoming an appealing alternative because it eliminates the need for a standard curve, which takes up valuable sample space on a PCR plate, and is less affected by inhibitors (Doi et al., 2015; Hunter et al., 2017; Baker et al., 2018; Hunter et al., 2018). Further, ddPCR has been shown to outperform qPCR (Doi et al., 2015) and metabarcoding in the detection of a single-target species (Wood et al., 2019). A subset of positive samples are often sequenced using

conventional approaches for DNA sequencing of gene fragments (Sanger sequencing, Deiner et al., 2017a) and comparison to publicly available DNA sequence databases[3] to confirm that the positives were, in fact, the target species.

NGS allows for simultaneous amplification of all DNA fragments in a mixed-species sample. Samples from the environment are treated to the same DNA isolation and purification processes as for a single-species assay. However, the next steps involve preparing DNA into fragments with additional portions of adapters and tagged primers added to generate 'libraries' of the field obtained eDNA (this process is detailed in Chapter 7, Taberlet et al., 2018). This is a time-consuming process requiring technical expertise but is regularly performed in most genetics laboratories and can be easily applied to biodiversity conservation studies through collaboration or pay-for-service.

8.2.6 Data analyses

Single-species targeted eDNA assays primarily provide presence/absence results (Goldberg et al., 2016). However, the quantitation of DNA amplification through time or DNA molecules per microliter from positive results can be used to estimate the relative abundance of a species (Doi et al., 2017). Interpretation of results need to be carefully considered as eDNA captures DNA that could have been incorporated into the water system through run-off of water crossing land and thus may include DNA from faecal samples or carcasses, which could lead to inflated abundance estimates (Merkes et al., 2014; Sepulveda et al., 2019). Further, one must decide if, given the number of replicates run for a single sample, whether a single positive among the replicates is enough to expend resources to deploy field confirmation efforts (Figure 8.1).

Bioinformatic tools are required to analyse NGS output. These analyses involve cleaning up spurious sequences generated in the process of amplification, trimming the adapters and primers off of sequences, and other data processing. The final

[3] https://www.ncbi.nlm.nih.gov/genbank/
https://www.ebi.ac.uk/ena

step is matching generated sequences to either those available in publicly available databases or a laboratory-generated database for taxonomic identification (detailed in Taberlet et al., 2018; Ruppert et al., 2019). This effort can be computationally expensive and requires technical expertise, yet it can return an expansive data set of biodiversity from a simple sample of water, soil, air, etc.. In an effort to simplify data processing and analyses and make generating results more accessible to non-experts, the Anacapa Toolkit (Curd et al., 2019) was developed. This software module takes raw data and processes it, helps generate a database for taxonomic identifications, and provides statistics for biodiversity and genetic diversity assessments. As more conservation practitioners turn to eDNA for biodiversity assessments, the increasing need for user-friendly software will be further addressed.

8.3 Wider application—review of what has been done

In the realm of conservation, eDNA has become a fairly well-accepted tool in recent years (Lodge et al., 2012; Bohmann et al., 2014; Thomsen & Willerslev, 2015; Barnes & Turner, 2016; Deiner et al., 2017a; Harper et al., 2019). Environmental DNA as a tool for detection and monitoring of target species and biodiversity surveillance has largely been applied in aquatic systems (Deiner et al., 2015; Deiner et al., 2017a; Cristescu & Hebert, 2018).

Environmental DNA has been used extensively for the detection of fish species and fish biodiversity assessments across aquatic ecosystems for nearly a decade (see review, Thomsen et al., 2012b; Shu et al., 2020). This tool is capable of detecting a wide range of marine taxa from plankton and phytoplankton, fish, and mammals (Thomsen et al., 2012a; Thomsen et al., 2012b, Westfall et al., 2017; Hansen et al., 2018; Stoeckle et al., 2018; Shu et al., 2020). Environmental DNA has been used to monitor marine species of concern (Costello et al., 2017; Boussarie et al., 2018; Hansen et al., 2018) and invasive species (Westfall et al., 2017), including invasive species from ship's ballast water (Gerhard & Gunsch, 2019). Further, it has been shown to be quite sensitive;

for example, detection of killer whales was possible for up to 2 hours after animals left the area (Baker et al., 2018), and detection of species richness of shark species was increased by metabarcoding of eDNA over traditional methods (Boussarie et al., 2018). When compared to other methods of marine diversity measurements, eDNA appears to be comparable or more sensitive at times but may be limited in other cases (Boussarie et al., 2018; Jerde et al., 2019). Even so, a complete understanding of the strengths and limitations of eDNA in varying marine ecosystems and across species is still being worked out (Cristescu & Hebert, 2018; Hansen et al., 2018; Jerde et al., 2019). Environmental DNA sampling in marine ecosystems has been limited by the necessity of humans to collect water samples, particularly in deep waters. Autonomous robots have been tested and compared with the peristaltic pumps and freezing protocols used during manual capture for effective DNA capture and preservation of samples (Yamahara et al., 2019). This work demonstrates that the use of robots for collecting eDNA and assessing biodiversity has its own challenges, but also advantages, and promises to be another viable method for collecting eDNA from places where humans have difficulty accessing marine environments (Yamahara et al., 2019) and almost certainly in other systems as well.

Freshwater systems were testbeds of eDNA and many of the earliest studies focused on detecting invasive species (Ficetola et al., 2008b; Piaggio et al., 2014; Doi et al., 2015). More recently, eDNA continues to be a critical tool for detecting invasive species. For example, eDNA has been used to detect invasive dreissenid mussels in the United States (e.g. Amberg et al., 2019; Sepulveda et al., 2019; Shogren et al., 2019) and the United Kingdom (Blackman et al., 2020). Exploiting eDNA in a reservoir in Montana demonstrated that this tool was more effective at detecting invasive mussels at low densities than plankton tow samples and thus extended the seasons in which they could be detected (Sepulveda et al., 2019). Further, the ongoing novel development of eDNA assays for things such as invasive nutria (Myocastor coypus; Akamatsu et al., 2018), fish ectoparasites (Fossøy et al., 2020), and aquatic plants (Anglès d'Auriac et al., 2019) demonstrate the utility of this tool in the detection of early stages of invasion and thus provide a critical alarm system for implementing management actions (see review in Mahon & Jerde, 2016). The study of vector-borne pathogens is important to predicting disease outbreak and managing pathogens to protect humans, wildlife, and domestic animals. The utility of eDNA to understand the movements of hosts that cannot otherwise be easily detected is highlighted by a study of the presence of freshwater snails (*Bulinus truncates*) that are host to a human pathogen (*Schistosoma* parasites) (Mulero et al., 2020). The development of this eDNA assay will contribute to the documentation of the distribution of the snail and thus the potential distribution of a pathogen of human health concern and is a good model for other vector-borne diseases (Mulero et al., 2020). Environmental DNA in freshwater systems has also been used to detect species of concern (Barnes & Turner, 2016; Goldberg et al., 2016; Cristescu & Hebert, 2018; Cilleros et al., 2019; Cristescu, 2019; Huerlimann et al., 2020) and is still proving its utility in finding rare target species. In Japan, a new population of an endangered salamander, Yamato salamander (*Hynobius vandenburghi*) was predicted, through habitat modelling using Geographic Information Systems (GIS), and confirmatory presence was assessed using eDNA (Sakai et al., 2019). Environmental DNA used to monitor or detect endangered species is a particularly useful tool because it does not require the capture or handling of sensitive species, which can be vulnerable to the stresses of trapping, vulnerable to pathogens transmitted by humans, or difficult to detect through trapping (Adams et al., 2019). Importantly, beyond continuing to develop the ability to quantify biodiversity using NGS approaches, the inference of genetic diversity and other population genetic parameters among populations of species of concern will greatly expand the utility of eDNA in biodiversity conservation efforts (Adams et al., 2019; Sigsgaard et al., 2020).

An emerging utility of eDNA is identifying biodiversity of communities that are too small to otherwise characterize (termed 'Biomonitoring 2.0', Baird and Hajibabaei, 2012); these include terrestrial microbial communities (e.g. water quality, Gerhard & Gunsch, 2019), soil bacterial communities (e.g. Fierer & Jackson, 2006; Warren-Rhodes et al., 2019), pathogens (e.g. avian influenza, Koçer, 2010), and pollen grain diversity (Bell et al., 2016) in air samples (Kraaijeveld et al., 2015), and on insects (Widmer et al., 2000). Samples from soil have provided an avenue to a more robust understanding of flora and fauna over the past 400,000 years (Hofreiter et al., 2003; Willerslev et al., 2003; Thomsen & Willerslev, 2015). The study of DNA of ancient ecosystems has also been greatly advanced by the ability to isolate and amplify DNA from ice cores (Willerslev et al., 2004; Willerslev et al., 2007). Recently, eDNA was used for understanding insect diversity visiting flowers. By collecting flowers and extracting DNA from them 135 arthropod species were detected, and clusters of species had associations with particular plant species (Thomsen & Sigsgaard, 2019). The pursuit for global biomonitoring through eDNA holds much promise for documenting the biodiversity of species that are easily seen, as well as ones that are not (Porter & Hajibabaei, 2018; McGee et al., 2019).

Methods for detecting terrestrial organisms using eDNA are less common for many species and ecosystems (Harper et al., 2019). Methods developed for collection of eDNA from terrestrial wildlife have included sampling ephemeral aquatic sources such as snow for vertebrate DNA from tracks (Franklin et al., 2019; Kinoshita et al., 2019), soil samples to assess vertebrate diversity (zoo animals; Andersen et al., 2012), watering holes for coyote detection (Rodgers & Mock, 2015), and freshwater sources for invasive species such as Burmese pythons (Piaggio et al., 2014), feral swine (Williams et al., 2017; Williams et al., 2018), and nutria (Akamatsu et al., 2018). Environmental DNA has also been collected from crops (Saito et al., 2008) and depredated sage-grouse eggs (Hopken et al., 2016) to detect predator species involved in damage or predation events. Vertebrate faunal diversity has been determined when DNA was collected from blood- or carrion-feeding insects (Calvignac-Spencer et al., 2013; Lynggaard et al., 2019), thus exploiting these species as ecosystem samplers. Plants have been swabbed for eDNA of herbivores such as deer and endangered mice (Nichols et al., 2012; Nichols & Spong, 2014; Nichols et al., 2015; Lyman et al., 2019) and to assess arthropod diversity

(Bittleston et al., 2016; Thomsen & Sigsgaard, 2019). As an effort to implement a sensitive tool to elucidate the influence of ungulate grazing on plant communities, the application of eDNA was tested (Nichols et al., 2012). It was demonstrated that buccal cells from saliva left behind on browsed twigs could amplify eDNA from various ungulate species and that it could be detected for up to 12 weeks after deposition (Nichols et al., 2012). Once the method was developed, it was then applied to a broader ecosystem in Sweden where ungulate species causing damage to tree plantations and browsing preferences among ungulate species were detected all using eDNA (Nichols & Spong, 2014; Nichols et al., 2015). Such studies have demonstrated that the development and application of the eDNA assay require extensive work but provide critical conservation and management results (Williams et al., 2016; Williams et al., 2017; Williams et al., 2018). The utility of eDNA for conservation and management of terrestrial species is only now becoming realized and, with more interest, it will become as routinely applied as it is in aquatic systems.

8.4 Case study

One of the largest programmes to implement eDNA for conservation purposes is for the management of a highly invasive fish species (Asian carp: Bighead Carp, *Hypophthalmichthys nobilis* and Silver Carp, *H. molitix*). The United States Fish and Wildlife Service (USFWS) uses water samples to monitor the progress of invasion by Asian carp[4] (USFWS, 2018), with the goal of detecting, managing, and eventually arresting the spread of these species in the Great Lakes, where they would decimate native fisheries and recreational value of waterways as they have done in other invaded areas (Darling & Mahon, 2011; Jerde et al., 2011). The need for eDNA was identified as a critical monitoring tool and species-specific primers were developed and tested in 2009–2010 (Jerde et al., 2011). The developed eDNA assay was applied to the monitoring programme alongside electrofishing and other methods. Environmental DNA results identified

[4] https://www.fws.gov/midwest/fisheries/eDNA.html

that both carp species had moved north of the electric barriers set up to block their northward dispersal. Further, eDNA was more effective at documenting the invasive front than electrofishing implemented simultaneously for comparison between the methods (Jerde et al., 2011). Later work focused on sequencing the mitochondrial genome of both carp species and developing a suite of markers that could be used together to increase the probability of detection (Farrington et al., 2015). Current eDNA surveillance uses a set of primers that capture both species at once, and then qPCR positives are further tested with species-specific primers to assess which species were detected (Farrington et al., 2015; USFWS, 2018).

Some of this programme's critical components are collaborations with state agencies within the affected and potentially affected areas, and the carefully developed Quality Assurance Project Plan (Woldt et al., 2015; USFWS, 2018). This plan outlines the field and laboratory methods, laboratory validation and calibration results, quality control and assurance procedures, and criteria for declaring positive detections. It also describes the responsibilities of each participating partner agency. This project continues to refine its sampling strategy and its understanding of detection probabilities given the freshwater ecosystems where they are monitoring. Using their vast sampling and processing capabilities, they have used occurrence modelling (similar to occupancy models) and simulations to assess their probability of detection across time, habitats, and biotic factors (Mize et al., 2019). This work identified the ideal sample size per site, the best time of year, and types of water bodies to sample to accomplish the highest probability of detection of both invasive species of carp in the target freshwater ecosystem. The work to monitor these invasive species and manage the invasion using eDNA as a primary tool has not been without controversy (Darling and Mahon, 2011), but the USFWS has aimed to address the concerns through ongoing research (Jerde et al., 2011; Klymus et al., 2015; Erickson et al., 2019; Mize et al., 2019) and the Quality Assurance Project Plan (QAPP), which is regularly adapted and updated (USFWS, 2018). In many ways, this project leads the way as an example of how eDNA can be implemented on a broad scale to aid and inform conservation management of native biodiversity and long-term monitoring of invasive species.

8.5 Limitations/constraints

Different organisms shed DNA differently and at different rates. DNA persistence in the environment is affected by numerous environmental variables (e.g. pH, temperature, UV, microbial organisms), and these may change across time (Taberlet et al., 2018). Further, eDNA is not distributed homogeneously in the environment and is affected by dispersion and diffusion in an aquatic habitat (Furlan et al., 2016). The co-occurrence of DNA inhibitors in the environment may reduce detection probability (McKee et al., 2015). Currently, eDNA cannot be used to assess a species' population size, but NGS allows relative comparison of signal across samples and some labs are successfully using the technology to provide abundance and biodiversity data. The attempts to address the limitations and constraint of eDNA application is addressed in more detail under future directions later in the chapter.

8.6 Social impact/privacy

There have thus far not been any privacy concerns documented associated with the use of eDNA. However, one might imagine an instance where a private landowner will not allow access to their property for a resource manager or researcher to monitor for an endangered species or a potentially invasive species on their property. However, the waters downstream can be sampled and successful detection could lead to further action or enforcement against the property owner. This could potentially lead to concerns about privacy, but this has yet to become an issue.

As with other tools that allow us to better understand biodiversity and associated anthropogenic impact, it is hard to account fully for social impacts. Environmental DNA has not been widely used to make management decisions without confirmation of eDNA results with another tool (Sepulveda et al., 2020). However, using the Daubert standard, which is used to evaluate reliability in United States Federal Courts, eDNA was determined to be a viable, stand-alone method for rigorous decision-making under law (Sepulveda et al., 2020). The use of eDNA data in a structured decision-making framework could rigorously inform management actions (Sepulveda et al., 2020). With the application of eDNA to conservation management, the contributions of data from this tool to biodiversity conservation would be an important social impact. Environmental DNA has been shown to be a viable tool in detecting rare and elusive or small organisms, and now is proven to be valid data in a legal arena (Sepulveda et al., 2020). Thus the impacts to society through policy actions based on eDNA data will increase.

8.7 Future directions

Largely, those using eDNA are processing samples in the laboratory on benchtop instruments (e.g. rtPCR instruments, MiSeq, PacBio, etc.), but some are testing the applicability of handheld devices directly in the field (MinIon, NGS device, Jain et al., 2016; Deiner et al., 2017a) (Biomeme, rtPCR, Sepulveda et al., 2018). The promise of this advancing technology is that one can collect environmental samples and process them within an hour or two from a laptop and an instrument the size of a USB flash drive. The handheld NGS device, MinIon, has been used effectively in the field in Antarctica to obtain long-read sequences from ancient and modern microbial mats found in the environment (Johnson et al., 2017) and sequence ray-finned fish mitogenomes from water (Deiner et al., 2017a). The MinIon introduces more sequencing error than desirable for most eDNA applications, and challenges with sensitivity and susceptibility to PCR inhibition of the Biomeme platform means they are not yet ready to become the standard for eDNA studies yet (Sepulveda et al., 2018; Thomas et al., 2020). However, handheld field-deployable platforms are clearly the next advancement to come to eDNA applications. These developments aim to minimize sample contamination by having all collection steps completely contained within the instrument, making field-based sampling easier and detection more rapid (e.g. Sepulveda et al., 2018; Thomas et al., 2018; Thomas et al., 2019). This will further facilitate the implementation of eDNA monitoring programmes to detect rare and endangered species or invasive species early in the invasion process when they might be halted.

The current state of using genomic approaches does not yet deliver on the promise of sequencing every single fragment of DNA from a mixed-species, environmental sample. Biases introduced from primers and other reagents, steps in the laboratory, or analytical process of NGS, which includes metabarcoding and metagenomics (whole-genome sequencing), do not allow quantification of the number of individuals or reliable detection of rare species quite yet. However, recent studies are advancing methods that allow for quantification from NGS data and it is highly likely in the near future that we will be able to use genomics to regularly assess abundance and reliably detect rare individuals (Bush et al., 2019; Doi et al., 2019; Shelton et al., 2019; Singer et al., 2019; Zinger et al., 2019). The recent application of a part of the bacterial immune system, clustered regularly interspaced short palindromic repeats (CRISPR) paired with Cas enzymes, has altered (Knott & Doudna, 2018), and will continue to provide innovation for, genomic sequencing applications that are directly relevant to eDNA as a biodiversity conservation tool (Phelps et al., 2020). As progress towards a single-species biosensor for environmental deployment, a proof-of-concept study demonstrated the use of CRISPR/Cas to induce fluorescence in the presence of a fish species (*Salmo salar*, Williams et al., 2019). This study is the first to apply CRISPR/Cas technology to eDNA and hints at the future promise of emerging technologies coupled with environmental sampling.

The collection of RNA from the environment may be another avenue for exploring biodiversity as well as pathogens (Cristescu, 2019). The use of eRNA would allow for a better understanding of gene expression in different habitats, but RNA is a very unstable molecule and thus hard to capture (Cristescu, 2019). Despite this, RNA has been captured and isolated from the environment, including from various fossil types (Cristescu, 2019). There is evidence from one study using metabarcoding that RNA for biodiversity studies may provide more

agreement to morphological surveys than DNA (Pochon et al., 2017). The isolation of RNA from the environment allows the detection of RNA viruses, such as the detection of SARS-CoV-2 in human wastewater in the Netherlands (Medema et al., 2020). The detection of this virus, which is shed in human faeces, occurred before human cases of the COVID-19 disease caused by the virus were reported (Medema et al., 2020). Thus, eDNA may serve as an early warning signal of an upcoming outbreak for this and other zoonotic pathogens. Studies have demonstrated the persistence of Avian Influenza Viruses (AIV), an RNA virus, in wild waters (see review in Stallknecht et al., 2010). These AIV viruses in water can infect wild birds, and thus the environment can serve as a transmission pathway (VanDalen et al., 2010). Highly pathogenic AIV subtypes were detected from eDNA near a poultry farm outbreak site (Borchardt et al., 2017). As AIV affect wildlife, livestock, and human health, it is an important model for the utility of eDNA in detecting zoonotic pathogens in the environment. It will be important for conservation scientists to continue to pursue the application of eDNA for pathogen detection for early warning systems and elucidating transmission pathways.

Loop-mediated isothermal amplification (LAMP) method is not a traditional PCR (Lee, 2017) and thus does not require expensive instrumentation. The isothermal amplification produces a highly specific and efficient amplification product if the target is available in a sample. This method is simple to apply in the field and has been used to identify bacterial, viral, parasitic, and fungal pathogens (Fu et al., 2011), yet it has not been widely applied to environmental detection or monitoring for conservation purposes. One advantage of a LAMP assay is it can easily be deployed as part of a biosensor through technology that already exists (Jones, 2016). Biosensors utilizing a LAMP assay and technology such as handheld dipsticks for single-species detection have already been developed for human pathogen detection (Prompamorn et al., 2011) but can produce false positives and require further development to achieve the promise they hold for the future of eDNA (Zasada et al., 2018). An ongoing effort to apply LAMP to eDNA and develop a handheld dipstick to detect invasive carp[5] has been in development (Merkes, 2020), but this is currently the only LAMP assay being tested for eDNA.

Recently, the University of California Conservation Genomics Consortium started an eDNA programme involving citizen scientists called CALeD-NA (Meyer et al., 2019). This focuses on sampling soils and sediments within the University of California Reserve System and posting results online. This effort is a tremendous undertaking and offers a huge promise of providing a better understanding of biodiversity across California, documenting what is currently there, and monitoring biodiversity changes through time (Meyer et al., 2019). If successful and popular, it is likely that other groups will aim to undertake similar large-scale eDNA landscape biodiversity research approaches. On a national level, Canada has made efforts to develop and evaluate the utility of eDNA for fisheries and develop national standards for its use in the management of biodiversity in aquatic ecosystems (Baillie et al., 2019). This effort has led to the creation of a national technical working group charged with integrating eDNA into federal scientific research programmes that guide resource management decisions (Baillie et al., 2019). These types of regional or federal efforts to standardize the use of eDNA are important to the integration of this tool in biodiversity conservation and management of invasive species.

In this world of fast-paced technological advances, not all new methods prove useful in an applied context. Although eDNA has not been used regularly in biodiversity conservation for more than a decade, it has proven to be an extremely practical and informative tool. The utility of eDNA is supported by ongoing advancements and development of novel applications. There is no easy way to standardize the application or methods of eDNA as the conservation question, and the target system must drive the selection of a range of options at every step. However, guidelines now exist for the best practices of optimizing a sampling scheme and sample processing for eDNA applications (Goldberg et al., 2016; Jeunen et al., 2019; Klymus et al., 2020; The eDNA Society, 2019; Shu et al., 2020). Further, the ranks of experienced eDNA practitioners have expanded globally; thus, it is fairly easy to find expert consultation. Therefore, it is now practical and prudent to adopt eDNA in the service of biodiversity conservation efforts.

References

Acevedo-whitehouse, K., Rocha-Gosselin, A., & Gendron, D. (2010). A novel non-invasive tool for disease surveillance of free-ranging whales and its relevance to conservation programs. *Animal Conservation*, 13, 217–225.

Adams, C. I., Knapp, M., Gemmell, N. J., Jeunen, G.-J., Bunce, M., Lamare, M. D., & Taylor, H. R. (2019). Beyond biodiversity: can environmental DNA (eDNA) cut it as a population genetics tool? *Genes*, 10, 192.

Akamatsu, Y., Goto, M., Inui, R., Yamanaka, H., Komuro, T., & Kono, Y. (2018). Monitoring of *Myocastor coypus* using environmental DNA and estimation of the potential habitat in the Yamaguchi prefecture. *Ecology and Civil Engineering*, 21, 1–8.

Amberg, J. J., Merkes, C. M., Stott, W., Rees, C. B., & Erickson, R. A. (2019). Environmental DNA as a tool to help inform zebra mussel, *Dreissena polymorpha*, management in inland lakes. *Management of Biological Invasions*, 10, 96.

Andersen, K., Bird, K. L., Rasmussen, M., et al. (2012). Meta-barcoding of 'dirt' DNA from soil reflects vertebrate biodiversity. *Molecular Ecology*, 21, 1966–1979.

Andruszkiewicz, E. A., Koseff, J. R., Fringer, O. B., et al. (2019). Modeling environmental DNA transport in the coastal ocean using lagrangian particle tracking. *Frontiers in Marine Science*, 6, 477.

Anglès d'Auriac, M. B., Strand, D. A., Mjelde, M., Demars, B. O. L., & Thaulow, J. (2019). Detection of an invasive aquatic plant in natural water bodies using environmental DNA. *PLoS One*, 14, e0219700.

Baillie, S. M., Mcgowan, C., May-mcnally, S., Leggatt, R., Sutherland, B. J., & Robinson, S. (2019). *Environmental DNA and Its Applications to Fisheries and Oceans Canada: National Needs and Priorities*. Fisheries and Oceans Canada (Pêches et Oceans Canada), Canada.

Baird, D. J. & Hajibabaei, M. (2012). Biomonitoring 2.0: a new paradigm in ecosystem assessment made possible by next-generation DNA sequencing. *Molecular Ecology*, 21, 2039–2044.

---

[5] https://www.usgs.gov/centers/umesc/science/developing-a-portable-lamp-assay-detecting-grass-and-black-carp

Baker, C. S., Steel, D., Nieukirk, S., & Klinck, H. (2018). Environmental DNA (eDNA) from the wake of the whales: droplet digital PCR for detection and species identification. *Frontiers in Marine Science*, 5, 133.

Barnes, M. A. & Turner, C. R. (2016). The ecology of environmental DNA and implications for conservation genetics. *Conservation Genetics*, 17, 1–17.

Barnes, M. A., Turner, C. R., Jerde, C. L., Renshaw, M. A., Chadderton, W. L., & Lodge, D. M. (2014). Environmental conditions influence eDNA persistence in aquatic systems. *Environmental Science & Technology*, 48, 1819–1827.

Bell, K. L., De Vere, N., Keller, A., et al. (2016). Pollen DNA barcoding: current applications and future prospects. *Genome*, 59, 629–640.

Bittleston, L. S., Baker, C. C. M., Strominger, L. B., Pringle, A., & Pierce, N. E. (2016). Metabarcoding as a tool for investigating arthropod diversity in Nepenthes pitcher plants. *Austral Ecology*, 41, 120–132.

Blackman, R., Benucci, M., Donnelly, R., et al. (2020). Simple, sensitive and species-specific assays for detecting quagga and zebra mussels (*Dreissena rostriformis bugensis* and *D. polymorpha*) using environmental DNA. *Management of Biological Invasions*, 11, 218–236.

Blaxter, M., Mann, J., Chapman, T., Thomas, F., Whitton, C., Floyd, R., & Abebe, E. (2005). Defining operational taxonomic units using DNA barcode data. *Philosophical Transactions of the Royal Society B: Biological Sciences*, 360, 1935–1943.

Bohmann, K., Evans, A., Gilbert, M. T. P., et al. (2014). Environmental DNA for wildlife biology and biodiversity monitoring. *Trends in Ecology & Evolution*, 29, 358–367.

Borchardt, M. A., Spencer, S. K., Hubbard, L. E., Firnstahl, A. D., Stokdyk, J. P. & Kolpin, D. W. (2017). Avian influenza virus RNA in groundwater wells supplying poultry farms affected by the 2015 influenza outbreak. *Environmental Science & Technology Letters*, 4, 268–272.

Boussarie, G., Bakker, J., Wangensteen, O. S., et al. (2018). Environmental DNA illuminates the dark diversity of sharks. *Science Advances*, 4, eaap9661.

Burton, A. C., Neilson, E., Moreira, D., et al. (2015). Wildlife camera trapping: a review and recommendations for linking surveys to ecological processes. *Journal of Applied Ecology*, 52, 675–685.

Bush, A., Compson, Z., Monk, W., et al. (2019). Studying ecosystems with DNA metabarcoding: lessons from aquatic biomonitoring. *Frontiers of Ecological Evolution*, 7, 434.

Calvignac-Spencer, S., Merkel, K., Kutzner, N., et al. (2013). Carrion fly-derived DNA as a tool for comprehensive and cost-effective assessment of mammalian biodiversity. *Molecular Ecology*, 22, 915–924.

Caravaggi, A., Banks, P. B., Burton, A. C., et al. (2017). A review of camera trapping for conservation behaviour research. *Remote Sensing in Ecology and Conservation*, 3, 109–122.

Cilleros, K., Valentini, A., Allard, L., et al. (2019). Unlocking biodiversity and conservation studies in high-diversity environments using environmental DNA (eDNA): a test with Guianese freshwater fishes. *Molecular Ecology Resources*, 19, 27–46.

Conrad, K. F., Fox, R., & Woiwod, I. P. (2007). Monitoring biodiversity: measuring long-term changes in insect abundance. In Stewart, I. J. A., New, T. R., & Lewis, O. T. (eds.). Insect Conservation Biology. Royal Entomological Society, St Albans, UK (pp. 203–225).

Costello, M. J., Basher, Z., Mcleod, L., et al. (2017). Methods for the study of marine biodiversity. In Walters, M., & Scholes, R. J. (eds.). *The GEO Handbook on Biodiversity Observation Networks*. Springer, Cham, Switzerland (pp. 129–164).

Cristescu, M. E. (2019). Can environmental RNA revolutionize biodiversity science? *Trends in Ecology & Evolution*, 34, 694–697.

Cristescu, M. E., & Hebert, P. D. N. (2018). Uses and misuses of environmental DNA in biodiversity science and conservation. *Annual Review of Ecology, Evolution, and Systematics*, 49, 209–230.

Curd, E. E., Gold, Z., Kandlikar, G. S., et al. (2019). Anacapa Toolkit: an environmental DNA toolkit for processing multilocus metabarcode datasets. *Methods in Ecology and Evolution*, 10, 1469–1475.

Darling, J. A. (2019). How to learn to stop worrying and love environmental DNA monitoring. *Aquatic Ecosystem Health & Management*, 1–13.

Darling, J. A., & Mahon, A. R. (2011). From molecules to management: adopting DNA-based methods for monitoring biological invasions in aquatic environments. *Environmental Research*, 111, 978–988.

Davis, A. J., Keiter, D. A., Kierepka, E. M., et al. (2020). A comparison of cost and quality of three methods for estimating density for wild pigs (*Sus scrofa*). *Scientific Reports*, 10, 2047.

Davis, A. J., Williams, K. E., Snow, N. P., Pepin, K. M., & Piaggio, A. J. (2018). Accounting for observation processes across multiple levels of uncertainty improves inference of species distributions and guides adaptive sampling of environmental DNA. *Ecology and Evolution*, 8, 10879–10892.

Deiner, K., Bik, H. M., Mächler, E., et al. (2017a). Environmental DNA metabarcoding: transforming how we survey animal and plant communities. *Molecular Ecology*, 26, 5872–5895.

Deiner, K., Renshaw, M. A., Li, Y., Olds, B. P., Lodge, D. M., & Pfrender, M. E. (2017b). Long-range PCR allows sequencing of mitochondrial genomes from environmental DNA. *Methods in Ecology and Evolution*, 8, 1888–1898.

Deiner, K., Walser, J.-C., Mächler, E., & Altermatt, F. (2015). Choice of capture and extraction methods affect detection of freshwater biodiversity from environmental DNA. *Biological Conservation*, 183, 53–63.

Doi, H., Fukaya, K., Oka, S.-I., Sato, K., Kondoh, M., & Miya, M. (2019). Evaluation of detection probabilities at the water-filtering and initial PCR steps in environmental DNA metabarcoding using a multispecies site occupancy model. *Scientific Reports*, 9, 3581.

Doi, H., Inui, R., Akamatsu, Y., et al. (2017). Environmental DNA analysis for estimating the abundance and biomass of stream fish. *Freshwater Biology*, 62, 30–39.

Doi, H., Takahara, T., Minamoto, T., Matsuhashi, S., Uchii, K., & Yamanaka, H. (2015). Droplet digital polymerase chain reaction (PCR) outperforms real-time PCR in the detection of environmental DNA from an invasive fish species. *Environmental Science & Technology*, 49, 5601–5608.

The eDNA Society (2019). *Environmental DNA Sampling and Experiment Manual Version 2.1*. The eDNA Society, Otsu, Japan.

Ellison, S. L. R., English, C. A., Burns, M. J., & Keer, J. T. (2006). Routes to improving the reliability of low level DNA analysis using real-time PCR. *BMC Biotechnology*, 6, 33.

Erickson, R. A., Merkes, C. M., & Mize, E. L. (2019). Sampling designs for landscape-level eDNA monitoring programs. *Integrated Environmental Assessment and Management*, 15, 760–771.

Farrington, H. L., Edwards, C. E., Guan, X., Carr, M. R., Baerwaldt, K., & Lance, R. F. (2015). Mitochondrial genome sequencing and development of genetic markers for the detection of DNA of invasive bighead and silver carp (*Hypophthalmichthys nobilis* and *H. molitrix*) in environmental water samples from the United States. *PLoS One*, 10, e0117803.

Ficetola, G. F., Bonin, A., & Miaud, C. (2008a). Population genetics reveals origin and number of founders in a biological invasion. *Molecular Ecology*, 17, 773–782.

Ficetola, G. F., Miaud, C., Pompanon, F., & Taberlet, P. (2008b). Species detection using environmental DNA from water samples. *Biology Letters*, 4, 423–425.

Fierer, N., & Jackson, R. B. (2006). The diversity and biogeography of soil bacterial communities. *Proceedings of the National Academy of Sciences of the United States of America*, 103, 626.

Fossøy, F., Brandsegg, H., Sivertsgård, R., et al. (2020). Monitoring presence and abundance of two gyrodactylid ectoparasites and their salmonid hosts using environmental DNA. *Environmental DNA*, 2, 53–62.

Franklin, T. W., Mckelvey, K. S., Golding, J. D., et al. (2019). Using environmental DNA methods to improve winter surveys for rare carnivores: DNA from snow and improved noninvasive techniques. *Biological Conservation*, 229, 50–58.

Fu, S., Qu, G., Guo, S., et al. (2011). Applications of loop-mediated isothermal DNA amplification. *Applied Biochemistry and Biotechnology*, 163, 845–850.

Furlan, E. M., Gleeson, D., Hardy, C. M., & Duncan, R. P. (2016). A framework for estimating the sensitivity of eDNA surveys. *Molecular ecology resources*, 16, 641–654.

Garden, J. G., Mcalpine, C. A., Possingham, H. P., & Jones, D. N. (2007). Using multiple survey methods to detect terrestrial reptiles and mammals: what are the most successful and cost-efficient combinations? *Wildlife Research*, 34, 218–227.

Gerhard, W. A., & Gunsch, C. K. (2019). Metabarcoding and machine learning analysis of environmental DNA in ballast water arriving to hub ports. *Environment International*, 124, 312–319.

Gogarten, J. F., Hoffmann, C., Arandjelovic, M., et al. (2020). Fly-derived DNA and camera traps are complementary tools for assessing mammalian biodiversity. *Environmental DNA*, 2, 63–76.

Goldberg, C. S., Strickler, K. M., & Fremier, A. K. (2017). Environmental DNA as a tool for inventory and monitoring of aquatic vertebrates. In Fremier, D. A. (ed.) *Final Report United States Department of Defense Environmental Security Technology Certification Program Project RC-201204*. Washington State University, Washington, DC (pp. 1–176).

Goldberg, C. S., Turner, C. R., Deiner, K., et al. (2016). Critical considerations for the application of environmental DNA methods to detect aquatic species. *Methods in Ecology and Evolution*, 7, 1299–1307.

Hansen, B. K., Bekkevold, D., Clausen, L. W., & Nielsen, E. E. (2018). The sceptical optimist: challenges and perspectives for the application of environmental DNA in marine fisheries. *Fish and Fisheries*, 19, 751–768.

Hansen, H., Hess, S. C., Cole, D., & Banko, P. C. (2007). Using population genetic tools to develop a control strategy for feral cats (*Felis catus*) in Hawai'i. *Wildlife Research*, 34, 587–596.

Harper, L. R., Buxton, A. S., Rees, H. C., et al. (2019). Prospects and challenges of environmental DNA (eDNA) monitoring in freshwater ponds. *Hydrobiologia*, 826, 25–41.

Harper, L. R., Handley, L. L., Carpenter, A. I., et al. (2019). Environmental DNA (eDNA) metabarcoding of pond water as a tool to survey conservation and management priority mammals. *Biological Conservation*, 238, 108225.

Heino, M., Porteiro, F. M., Sutton, T. T., Falkenhaug, T., Godø, O. R., & Piatkowski, U. (2011). Catchability of pelagic trawls for sampling deep-living nekton in the mid-North Atlantic. *ICES Journal of Marine Science*, 68, 377–389.

Hofreiter, M., Mead, J., Martin, P., & Poinar, H. (2003). Molecular caving. *Current Biology*, 13, 693–695.

Hopken, M. W., Orning, E. K., Young, J. K., & Piaggio, A. J. (2016). Molecular forensics in avian conservation: a DNA-based approach for identifying mammalian predators of ground-nesting birds and eggs. *BMC Research Notes*, 9, 14.

Huerlimann, R., Cooper, M., Edmunds, R., et al. (2020). Enhancing tropical conservation and ecology research with aquatic environmental DNA methods: an introduction for non-environmental DNA specialists. *Animal Conservation*, 23, 632–645.

Hunter, M. E., Dorazio, R. M., Butterfield, J. S. S., Meigs-friend, G., Nico, L. G., & Ferrante, J. A. (2017). Detection limits of quantitative and digital PCR assays and their influence in presence–absence surveys of environmental DNA. *Molecular Ecology Resources*, 17, 221–229.

Hunter, M. E., Meigs-friend, G., Ferrante, J. A., et al. (2018). Surveys of environmental DNA (eDNA): a new approach to estimate occurrence in vulnerable manatee populations. *Endangered Species Research*, 35, 101–111.

Jain, M., Olsen, H. E., Paten, B., & Akeson, M. (2016). The Oxford Nanopore MinION: delivery of nanopore sequencing to the genomics community. *Genome Biology*, 17, 239.

Jerde, C. L., Mahon, A. R., Chadderton, W. L., & Lodge, D. M. (2011). Sight-unseen detection of rare aquatic species using environmental DNA. *Conservation Letters*, 4, 150–157.

Jerde, C. L., Wilson, E. A., & Dressler, T. L. (2019). Measuring global fish species richness with eDNA metabarcoding. *Molecular Ecology Resources*, 19, 19–22.

Jeunen, G.-J., Knapp, M., Spencer, H. G., et al. (2019). Species-level biodiversity assessment using marine environmental DNA metabarcoding requires protocol optimization and standardization. *Ecology and Evolution*, 9, 1323–1335.

Jo, T., Murakami, H., Masuda, R., Sakata, M. K., Yamamoto, S., & Minamoto, T. (2017). Rapid degradation of longer DNA fragments enables the improved estimation of distribution and biomass using environmental DNA. *Molecular Ecology Resources*, 17, e25-e33.

Johnson, S. S., Zaikova, E., Goerlitz, D. S., Bai, Y., & Tighe, S. W. (2017). Real-Time DNA sequencing in the antarctic dry valleys using the Oxford nanopore sequencer. *Journal of Biomolecular Techniques: JBT*, 28, 2–7.

Jones, B. C. (2016). Development of a *Simple Hanheld Biosensor* for *Waterborne Pathogens*. Undergraduate thesis, University of Arkansas, Fayetteville, NC.

Jones, J. (1992). Environmental impact of trawling on the seabed: a review. *New Zealand Journal of Marine and Freshwater Research*, 26, 59–67.

Kandlikar, G., Gold, Z., Cowen, M., et al. (2018). ranacapa: an R package and Shiny web app to explore environmental DNA data with exploratory statistics and interactive visualizations [version 1; peer review: 1 approved, 2 approved with reservations]. *F1000Research*, 7.

Kinoshita, G., Yonezawa, S., Murakami, S., & Isagi, Y. (2019). Environmental DNA collected from snow tracks is useful for identification of mammalian species. *Zoological Science*, 36, 198–207.

Klymus, K. E., Merkes, C. M., Allison, M. J., et al. (2020). Reporting the limits of detection and quantification for environmental DNA assays. *Environmental DNA*, 2, 271–283.

Klymus, K. E., Richter, C. A., Chapman, D. C., & Paukert, C. (2015). Quantification of eDNA shedding rates from invasive bighead carp Hypophthalmichthys nobilis and silver carp Hypophthalmichthys molitrix. *Biological Conservation*, 183, 77–84.

Knott, G. J., & Doudna, J. A. (2018). CRISPR-Cas guides the future of genetic engineering. *Science*, 361, 866.

Knudsen, S. W., Ebert, R. B., Hesselsøe, M., et al. (2019). Species-specific detection and quantification of environmental DNA from marine fishes in the Baltic Sea. *Journal of Experimental Marine Biology and Ecology*, 510, 31–45.

Koçer, Z. A. (2010). *Detection of Influenza A Viruses from Environmental Lake and Pond Ice*. Bowling Green State University, Bowling Green, OH.

Kraaijeveld, K., De Weger, L. A., Ventayol García, M., et al. (2015). Efficient and sensitive identification and quantification of airborne pollen using next-generation DNA sequencing. *Molecular Ecology Resources*, 15, 8–16.

Kumar, G., Eble, J. E., & Gaither, M. R. (2020). A practical guide to sample preservation and pre-PCR processing of aquatic environmental DNA. *Molecular Ecology Resources*, 20, 29–39.

Lacoursière-roussel, A., Côté, G., Leclerc, V., & Bernatchez, L. (2016). Quantifying relative fish abundance with eDNA: a promising tool for fisheries management. *Journal of Applied Ecology*, 53, 1148–1157.

Laramie, M. B., Pilliod, D. S., Goldberg, C. S., & Strickler, K. M. (2015). *Environmental DNA Sampling Protocol—Filtering Water to Capture DNA from Aquatic Organisms*. Techniques and Methods. USGS, Reston, VA.

Leduc, N., Lacoursière-roussel, A., Howland, K. L., et al. (2019). Comparing eDNA metabarcoding and species collection for documenting Arctic metazoan biodiversity. *Environmental DNA*, 1, 342–358.

Lee, P. L. M. (2017). DNA amplification in the field: move over PCR, here comes LAMP. *Molecular Ecology Resources*, 17, 138–141.

Lodge, D. M., Turner, C. R., Jerde, C. L., et al. (2012). Conservation in a cup of water: estimating biodiversity and population abundance from environmental DNA. *Molecular Ecology*, 21, 2555–2558.

Lonsinger, R. C., Gese, E. M., Dempsey, S. J., Kluever, B. M., Johnson, T. R., & Waits, L. P. (2015). Balancing sample accumulation and DNA degradation rates to optimize noninvasive genetic sampling of sympatric carnivores. *Molecular Ecology Resources*, 15, 831–842.

Lyman, J., Sanchez, D., Sobek, C., et al. (2019). Environmental DNA for monitoring. Abstract for The Wildlife Society Conference 2019.

Lynggaard, C., Nielsen, M., Santos-bay, L., Gastauer, M., Oliveira, G., & Bohmann, K. (2019). Vertebrate diversity revealed by metabarcoding of bulk arthropod samples from tropical forests. *Environmental DNA*, 00, 1–13.

Mahon, A. R., & Jerde, C. L. (2016). Using environmental DNA for invasive species surveillance and monitoring. *Methods in Molecular Biology*, 1452, 131–142.

Mallet, D., & Pelletier, D. (2014). Underwater video techniques for observing coastal marine biodiversity: a review of sixty years of publications (1952–2012). *Fisheries Research*, 154, 44–62.

Mariani, S., Baillie, C., Colosimo, G., & Riesgo, A. (2019). Sponges as natural environmental DNA samplers. *Current Biology*, 29, R401-R402.

Mcgee, K. M., Robinson, C. V., & Hajibabaei, M. (2019). Gaps in DNA-based biomonitoring across the globe. *Frontiers in Ecology and Evolution*, 7, 337.

Mckee, A. M., Spear, S. F., & Pierson, T. W. (2015). The effect of dilution and the use of a post-extraction nucleic acid purification column on the accuracy, precision, and inhibition of environmental DNA samples. *Biological Conservation*, 183, 70–76.

Mcleod, L. E., & Costello, M. J. (2017). Light traps for sampling marine biodiversity. *Helgoland Marine Research*, 71, 2.

Medema, G., Heijnen, L., Elsinga, G., Italiaander, R., & Brouwer, A. (2020). Presence of SARS-Coronavirus-2 in sewage. *medRxiv*, 2020.03.29.20045880.

Merkes, C. (2020). In situ detection of multiple predominant species of Asian carp. US Patent App. 16/157,565.

Merkes, C. M., Mccalla, S. G., Jensen, N. R., Gaikowski, M. P., & Amberg, J. J. (2014). Persistence of DNA in carcasses, slime and avian feces may affect interpretation of environmental DNA data. *PLoS One*, 9, e113346.

Meyer, R. S., Curd, E. E., Schweizer, T., et al. (2019). The California environmental DNA 'CALeDNA' program. *bioRxiv*, 503383.

Mize, E. L., Erickson, R. A., Merkes, C. M., et al. (2019). Refinement of eDNA as an early monitoring tool at the landscape-level: study design considerations. *Ecological Applications*, 29, e01951.

Mulero, S., Boissier, J., Allienne, J.-F., Quilichini, Y., Foata, J., Pointier, J.-P., & Rey, O. (2020). Environmental DNA for detecting Bulinus truncatus: a new environmental surveillance tool for schistosomiasis emergence risk assessment. *Environmental DNA*, 2, 161–174.

Nichols, R. V., Cromsigt, J. P., & Spong, G. (2015). Using eDNA to experimentally test ungulate browsing preferences. *SpringerPlus*, 4, 489.

Nichols, R. V., Koenigsson, H., Danell, K., & Spong, G. (2012). Browsed twig environmental DNA: diagnostic PCR to identify ungulate species. *Molecular Ecology Resources*, 12, 983–989.

Nichols, R. V., & Spong, G. (2014). Ungulate browsing on conifers during summer as revealed by DNA. *Scandinavian Journal of Forest Research*, 29, 650–652.

Phelps, M. P., Seeb, L. W., & Seeb, J. E. (2020). Transforming ecology and conservation biology through genome editing. *Conservation Biology*, 34, 54–65.

Piaggio, A. J., Engeman, R. M., Hopken, M. W., et al. (2014). Detecting an elusive invasive species: a diagnostic PCR to detect Burmese python in Florida waters and an assessment of persistence of environmental DNA. *Molecular Ecology Resources*, 14, 374–380.

Pilliod, D. S., Goldberg, C. S., Arkle, R. S., & Waits, L. P. (2013). Estimating occupancy and abundance of stream amphibians using environmental DNA from filtered water samples. *Canadian Journal of Fisheries and Aquatic Sciences*, 70, 1123–1130.

Pochon, X., Zaiko, A., Fletcher, L. M., Laroche, O., & Wood, S. A. (2017). Wanted dead or alive? Using metabarcoding of environmental DNA and RNA to distinguish living assemblages for biosecurity applications. *PLoS One*, 12, e0187636.

Port, J. A., O'Donnell, J. L., Romero-Maraccini, O. C., et al. (2016). Assessing vertebrate biodiversity in a kelp forest ecosystem using environmental DNA. *Molecular Ecology*, 25, 527–541.

Porter, T. M., & Hajibabaei, M. (2018). Scaling up: a guide to high-throughput genomic approaches for biodiversity analysis. *Molecular Ecology*, 27, 313–338.

Prompamorn, P., Sithigorngul, P., Rukpratanporn, S., Longyant, S., Sridulyakul, P., & Chaivisuthangkura, P. (2011). The development of loop-mediated isothermal amplification combined with lateral flow dipstick for detection of Vibrio parahaemolyticus. *Letters in Applied Microbiology*, 52, 344–351.

Rees, H. C., Maddison, B. C., Middleditch, D. J., Patmore, J. R. M., & Gough, K. C. (2014). REVIEW: the detection of aquatic animal species using environmental DNA – a review of eDNA as a survey tool in ecology. *Journal of Applied Ecology*, 51, 1450–1459.

Renshaw, M. A., Olds, B. P., Jerde, C. L., Mcveigh, M. M., & Lodge, D. M. (2015). The room temperature preservation of filtered environmental DNA samples and assimilation into a phenol–chloroform–isoamyl alcohol DNA extraction. *Molecular Ecology Resources*, 15, 168–176.

Richardson, A. J., Walne, A. W., John, A. W. G., et al. (2006). Using continuous plankton recorder data. *Progress in Oceanography*, 68, 27–74.

Rodgers, T. W., & Mock, K. E. (2015). Drinking water as a source of environmental DNA for the detection of terrestrial wildlife species. *Conservation Genetics Resources*, 7, 693–696.

Ruppert, K. M., Kline, R. J., & Rahman, M. S. (2019). Past, present, and future perspectives of environmental DNA (eDNA) metabarcoding: a systematic review in methods, monitoring, and applications of global eDNA. Global Ecology and Conservation, 17, e00547.

Saito, M., Yamauchi, K., & Aoi, T. (2008). Individual identification of Asiatic black bears using extracted DNA from damaged crops. *Ursus*, 19, 162–167.

Sakai, Y., Kusakabe, A., Tsuchida, K., et al. (2019). Discovery of an unrecorded population of Yamato salamander (*Hynobius vandenburghi*) by GIS and eDNA analysis. *Environmental DNA*, 1, 281–289.

Schnell, I. B., Thomsen, P. F., Wilkinson, N., et al. (2012). Screening mammal biodiversity using DNA from leeches. *Current Biology*, 22, R262–R263.

Schwartz, M. K., Luikart, G., & Waples, R. S. (2007). Genetic monitoring as a promising tool for conservation and management. *Trends in Ecology & Evolution*, 22, 25–33.

Sepulveda, A. J., Amberg, J. J., & Hanson, E. (2019). Using environmental DNA to extend the window of early detection for dreissenid mussels. *Management of Biological Invasions*, 10, 342.

Sepulveda, A. J., Hutchins, P. R., Massengill, R. L., Dunker, K. J., & Barnes, M. (2018). Tradeoffs of a portable, field-based environmental DNA platform for detecting invasive northern pike (*Esox lucius*) in Alaska. *Management of Biological Invasions*, 9, 253–258.

Sepulveda, A. J., Nelson, N. M., Jerde, C. L., & Luikart, G. (2020). Are environmental DNA methods ready for aquatic invasive species management? *Trends in Ecology & Evolution*, 35(8), 668–678.

Shardlow, T. F., & Hyatt, K. D. (2013). Quantifying associations of large vertebrates with salmon in riparian areas of British Columbia streams by means of camera-traps, bait stations, and hair samples. *Ecological Indicators*, 27, 97–107.

Shelton, A. O., Kelly, R. P., O'Donnell, J. L., et al. (2019). Environmental DNA provides quantitative estimates of a threatened salmon species. *Biological Conservation*, 237, 383–391.

Shogren, A. J., Tank, J. L., Egan, S. P., Bolster, D., & Riis, T. (2019). Riverine distribution of mussel environmental DNA reflects a balance among density, transport, and removal processes. *Freshwater Biology*, 64, 1467–1479.

Shu, L., Ludwig, A., & Peng, Z. (2020). Standards for methods utilizing environmental DNA for detection of fish species. *Genes*, 11, 296.

Sigsgaard, E. E., Jensen, M. R., Winkelmann, I. E., Møller, P. R., Hansen, M. M., & Thomsen, P. F. (2020). Population-level inferences from environmental DNA—current status and future perspectives. *Evolutionary Applications*, 13, 245–262.

Singer, G., Fahner, N. A., Barnes, J., McCarthy, A., & Hajibabaei, M. (2019). Comprehensive biodiversity analysis via ultra-deep patterned flow cell technology: a case study of eDNA metabarcoding seawater. *Scientific Reports*, 9, 5991.

Stallknecht, D. E., Goekjian, V. H., Wilcox, B. R., Poulson, R. L., & Brown, J. D. (2010). Avian influenza virus in aquatic habitats: what do we need to learn? *Avian Diseases*, 54, 461–465.

Stoeckle, M. Y., Mishu, M. D., & Charlop-Powers, Z. (2018). GoFish: a streamlined environmental DNA presence/absence assay for marine vertebrates. *BioRxiv*, 2, 331322.

Taberlet, P., Bonin, A., Zinger, L., & Coissac, E. (2018). *Environmental DNA: For Biodiversity Research and Monitoring*. Oxford University Press, Oxford, UK.

Thomas, A. C., Howard, J., Nguyen, P. L., Seimon, T. A., & Goldberg, C. S. (2018). ANDe™: a fully integrated environmental DNA sampling system. *Methods in Ecology and Evolution*, 9, 1379–1385.

Thomas, A. C., Tank, S., Nguyen, P. L., Ponce, J., Sinnesael, M., & Goldberg, C. S. (2020). A system for rapid eDNA detection of aquatic invasive species. *Environmental DNA*, 2, 261–270.

Thompson, W. (2013). *Sampling Rare or Elusive Species: Concepts, Designs, and Techniques for estimating population parameters*. Island Press, Washington, DC.

Thomsen, P. F., Kielgast, J., Iversen, L. L., et al. (2012a). Monitoring endangered freshwater biodiversity using environmental DNA. *Molecular Ecology*, 21, 2565–2573.

Thomsen, P. F., Kielgast, J., Iversen, L. L., Moller, P. R., Rasmussen, M., & Willerslev, E. (2012b). Detection of a diverse marine fish fauna using environmental DNA from seawater samples. *PLoS One*, 7, e41732.

Thomsen, P. F., & Willerslev, E. (2015). Environmental DNA—an emerging tool in conservation for monitoring past and present biodiversity. *Biological Conservation*, 183, 4–18.

Thomsen, P. F., & Sigsgaard, E. E. (2019). Environmental DNA metabarcoding of wild flowers reveals diverse communities of terrestrial arthropods. *Ecology and Evolution*, 9, 1665–1679.

USFWS (2018). *Quality Assurance Project Plan (QAPP) eDNA Monitoring of Bighead and Silver Carps*. U. S. Fish and Wildlife Service, Midwest Region Bloomington, MN.

Valentini, A., Pompanon, F., & Taberlet, P. (2009). DNA barcoding for ecologists. *Trends in Ecology & Evolution*, 24, 110–117.

Valentini, A., Taberlet, P., Miaud, C., et al. (2016). Next-generation monitoring of aquatic biodiversity using environmental DNA metabarcoding. *Molecular Ecology*, 25, 929–942.

Vandalen, K. K., Franklin, A. B., Mooers, N. L., Sullivan, H. J., & Shriner, S. A. (2010). Shedding light on avian influenza H4N6 infection in mallards: modes of transmission and implications for surveillance. *PLoS One*, 5, e12851.

Von Duyke, A. L., Kruger, E., Wijkmark, N., Näslund, J., Hellström, P., & Hellström, M. (2019). Evaluation of environmental DNA (eDNA) collected from tracks in the snow as a means to monitor individual polar bears (Ursus maritimus) in the Chukchi and Beaufort Seas. Available at: http://www.north-slope.org/assets/images/uploads/NSB-DWM_PRR-2019-01_Preliminary_Research_Report_Polar_Bear_eDNA_2019.10.17.pdf

Waits, L. P., & Paetkau, D. (2005). Noninvasive genetic sampling tools for wildlife biologists: a review of applications and recommendations for accurate data collection. *Journal of Wildlife Management*, 69, 1419–1433.

Warren-Rhodes, K. A., Lee, K. C., Archer, S. D. J., et al. (2019). Subsurface microbial habitats in an extreme desert mars-analog environment. *Frontiers in Microbiology*, 10, 69.

Wearn, O. R., & Glover-Kapfer, P. (2019). Snap happy: camera traps are an effective sampling tool when compared with alternative methods. *Royal Society Open Science*, 6, 181748.

Westfall, K., Therriault, T., Cristescu, M., Miller, K., & Abbott, C. (2017). A new biosurveillance tool for a global problem: metabarcoding of environmental DNA to identify marine invasive species. *Genome*, 60, 1009–1011.

Widmer, A., Cozzolino, S., Pellegrino, G., Soliva, M., & Dafni, A. (2000). Molecular analysis of orchid pollinaria and pollinaria-remains found on insects. *Molecular Ecology*, 9, 1911–1914.

Wilcox, T. M., Carim, K. J., Young, M. K., Mckelvey, K. S., Franklin, T. W., & Schwartz, M. K. (2018). Comment: the Importance of Sound Methodology in Environmental DNA Sampling. *North American Journal of Fisheries Management*, 38, 592–596.

Willerslev, E., Cappellini, E., Boomsma, W., et al. (2007). Ancient biomolecules from deep ice cores reveal a forested southern Greenland. *Science*, 317, 111.

Willerslev, E., Hansen, A. J., Binladen, J., et al. (2003). Diverse plant and animal genetic records from holocene and pleistocene sediments. *Science*, 300, 791.

Willerslev, E., Hansen, A. J., & Poinar, H. N. (2004). Isolation of nucleic acids and cultures from fossil ice and permafrost. *Trends in Ecology & Evolution*, 19, 141–147.

Williams, K. E., Huyvaert, K. P., & Piaggio, A. J. (2016). No filters, no fridges: a method for preservation of water samples for eDNA analysis. *BMC Research Notes*, 9, 298.

Williams, K. E., Huyvaert, K. P., & Piaggio, A. J. (2017). Clearing muddied waters: capture of environmental DNA from turbid waters. *PLoS One*, 12, e0179282.

Williams, K. E., Huyvaert, K. P., Vercauteren, K. C., Davis, A. J., & Piaggio, A. J. (2018). Detection and persistence of environmental DNA from an invasive, terrestrial mammal. *Ecology and Evolution*, 8, 688–695.

Williams, M.-A., O'Grady, J., Ball, B., et al. (2019). The application of CRISPR-Cas for single species identification from environmental DNA. *Molecular Ecology Resources*, 19, 1106–1114.

Woldt, A., Baerwaldt, K., Monroe, E., et al. (2015). *Quality Assurance Project Plan: eDNA Monitoring of Bighead and Silver Carps*. U.S. Fish and Wildlife Service USFWS Great Lakes Region 3, Bloomington, MN.

Wood, S. A., Pochon, X., Laroche, O., Von Ammon, U., Adamson, J., & Zaiko, A. (2019). A comparison of droplet digital polymerase chain reaction (PCR), quantitative PCR and metabarcoding for species-specific detection in environmental DNA. *Molecular Ecology Resources*, 19, 1407–1419.

Yamahara, K. M., Preston, C. M., Birch, J., et al. (2019). In situ autonomous acquisition and preservation of marine environmental DNA using an autonomous underwater vehicle. *Frontiers in Marine Science*, 6, 373.

Zasada, A. A., Zacharczuk, K., Formińska, K., Wiatrzyk, A., Ziółkowski, R., & Malinowska, E. (2018). Isothermal DNA amplification combined with lateral flow dipsticks for detection of biothreat agents. *Analytical Biochemistry*, 560, 60–66.

Zinger, L., Bonin, A., Alsos, I. G., et al. (2019). DNA metabarcoding—Need for robust experimental designs to draw sound ecological conclusions. *Molecular Ecology*, 28, 1857–1862.

CHAPTER 9

Mobile data collection apps

Edward McLester and Alex K. Piel

9.1 Introduction

To effectively inform conservation strategies, it is of fundamental importance to understand the biology of the organism or system that is to be conserved. Given the pervasiveness of biodiversity crises around the globe, there is a need for large data sets across wide spatial and temporal scales (Sutherland et al., 2004; Schwartz et al., 2018). Building larger data sets and efficiently transforming data to conservation action increasingly requires innovation of data collection methods (Marvin et al., 2016; Marshall et al., 2018). How, then, can we make use of modern technology to quickly and reliably convert our observations to an analysable, storable format when faced with the logistical, financial, and environmental constraints that field researchers encounter?

At their most basic, traditional methods of data collection by hand rely on pen and paper. This simplicity is not a fault in itself—rather, recording data by hand allows unlimited flexibility to change data collection protocols on the fly, to include any number of ad libitum observations, and is universally implementable across all study sites and budgets. Pen and paper are inexpensive and can be complemented by any number of other tools (e.g. Global Navigation Satellite System [GNSS] units; cameras; audio and video recorders) as needed. Nonetheless, the most obvious disadvantages to data collection by hand are the lack of (1) standardization, (2) automation, and (3) integration, in the workflow (Marshall et al., 2018). First, variability in legibility may reduce data reliability.

Moreover, handwritten data may be harder to standardize and maintain interindividual reliability given propensity for spelling mistakes or user-specific languages (Verma et al., 2016). Even with training, data collectors may vary in their coding or abbreviations, especially when recording unusual or rare events that are not clearly specified in the data collection protocol. Second, transcribing and storing the data for analysis must be done manually—or at least, using an additional step such as processing with handwriting recognition software—which is a labour-intensive process and may introduce errors that require more time to correct (McDonald & Johnson, 2014). Similarly, recording observations is entirely dependent on the observer's writing speed, which will also vary substantially depending on the environment (e.g. in inclement weather). Third, pen and paper can be difficult to integrate into workflows that require other devices—for example, GNSS coordinates and image filenames or timestamps must be copied or coded from the corresponding device at the point of capture. As well as being slower, each such additional step in the workflow introduces a greater susceptibility to unintentional errors by the observer. Depending on observer training and the protocol complexity, this may lead to substantial interobserver variation in the order of steps taken to record each observation (Figure 9.1).

Technological methods are designed to automate some or all parts of the data collection workflow, thereby increasing standardization of data collection across users and reducing opportunities for

Edward McLester and Alex K. Piel, *Mobile data collection apps*. In: *Conservation Technology*.
Edited by Serge A. Wich and Alex K. Piel, Oxford University Press.
© Oxford University Press 2021. DOI: 10.1093/oso/9780198850243.003.0009

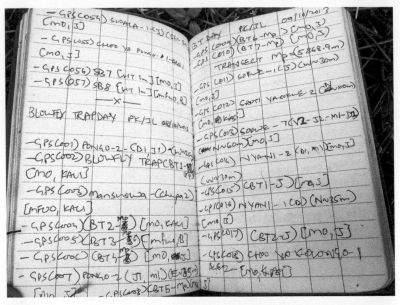

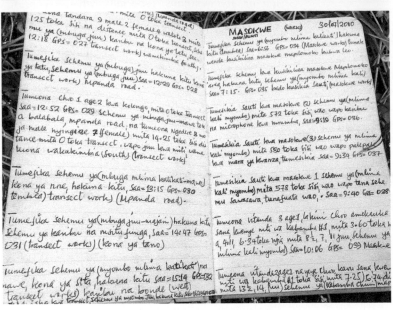

Figure 9.1 Examples of handwritten notebooks used to record behavioural and ecological data (see 9.4 Case Study). Note contrasting handwriting styles and legibility between the two users. The small and finite amount of writing space also means long details or corrections to mistakes do not always fit into predefined columns, giving potential for confusion during transcription. Longer notes are written out in full (rather than abbreviated or as shorthand) and may require time to translate if data are to be analysed as part of international collaborations (credits: AP).

user error (Newman et al., 2012). In the early 2000s, personal digital assistants (PDA) running proprietary operating systems (e.g. Palm OS or Windows Mobile-based devices) provided alternative hardware to collecting data by hand. As discussed by Waddle et al. (2003), however, these devices were generally expensive, slow to process large data sets, and could only back up data by downloading to a computer—leaving them susceptible to data loss

if damaged beforehand. The inclusion of sensors such as GNSS or cameras was also limited, meaning expensive, bulky setups comprising multiple pieces of equipment were usually required for specific tasks beyond entering written observations (e.g. see the PDA-GPS combinations described by Diefenbach et al., 2002 and Marshall et al., 2018).

The emergence of the smartphone in the last decade has provided a considerable increase in the accessibility of alternatives to collecting data by hand. Mobile phones have transitioned from devices designed primarily for calling and messaging to all-in-one computing solutions, complete with multiple cameras; WiFi, GNSS, and Bluetooth capability, accelerometers, and processors and onboard storage capable of running software in line with desktop computers (Aanensen et al., 2009; Snaddon et al., 2013; Berger-Tal & Lahoz-Monfort, 2018). A similar trend can be seen with the (re)introduction of the tablet computer in 2010, with tablet models including both low-end media consumption devices, and high-end devices marketed as replacements for traditional laptop computers. USB charging is ubiquitous in smartphones and tablets, allowing devices to be powered from a range of sources (e.g. AC and DC power systems; portable power banks; laptops and desktop computers).

The popularity of these devices has led to a substantial demand for both hardware and software. Their expansion has been especially significant in developing countries where, in addition to communication, these devices provide solutions for otherwise missing or inaccessible infrastructure (e.g. banking; healthcare—Pimm et al., 2015). Most importantly for financially constrained researchers, this user uptake has led to a range of models and prices suitable for almost all budgets. While these starting costs can still appear prohibitive compared to the cost of pen and paper, evidence indicates short-term costs from purchasing devices and apps, as well as time costs from training team members, can be offset by savings in paid time once researchers have been familiarized with a digitized data collection protocol. For example, Leisher (2014) reviewed three surveys of local communities overseen by The Nature Conservancy in South Africa and Tanzania (20102011; paper-based) and Kenya (2013; iPad-based). The authors found that the time taken to complete a survey was also significantly shorter when using tablets, allowing data collectors to complete more surveys per work hour.

Moreover, the cost per survey (i.e. survey materials and paid working hours for pre-analysis data cleaning) was significantly lower for the tablet-based study, despite a higher initial cost of purchasing six iPads compared to printing paper forms, in line with two similar studies (Zhang et al., 2012; King et al., 2013; reviewed in Leisher, 2014). Specifically, the tablet-based survey exhibited a substantially lower number of data entry errors that required time to correct compared with the handwritten surveys. In a comparison of handwritten and app-based capture-mark-recapture protocols, Bateman et al., (2013) also found that the data entry was significantly quicker (up to twice as fast) and the number of data entry errors was significantly lower (by almost a third) in the app-based protocol, regardless of which protocol researchers were given to use first.

The technological advances that have made their way into current smartphones and tablets mean that the potential for these consumer devices to be powerful and accessible tools for scientific researchers has now been realized in just a matter of years. As such, the rapid pace at which mobile computing has and continues to evolve means updates on available hardware and software, as well as applications of these devices in conservation science, are warranted.

9.2 New technology

9.2.1 Data collection hardware

Given that efficient implementation of data collection software depends equally on the device it is running on as the software itself, it is worth briefly reviewing current mobile hardware that supports data collection software applications (usually shortened to 'apps' in the context of mobile devices). The sheer range of devices currently available and the continuous rates at which new devices are released mean that a review of all smartphone and tablet models suitable for data collection is beyond this chapter's scope. Instead, here we briefly summarize key features that are likely to be of importance when choosing a data collection device. This is especially pertinent when discussing the use of consumer-orientated devices for data collection, given that the features that are likely to be of interest for field researchers are often not advertised by manufacturers (or at least, not with scientific research in mind).

9.2.1.1 WiFi, cellular, and satellite (GNSS) connectivity

WiFi (wireless LAN) connectivity is ubiquitous among modern smartphones and tablets, with higher-end models supporting 5 GHz band connectivity from some routers that offer greater bandwidth than the usual 2.4 GHz band. WiFi is a universally compatible standard worldwide and includes compatibility with mobile satellite internet hotspots (e.g. see services from Iridium, Inmarsat, and Thuraya). In addition to internet connectivity via a wireless router, WiFi compatibility also enables local data transfer from WiFi-enabled peripherals such as digital cameras, camera traps, and memory cards without the need for cables or additional accessories.

Unlike smartphones, cellular connectivity is a key difference between tablet models that otherwise resemble smartphones concerning usability and functionality, save for the larger form factor. Most tablets are offered in cellular (accepting SIM cards) and non-cellular ('WiFi only') models. Cellular connectivity in either a smartphone or a tablet may be useful to take advantage of developing projects such as Loon, which aims to expand cellular internet access to otherwise inaccessible locations using networks of balloons that provide internet connectivity through either direct cellular service signals or via proprietary receivers by users on the ground. In Kenya, Loon was first deployed commercially in 2020, with an initial service range spanning 50,000 km^2 which provided network access through cellular 4G service to thousands of households that previously had only sporadic cellular internet access[1]. Otherwise, non-cellular tablets models are usually substantially cheaper than equivalent cellular models. Given that cellular internet service can usually be shared relatively easily from a single internet-connected device using mobile tethering or WiFi hotspots, cost savings can be made by choosing non-cellular tablets, compared with either cellular tablets or smartphones.

GNSS coordinates often comprise an important element of field observations. Integration of GNSS antennae into small form factor mobile devices is common, given everyday applications (e.g. driving navigation aids), and provides a useful alternative to bulkier external USB-connected GNSS peripherals for laptops, for example. While 'GPS' usually refers to the United States owned and operated NAVSTAR GPS network that offers global coverage, some devices support supplementation from additional regional GNSS networks (satellite-based augmentation system—SBAS) for more reliable fixes. Examples of SBAS support on consumer smartphones and tablets include compatibility with the GLONASS (Russia), QZSS (Japan), BeiDou (China), GAGAN (India), and Galileo (European Union) systems. In addition, support for assisted GNSS (e.g. A-GPS) means using a cellular signal to assist with GNSS triangulation, which can reduce battery use. First introduced in Android smartphones in 2018, dual-frequency GNSS support, which increases triangulation accuracy (e.g. minimum 30 cm cf. typical ~5 m for single frequency) by supporting connectivity to two frequencies rather than a single frequency from each satellite, should become more widespread in future devices. Unfortunately, predicting GNSS signal accuracy and reliability from a device's specifications alone can be difficult, not least because information concerning GNSS connectivity is often vague or missing in manufacturer descriptions. Instead, several apps can report details of GNSS fixes, such as accuracy in metres, number and positions of triangulating satellites, and SBAS availability (e.g. GPS Status & Toolbox and GPSTest for Android; GPS Diagnostic: Satellite Test for iOS).

Bluetooth connectivity is also a common, if not universal, feature that enables both connection to external peripherals and wireless data transfer. In the first instance, peripherals can include GNSS receivers—which may supplement or provide more accurate fixes than the internal GNSS—keyboards, mice, and styluses for data entry, or speakers and headphones for playback experiments, for example. In the second instance, similar to local WiFi, Bluetooth data transfer also means that data can be extracted from compatible sources—e.g. Kestrel and HOBO weather stations—in the field without the

---

[1] https://www.nytimes.com/2020/07/07/world/africa/google-loon-balloon-kenya.html

need for additional expensive or bulky proprietary accessories.

9.2.1.2 Battery life

Device battery life (usually specified in milliamp hours [mAh] for smartphones and tablets) is almost always an important consideration for researchers in field environments where power is expensive or only available sporadically. Fortunately, as applications of mobile devices in everyday life have increased, similar demands from consumers have made battery life in mobile devices a competitive selling point. As such, many mobile devices come equipped with batteries designed to last for a full day's worth of work and media consumption. Most data collection apps require very little computing power to run. Besides, low-end devices with minimum-spec and low-power processors are usually sufficient. As such, the battery lives of these devices can match or exceed those advertised to consumers if multitasking is kept to a minimum. Battery life can also be increased by lower screen resolutions and brightness and limiting WiFi, cellular network, GNSS, and Bluetooth uptimes. As a rough starting point, in these author's experience, a fully charged battery of $\geq$3500mAh on a 7-inch Android tablet (ASUS ZenPad Z370C) is sufficient for 1–13 hours' data collection (15-minute interval form filling; regular map-viewing and note-taking; continuous GNSS signal; medium screen brightness at 1280×800 resolution; n = daily for 13 months).

Most mobile devices charge using USB connections via micro B and type C ports. Adapters for either port to AC or DC power outlets are readily available. Low amperage USB charging works particularly well with DC outputs from solar power systems often used in remote field environments because it does not require inefficient AC-DC inverters. Additionally, a wide range of small, portable rechargeable battery banks are available to easily pair with devices to extend battery life or provide extra recharges in the field. Battery banks with capacities ranging from 5000–30,000mAh are relatively inexpensive (<£50) to purchase. Portable battery banks are usually recharged by USB, although some variants can be charged directly by integrated solar panels. Larger battery banks (2–5 kg; 50,000–120,000 mAh; consider maximum mAh permitted onboard by airlines when purchasing) are available that include integrated AC-DC inverters and that can be recharged continuously using mains electricity or external solar panels.

Where solar power is not viable, thermoelectric generators may be an alternative. For example, the Hatsuden Nabe 'pan charger wonder pot' (Japan[2]) and PowerPot V (USA) are cooking pans that use residual energy from boiling water to power a DC or USB output (7–30 watts). Low wattage models can charge a smartphone in ca. 3–5 hours and may be useful in field environments where open fires are used for cooking. An early 2-watt Hatsuden Nabe was used successfully by Vitos et al. (2013b) to charge mobile devices during a community forest monitoring scheme in the Republic of Congo, where solar power was limited by dense canopy cover in the rainforest. Although the Hatsuden Nabe and PowerPot V appear to no longer be in manufacture, similar thermoelectric products or relatively inexpensive DIY equivalents may be worth exploring.

9.2.1.3 Ruggedness and protection in the field

The relative expense and fragility of electronic devices compared to pen and paper methods means minimizing risk of breakdown or damage is a primary concern for researchers. For mid- to high-end smartphone models, protection from accidental damage or weather is often a selling point for everyday consumers and therefore advertised clearly. Protection is usually quantified through Ingress Protection Ratings that comprise two digits (e.g. 'IP68'): the first indicating dust-proofing (scale 0–6 where 6 is completely dust-tight) and the second indicating water damage resistance (scale 0–9 where 9 is protected against high-pressure water or steam jets). Screens made of Corning's Gorilla Glass, or similar alternatives to regular glass, should afford better protection against cracks and scratches.

---

[2] Produced by Japan's TES New Energy (website: https://web.archive.org/web/20140927084216/ http://tes-ne.com/English/pot; since defunct).

Some budget and low-end devices are made of materials that may actually be more durable than more expensive fragile high-end devices (e.g. plastic cf. glass). Similarly, the popularity of many devices means that a large selection of manufacturer or cheaper third-party accessories are typically available, including soft covers, hard cases, screen protectors, and waterproof bags. Pairing cheaper, lower- or non-certified devices with appropriate accessories may afford a similar level of protection to certified devices at a cost saving.

Several manufacturers also offer specialized heavy-duty, 'rugged' smartphone and tablet models (e.g. Cat, Plum, AGM, Doogee, and Blackview for smartphones; Panasonic and Getac for tablets). These devices include outdoors-orientated features such as armoured casing for protection against drops, higher grades of shock-proofing and water-submersion, and larger batteries for extended use in the absence of power. Some of these devices are priced similarly to regular, non-rugged models, usually because the rugged features eschew otherwise expensive flagship features, such as screen and camera quality. For even more effective protection, but a typically higher cost than regular devices, some manufacturers (e.g. Panasonic; Getac; Trimble) offer tablets and laptops built around industry- and military-grade specifications. These specialist devices may also offer similarly high-specification features not otherwise found on consumer devices, such as high power GNSS antennae, external keyboards, barcode readers, and long-lasting, hot-swappable batteries. Note that such purpose-built devices may be restricted to old or otherwise deprecated versions of mobile operating systems to ensure compatibility with proprietary features, which may cause compatibility issues with data collection apps designed for modern consumer devices.

9.2.1.4 Operating system and app availability

The operating system is the software run by a device that performs baseline system functions, facilitates user input, and allows applications and other software to be installed. Operating systems may be strictly tied to hardware, usually in line with specific manufacturers. Among mobile devices, the two most common operating systems are Android[3] and Apple's iOS.[4] Android is an open-source, Linux-based software developed primarily by Google and found on devices from a range of manufacturers in addition to Google (e.g. Samsung, Huawei, Motorola). In contrast, iOS is proprietary software and is included only on Apple-manufactured devices (i.e. iPhones and iPads[5]).

While the features offered by Android and iOS are broadly comparable, a handful of caveats warrant mentioning. Most significantly, although many apps are available for both operating systems, some apps are only available for one or the other—therefore, app compatibility is extremely important to consider when investing in new devices. In both operating systems, apps are typically installed from the app stores (Google Play Store in Android, and the App Store in iOS[6]). In Android, however, apps obtained from other sources can also be installed manually using their respective standalone executable .apk files (the equivalent of a .msi file in Microsoft Windows), by copying them onto the device and navigating to them using a file explorer app. This allows for Android apps to be backed up or transferred locally and installed in the absence of an internet connection.

Being at least partially open-source software, many Android devices from manufacturers other than Google come preinstalled with the manufacturer's own apps, features, and aesthetics that add an extra layer of software to the underlying operating system. While occasionally useful, this so-called bloatware can often slow performance, reduce battery life, and add unnecessary distractions to what is usually otherwise required by researchers to be an efficient, functional device. It is generally worth

---

[3] Hereafter, 'Android' (a trademark of Google) refers to the most common consumer devices that run Android alongside Google's software suite, which includes the Google Play Store for downloading apps (in contrast to non-Google Android forks, such as Amazon's Fire OS, for which app availability and compatibility may vary).

[4] Microsoft's Windows 10 Mobile being deprecated as of 2019; but see Windows 10 and Windows 10X for tablets (e.g. Microsoft's own Surface line) and dual-screen devices, respectively. See also Huawei's Harmony OS, which may emerge as a large market share in the near future.

[5] Hereafter, 'iOS' refers to both iOS and iPadOS.

[6] Unreferenced apps named in this chapter are available from the corresponding app store at the time of writing.

checking the extent to which unnecessary apps can be hidden, disabled, or uninstalled to provide the most streamlined, foolproof, and distraction-free workflow for the end-user. This is particularly pertinent for devices that are to be used by users who may not have worked with mobile devices before (e.g. see Stevens et al., 2013 for an example of a streamlined Android user interface designed for citizen science data collection).

In contrast, devices sold directly by Google typically run more streamlined and recently released stock versions of Android .[7] iOS devices also avoid the risks of manufacturer customization because they are always produced by Apple, and therefore do not include any other manufacturer's own content. However, this restriction does mean that users who prefer apps compatible only with iOS will find a more limited range of devices and price points from which to choose. The cost of Apple and Google's own devices are usually significantly higher than third-party manufacturer Android devices, the latter of which can often provide similar enough features to suffice for conservation research at a substantial cost saving, depending on researcher needs.

9.2.2 Data collection software

A review of every mobile data collection app is a daunting endeavour. A cursory search online reveals an array of software designed to facilitate mobile data collection for a variety of devices, tasks, or study species. The number of useful and deployable apps decreases quickly, however, when deprecated apps (i.e. apps that are no longer regularly updated) are excluded. We consider deprecation to be an important factor because the fast rate at which new devices and operating system versions are released means that loss of compatibility can quickly become an issue if apps are not regularly updated at the same rate (Teacher et al., 2013). A lack of recent development does not necessarily eliminate the usefulness of an app or exclude functionality, but it does mean that without developer support, functionality and troubleshooting are not

guaranteed and may require trial-and-error to discover whether app features do or do not work on certain devices (see 9.7).

As such, given our intent for this chapter to retain relevance for current and near-future applications, we present a non-exhaustive list of data collection apps in active development with applications across conservation science. Specifically, we focused on apps that: (1) support broad functionality for use in data collection in conservation science; (2) are in active development at the time of writing; (3) run on either Android and/or iOS, given these will most likely continue to be the market-leading operating systems;[8] and (4) do not require additional specialized hardware. We also did not test apps that fit these criteria but only offer commercial price tiers that are unlikely to be accessible to solo users (i.e. >$100 USD per month[9]).

9.2.2.1 App accessibility

We identified 11 apps that fitted our criteria for recommendation in this chapter, with some caveats (Table 9.1). Of these, five apps run only on Android; two run only on iOS; and four are compatible with both operating systems. While most of these apps are proprietary software, some are built around an existing file format for data collection. For example, XLSForms[10] is an open-source standard for coding data collection questionnaires in Microsoft Excel .xls or .xlsx files or, if preferred, a graphical wrapper (e.g. ODK Build) that once converted into finalized XForm .xml files using online or offline tools are compatible with several different apps and services. Allowing multiple apps to be developed using a single standard means that users have a choice

---

[7] See also devices by HMD's Nokia and Lenovo's Motorola, two manufacturers currently producing devices with close-to-stock versions of Android.

[8] Examples of apps that did not fit our criteria: *Ant-App* (Ahmed et al., 2014); *DORIS* (cited in Teacher et al., 2013); *Mongoose 2000* (Marshall et al., 2018); *Prim8* (McDonald & Johnson, 2014); *The Observer XT* (Zimmerman et al., 2009). Further lists of apps for ecological data collection are curated by Emiolio Bruna at http://www.brunalab.org/apps (see Marvin et al., 2016) and reviewed in Andrachuk et al. (2019) and Aitkenhead et al. (2014—for examples of environmental monitoring using specific mobile device sensors).

[9] For example: *CommCare* (https://www.dimagi.com/commcare); *ArcGIS Survey123* (https://survey123. arcgis. com/); *SurveyCTO* (https://www.surveycto.com); *Secure Data Kit* (https://www.securedatakit.com).

[10] https://www.xlsform.org/en; https://www.opendata kit.org/xlsform

Table 9.1 A comparison of current mobile apps for general purpose data collection as of July 2020. Parentheses indicate where additional apps or software are required to make use of features not otherwise directly integrated. Data export options refer to .csv output file type, unless otherwise stated

| Author/ Service | Mobile app | Operating system | Form creation software | Cost | Data collection | | | | Data export | | | | Website |
					GNSS coordinates	Photo/ video	Audio	Drawing/ handwriting	Export offline (file type notes)	Export via SMS	Export to online cloud via internet	Cloud service	
Open Data Kit (ODK)	ODK Collect	Android [1]	Microsoft Excel (XLSForms standard)	Free	Yes	Yes	Yes	Yes	Yes (.xml; .csv via ODK Briefcase [7])	No	Yes (also including .kml; .json)	Google Drive; custom server (local or third-party including Google Compute Engine, Amazon Web Services) using ODK Aggregate	opendatakit. org
Ona	ODK Collect	Android [1]	Microsoft Excel (XLSForms standard)	Free; paid tiers (monthly subscription)	Yes	Yes	Yes	Yes	Yes (.xml; .csv via ODK Briefcase [7])	No	Yes (also including .xlsx; .kml; .json)	Fully integrated; Google Drive; custom server (local or third-party including Amazon Web Services) using proprietary service	ona.io
DataWinners	DataWinners	Android [2]	Microsoft Excel (XLSForms standard); Proprietary software (online via browser)	Free (one year); paid tiers (annual subscription)	Yes	Yes	Yes	Yes	Yes (.xml; .csv via ODK Briefcase [7])	Yes, to online cloud (not compatible with app)	Yes	Fully integrated	datawinners. com

Kobo Toolbox	KoBoCollect	Android[1]	Microsoft Excel (XLSForms standard); Proprietary software (online via browser)	Free	Yes	Yes	Yes	Yes (.xml; .csv via ODK Briefcase[7])	No	Yes (also .xlsx)	Fully integrated	kobotoolbox.org
CyberTracker	CyberTracker	Android[3]	Proprietary desktop software (offline)	Free	Yes	Yes	Yes	Yes (also including .xlsx; .kml; .xml; .html)	No	Yes.	Custom server (local or third-party) using FTP	cybertracker.org
Fulcrum	Fulcrum Mobile Data Collector	Android; iOS	Proprietary software (online via browser)	Paid tiers (monthly/annual subscription)	Yes	Yes	Yes	No	No	Yes (also including .xlsx; .sql; .sqlite)	Fully integrated	fulcrumapp.com
Imperial College, London	EpiCollect5	Android; iOS	Proprietary software (online via browser)	Free	Yes	Yes	No	No	No	Yes	Fully integrated	five.epicollect.net
HanDBase	HanDBase Database Manager	Android[2]; iOS	Proprietary desktop software (offline)[6]; In-app	One-off	Yes	No	Yes	Yes	No	No	-	ddhsoftware.com
FileMaker	FileMaker Go	iOS	Proprietary desktop software (offline)	One-off, paid cloud service (monthly/annual subscription)	Yes	Yes	Yes	Yes (also .xlsx)	No	Yes (via FileMaker Cloud or FileMaker Server)	Fully integrated using FileMaker Cloud; custom server (local or third-party) using FileMaker Server	filemaker.com/products/filemaker-go

continued

Table 9.1 *Continued*

Author/ Service	Mobile app	Operating system	Form creation software	Cost	Data collection				Export offline (file type notes)	Export via SMS	Data export		Website
					GNSS coordinates	Photo/ video	Audio	Drawing/ handwriting			Export to online cloud via internet	Cloud service	
Damien Caillaud/ Dian Fossey Gorilla Fund International	Animal Observer [8]	iOS [4]	Proprietary desktop software (offline via R)	Free	Yes	Yes	Yes	No	Yes (.csv via iTunes; R)	No	Yes (as .dat; .csv via R)	Custom server (local or third-party) using FTP	fosseyfund. github.io/ AOToolBox
Ross et al. (2016)	ZooMonitor [8]	Android; iOS [5]	Proprietary software (online via browser)	Free (for accredited organizations); paid tiers (annual subscription)	No (user-defined coordinates only)	No	No	No	No	No	Yes	Fully integrated	zoomonitor. org/home
Newton-Fisher (2012)	Animal Behaviour Pro	iOS	Microsoft Excel; In-app	One-off	No	No	No	No	Yes	No	No	-	kar.kent.ac. uk/44,969
Stevens et al. (2013); Extreme Citizen Science group, University College London	Sapelli Collector	Android	Proprietary desktop software (offline)	Free	Yes	Yes	Yes	No	Yes (also .xml)	Yes, to offline device also running app	Yes	Custom server (local or third-party) using GeoKey	sapelli. org

[1] iOS supported using Enketo via browser.
[2] Android app is deprecated.
[3] Manual (offline) install only (.apk installer provided by desktop software).
[4] iPad only.
[5] iOS and Android supported using web app via browser.
[6] Desktop software is deprecated.
[7] ODK Briefcase is Java-based desktop software for Windows, macOS, and Linux.
[8] Designed specifically for behavioural data collection.

of apps and price tiers that—thanks to inter-app compatibility—can be switched between as budgets allow: from free pricing for those willing to build and collate forms, apps, and optionally cloud servers; to paid subscriptions for services that conveniently integrate these features into a single, all-in-one solution.

9.2.2.2 Software features

The apps that we identified as having the most flexible customization, and therefore the broadest applications, generally follow a similar format. Users populate a form with questions in various formats (e.g. text boxes; multiple-choice; record GNSS coordinates; capture photo using device camera) and media that can be displayed as part of questions or answers (e.g. images; video; audio), which can encompass multiple pages, loops, and repeats. Apps designed more specifically for behavioural data (e.g. Animal Observer; Animal Behaviour Pro) generally had formats more closely divided into data collected as part of focal follows, scans, or ad libitum observations. In the case of Zoo Monitor, some aspects of data collection were more tightly constrained for use in captive environments (e.g. eschewing capturing GNSS coordinates for pinpointing a location on a predefined map by eye). However, we felt enough customization was possible in other questions for the app to be recommendable for broader use cases.

9.2.2.3 Data export and storage

Almost all reviewed apps (Table 9.1) support offline export of data as spreadsheets, although saving these spreadsheets directly onto the mobile device was only possible in some proprietary apps. Instead, for all of the apps using the XLSForms standard, local export requires using separate desktop software. Exporting data to a cloud server—either an existing server provided by the user, or one provided as a part of an integrated service—was a feature available in almost all of the apps we reviewed, and in the case of Fulcrum, is currently the only option for exporting data. Of particular note were the options included in DataWinners and Sapelli Collector service to transfer data using SMS. In the case of DataWinners, this method of sending data is not cross-compatible with forms filled in and

submitted using the app; however, we felt it noteworthy given the advantage of being able to exploit weak cellular service in remote areas that may support SMS service but not suffice for internet data service.

9.2.2.4 'DIY' versus integrated services

A distinction can be drawn between standalone apps—either free or requiring a one-time purchase—and apps included as part of more complete packages and mainly paid for by subscription. For example, commonly used apps such as ODK Collect and CyberTracker, or heavily animal behaviour-focused apps such as Animal Behaviour Pro and Animal Observer, provide a front-end for data collection. These apps may require other software or customization (and therefore potentially some technical knowledge) to set up an entire workflow that includes collating data from multiple users in a single (e.g. cloud) database, for example. As such, these apps may be especially useful in modular or 'DIY' workflows (Vitos et al., 2013a), in which flexibility and customization of software are desired or required (e.g. to meet prerequisites, such as the particular type of smartphone or tablet on which data will be collected, or cloud server/service to which data must be exported).

On the other hand, services such as FileMaker, Fulcrum, and Ona include data collection apps, cloud storage, and additional features, such as proprietary form-building software and data analytics, that are integrated into a single package. These services may suit researchers who require an 'oven-ready' workflow for data collection that can be quickly deployed to multiple users with little technical knowledge, and may be willing to pay subscription prices for the convenience of an all-in-one service. In addition, for users who require rapid analyses of data or prefer to eschew exporting data for analysis in separate software, some integrated services include some analytical features in their software. For example, FileMaker, Fulcrum, and Ona allow users to visualize heatmaps of collected form locations, generate PDF summary reports, and create various charts, respectively, directly within their software. These analytical features are generally missing or less fleshed-out in standalone apps designed foremost for data collection, which require users to export data for analysis in other software (e.g. Microsoft Excel; R; QGIS).

9.3 Applications

The variety of hardware and software features in current mobile devices can facilitate a wide range of conservation science applications. This section reviews several broad functionalities and discusses the advantages of digitizing workflows in each case.

Several of the apps we reviewed (Table 9.1) have been used in conservation projects, particularly those with the broadest customization and functionality (e.g. Open Data Kit- and XLSForms-based apps and CyberTracker). Mobile data collection apps are often adopted as direct substitutes for pen and paper methods, but the amount of data that can be recorded and the speed at which it can be organized and transmitted to a central database mean that data collection protocols can quickly be expanded to supersede what is possible by hand. For example, rangers at the Djelk Indigenous Protected Area, Australia, used Cyber-Tracker in place of paper forms and GNSS units to increase monitoring efficiency of feral wildlife, vegetation infestations, and prescribed vegetation burning (Ansell & Koenig, 2011; reviewed in Liebenberg et al., 2017). ODK Collect has been used for a similar purpose in several conservation projects. For example, The Jane Goodall Institute trained forest scouts to use the app to monitor chimpanzee presence and deforestation (Chapter 2). ODK Collect was used in a community forest monitoring scheme in Vietnam, whereby community members could complete questionnaires and record details, including photographic and video evidence, of illegal activities (Pratihast et al., 2012). The same app was also used in Brazil to run a community-based monitoring scheme of fisheries production to assess impacts of environmental change on catch quality (Oviedo & Bursztyn 2017). In the latter two cases, using a mobile app to delegate data collection to local communities resulted in data that were of comparable accuracy and reliability to expert or government data and were less expensive to collect.

The speed at which data can be transferred from individual devices for collation in a database, especially where the internet is available to update cloud databases online, means near real-time monitoring is possible. For example, KoBoCollect was used by community informants to report observations of human–wildlife conflict (e.g. crop-raiding and property damage by elephants in Tanzania) that allow authorities to identify problem individuals more quickly and mitigate future conflicts (Le Bel et al., 2016). The authors note that the app provides a more reliable and efficient upgrade to both pen and paper and a system of coded SMS messages that relayed observations to authorities (Le Bel et al., 2014). CyberTracker is used by rangers in South African national parks to monitor anti-poaching efforts based on data collected during patrols (Liebenberg et al., 2017). Similarly, ForestLink, an Open Data Kit-based app developed for the Rainforest Foundation UK, is used for real-time monitoring of illegal deforestation and community threats in several locations across Africa and the Americas. Data are uploaded to a central database using low-power (e.g. Rapsberry Pi-based) modems that transmits data using a satellite uplink, negating the need for cellular or internet connectivity. Alerts can then be quickly delegated to ground teams for investigation in person (Rainforest Foundation UK, 2019).

Mobile devices and apps provide opportunities for integration with other technological platforms. In Uganda, mobile devices are used during patrols by forest rangers in tandem with the Global Forest Watch platform, a system that analyses satellite imagery to detect forest loss and alert rangers on the ground who can investigate and verify breaches of regulations (Weisse et al., 2017; see Chapter 2).

Mobile devices are a core component of protected area management desktop software, such as Vulcan's EarthRanger and ESRI's ArcGIS for Protected Area Management (PAM), which collate data collected with mobile devices in order to provide rapid or real-time monitoring of events within protected areas (e.g. wildlife sightings, illegal activities, and enforcement). Both EarthRanger and

ArcGIS for PAM are compatible with proprietary and third-party data collection apps, such as Cyber-Tracker. In the case of EarthRanger, alerts can also be transmitted from a central computer to mobile devices (e.g. carried by rangers on patrol) using WhatsApp or SMS. Similarly, the Spatial Monitoring and Reporting Tool (SMART) mobile and desktop software is a widely used conservation monitoring and management tool. Data can be collected using mobile devices (using a third-party app, such as CyberTracker, or using SMART's own app, which is itself based on CyberTracker) and collated for overview and analysis in the SMART desktop program. SMART is detailed in full in Chapter 9.

9.3.1 Behavioural data collection

Behavioural data directly inform our understanding of species ecology and are therefore important considerations for conservation strategies that need to accurately reflect how animals use their environment and interact with other organisms, including humans (Sutherland, 1998). Behavioural data protocols usually comprise some combination of focal follows, scans, and ad libitum observations, which usually need to be collected alongside data from external devices such as location coordinates, timestamps, and image or audio captures. These data must be recorded relatively quickly to accurately reflect an animal's behaviour at a given point in time and to minimize wasting time that may be needed to collect data for other protocols (Marshall et al., 2018). Speed of data collection is likely to be limited by the nature of the study species (e.g. small, fast-moving, arboreal, or flying animals that are difficult to observe) and by adverse field conditions (e.g. poor weather or tough terrain). Furthermore, as discussed in Section 9.1, one of the major disadvantages of data collection with pen and paper is the lack of integration between a handwritten datasheet and other pieces of equipment such as GNSS units, cameras, and sound recorders, which can result in a slow, disjointed workflow, and more opportunities for user error.

Mobile apps are particularly well-suited for behavioural data collection because they support the fast input of complex protocols without compromising the amount or resolution of data recorded, which may occur if otherwise handwritten qualitative observations are reduced to code or shorthand (Bateman et al., 2013). For example, forms made with XLSForms support loops for repeated observations, conditional statements that display questions or pages depending on answers to previous questions or forms, and restrictions that prevent users forgetting to answer certain questions. Apps designed specifically for behavioural data (e.g. Animal Observer; Animal Behaviour Pro—Table 9.1) can also include useful features such as countdowns and reminders for scans, and pre-made templates to match group- and individual-level protocols that speed up form customization. Moreover, the use of onboard or external (e.g. via Bluetooth or WiFi) hardware features, including GNSS, cameras, and microphones functions, are integrated into almost all data collection apps (e.g. see Table 9.1). In many cases, timestamps or GNSS coordinates can be recorded automatically at a given stage in the protocol, avoiding any user input—and potentially mistaken or missing entries—altogether.

9.3.2 Citizen science and community engagement

In addition to data collected by researchers in the field, the popularity of smartphones as consumer devices means that collecting data through the general public ('citizen science'—reviewed in Graham et al., 2011) is now easier than ever and increasingly integrated into long-term projects (e.g. Rafiq et al., 2019). For fully integrated services, forms can be built using provided tools or templates and shared publicly with other users through a website or app (e.g. EpiCollect5—Table 9.1). For even wider dissemination, mobile apps can be built from the ground up and distributed through app marketplaces, where they can also be more easily discovered by or marketed to new users. By making apps publicly available to so many potential users through just a handful of standardized app marketplaces, apps can be adopted for extremely short-term projects with little prior notice needed for users. For example, wildlife tourism experiences can encourage users to download apps at the start of tours to report observations (e.g. see Whale Trails,[11] an Android and iOS app designed for whale watchers to record humpback whale GNSS tracks—Meynecke, 2014). For long-term citizen science projects that require a large number of users to collect sufficient data, mobile apps may be better suited to attracting and retaining users because they can be easily integrated into a format that appeals to a wider array of demographics and personal interests, such as interactive games (Bowser et al., 2013; Kim et al., 2013).

Mobile apps can also help facilitate data collection protocols that are designed with specific local communities in mind. Specifically, touch-screen devices with large screens can make user input more intuitive than handwritten methods because of the ease of displaying different media types to users, such as images, video, or audio to users. This can increase accessibility for a data collection protocol, even for users who may not have experience with smartphones or tablets. In turn, user enjoyment is likely to be greater, which is an important factor in motivating users to continue collecting data for citizen science projects (Kim et al., 2013). For example, Sapelli Collector was used in developing questionnaires with graphic interfaces based on input from local communities in Cameroon, which were suitable by communities with high rates of illiteracy for reporting observations of local wildlife crime (Stevens et al., 2013). CyberTracker and the XLSForms standard also support icons or pictures as replacements for text throughout questionnaires (e.g. multichoice answers). In the case of CyberTracker, a key use case has been the recording of indirect evidence of animal presence (e.g. prints), for which an icon-based interface is especially intuitive for non-literate users (reviewed in Liebenberg et al., 2017). Similarly, Vitos et al. (2013b) took advantage of the open-source code for ODK Collect to remove all text from the app to streamline the interface for non-literate users during community data collection in the Republic of Congo.

[11] Likely deprecated—see *Whale Track* for Android for a similar example in active development.

9.3.3 Mobile geographic information systems (GIS) and participatory mapping

Several GIS are available as mobile apps. Mobile-only GIS apps include SWMaps and Locus GIS for Android (free) and GIS Pro for iOS (paid). An Android port of the free and open-source QGIS was under development but is deprecated at the time of writing (as alternatives for viewing QGIS projects on mobile, see Input by Lutra Consulting for Android and iOS and QField for QGIS for Android). For direct spatial data collection, ODK Collect supports some spatial data question types (e.g. tracing of lines and polygons; see also GeoODK for Android and ArcGIS Survey123 included in ESRI's paid ArcGIS suite—footnote 7).

Mobile GIS apps provide similar functionality as desktop software for displaying, annotating, and in some cases analysing, vector (e.g. shapefiles) and raster data. An advantage of displaying spatial data on a mobile device is that the device GNSS location can be displayed in real time as an overlay, similar to conventional mapping apps (e.g. Google Maps; Apple Maps). As such, researchers can quickly and easily create fully customizable and navigable maps for any location for which pre-existing spatial data is available. These maps allow researchers to identify and plan routes to sites of interest (e.g. transect or sampling locations), avoid geographic hazards, and more quickly familiarize themselves with new study sites—especially important given that data collection is often tightly constrained by time available for fieldwork.

In addition to navigation, mobile GIS can be used to collect data directly. Location tracks may be recorded for a given time or distance intervals for later analysis, which is a similar functionality to that found in standalone GNSS units. We also note that running a GIS app that continuously acquires a GNSS fix even when minimized is one effective way of reducing GNSS triangulation times in other apps (e.g. data collection apps that record GNSS coordinates—Table 9.1) because the fix should already be available.

Mobile GIS are also useful tools in participatory mapping that collate spatially explicit data and knowledge from local communities (see Chapter 2). Issues such as land rights and vegetation usage may involve input from many community members, for which digital maps can provide useful sandboxes for drawing and discussing user-identified landscape features in relation to various spatial data, such as boundaries or disputed areas. Previous studies, such as those focusing on community practices and forest use in Ecuador (Delgado-Aguilar et al., 2017) and Suriname (Ramirez-Gomez et al., 2016), have used paper or hard copies of maps when interviewing or creating maps with local communities, which then require scanning and geo-referencing for analysis in desktop GIS. Mobile GIS may provide effective alternatives to paper maps, particularly in areas where spatial data are already available for creating custom maps that correspond to interview questions (McCall et al., 2016; see also Pacha, 2015).

9.3.4 Mobile devices as multipurpose tools

As mentioned in Section 9.1, the transition of smartphones and tablets into devices capable of rivalling desktop computer functionality means that there are many applications of these devices for field researchers in addition to recording observational data. It is impossible to review every use of the multitude of apps currently available for Android and iOS (which number in the millions for each operating system), but here we highlight some of the more common applications that can supplement primary data collection.

The popularity of mobile devices combined with the current operating system duopoly means that integration with an Android or iOS app over Bluetooth or WiFi has become a common standard for connectivity, as evidenced by the number of 'smart'-branded everyday products now available. App connectivity has also become standardized for certain equipment of use to conservation researchers. For example, several Kestrel and HOBO weather stations can be configured and data collected from using the free Kestrel LiNK and HOBOmobile apps, both of which are available for Android and iOS (Figure 9.2). Mobile devices are not only smaller and lighter to carry to remote locations than laptops—particularly specialized rugged models—that are usually used in conjunction with these stations,

but if already being used for data collection, can remove the need to purchase other devices specifically for this purpose. For digital cameras, some camera manufacturers (e.g. Canon Camera Connect and WirelessMobileUtility by Nikon, for Android and iOS) provide apps for certain digital camera models that add remote control functionality. Outside of proprietary apps, USB on-the-go technology is a connection standard for mobile devices that supports connectivity to a wide array of generic USB peripherals, such as flash drives and SD card readers that can benefit camera trap users (see also WiFi-enabled SD cards that enable wireless transfer of photos to mobile devices, camera compatibility permitting).

Many of the more generic apps bundled with most mobile devices are also useful for researchers as digital alternatives to pen and paper. Apps for recording text range from simple note-taking to mobile equivalents of desktop office suites (e.g. Microsoft Office, Polaris Office, and WPS Office for Android and iOS, all of which have free tiers). Some office suites also include PDF readers as alternatives to standalone equivalents such as Adobe Acrobat Reader—all of which can be used to refer to documents such as protocol instructions or published articles without the need for hard copies. Drawing or sketching apps can be useful where language or illiteracy can impede verbal communication. Image galleries provide a system to organize animal or plant IDs that can be examined at high resolutions in the field. While most variants of Android and iOS come bundled with basic gallery and file manager apps, third-party alternatives that allow detailed browsing by nested folder structures (e.g. Simple Gallery Pro; F-Stop Gallery Pro; FX File Explorer for Android) can be useful for organizing large photo catalogues by study group and subject, for example. Internet-permitting folders of images can also be synced with a single cloud service account to allow a manager or other user to update the database on multiple devices remotely and with one action (e.g. Microsoft OneDrive and Dropbox for Android and iOS, which offer 2–5 GB of free storage with paid tiers thereafter and allow files to be updated when online and subsequently accessed offline). Audio recorder apps allow for spoken data collection (see

Figure 9.2 Kestrel Drop 1 temperature logger connected wirelessly to a Motorola Moto E (2014) Android smartphone via Bluetooth and the Kestrel LiNK app. Universal standards for Bluetooth and WiFi wireless connectivity mean consumer smartphones and tablets are increasingly compatible with data recording loggers and sensors. In this case, wireless connectivity means the logger can be positioned several metres high in a tree while still allowing users to remotely adjust settings, observe measurements in real-time, and export data to comma separated spreadsheets (.csv files; credit: EM).

also speech-to-text apps, such as the free Otter Voice Notes for Android and iOS, that can transcribe spoken notes in real-time), while media players can be used for audio playback experiments using Bluetooth connectivity or through 3.5 mm wired output. Where the cellular network is available, SMS or free internet messaging apps (e.g. WhatsApp) can be cheaper and more intuitive replacements for two-way radio communication, especially in difficult field conditions where referring to a written message can be integral to avoiding miscommunication.

9.4 Case study: from paper to digital data collection for primate conservation at the Issa Valley, western Tanzania

Historically, behavioural data collection on wild animal presence, ecology, and behaviour has been relatively labour intensive, not just in data collection but also in storage, transfer, and transcription. Researchers at the Issa Valley, western Tanzania, began a long-term study of primate community ecology in 2008 (the Greater Mahale Ecosystem and Conservation [GMERC] Project—Piel et al., 2018). The project began with two foreign researchers and two local field staff. Field staff had varying literacy and writing abilities. Data collection was in Swahili and began by documenting all evidence of wildlife and human disturbance. With only four data collectors, attempts to standardize the documentation of these events were initially relatively straightforward. For example, data were collected in Rite in the Rain data books, which were converted to grid cells to remind assistants which information to record. Each staff member was allocated two days/month to transcribe data from books to data sheets, which could then be more easily transcribed (a second time) to an electronic format for eventual analysis.

Besides the obvious challenges of protecting paper against the elements, especially keeping data dry through rain, there were additional obstacles to handwritten data collection. Despite attempts at standardization, variability in data records persisted. For example, spellings of wildlife names changed both within and between data collectors. Thus, collectors began developing individually specific abbreviations for recording observations. Due to variation in literacy, important ad libitum observations of animal behaviour went either unrecorded or were reduced to short narratives with details omitted due to the time taken and space available on paper to write down observations. This issue was especially pertinent given the nature of such ecological work. Anticipating everything that occurs in the natural world is naturally impossible and often requires rapid or continuous recording of observations, otherwise key details may be missed. Finally, time spent transcribing data from books was considerable (ca. half a workday per week for each staff member).

Additional data were collected as the project expanded in scope, especially on animal behaviour. In a matter of months, researchers and local field staff had habituated two troops of baboons (*Papio cynocephalus*; Johnson et al., 2015) and another of red-tailed monkeys (*Cercopithecus ascanius*; McLester et al., 2019), which later fissioned into two daughter troops, while habituation of a focal chimpanzee (*Pan troglodytes schweinfurthii*) community continued. The project hired additional staff, and protocols quickly transformed from recording opportunistic encounters of wildlife and people to detailed behavioural observations (e.g. dawn to dusk 5-minute focal follows of individual animals). Researchers and assistants collected an increasingly wide range of data, including dietary diversity, social interactions, and activity budgets. As protocols expanded, data collection increasingly required large amounts of reference information to which researchers needed to refer, such as plant species names, individual identities of animal group members, and behavioural ethograms. In particular, mammal diversity monitoring conducted with line transects required research staff to be familiar with information on perpendicular and observational distances, flight responses, and the age, type, and location of snares. Increasingly, assistants would also encounter people who were illegally in the forest. A database on this growing threat was built, requiring data collection on human activity, the village of origin, and duration in the forest, which necessitated additional flexibility in the number and complexity of questions recorded per observation.

As the complexity of data collection protocols and the number of collectors grew, the time allocated for training, data transcription, and resolving confusion concerning spelling and nomenclature also increased. More personnel meant that more data could be collected, but also led to an increased need for quicker collation and transcription or digitization for eventual analysis. Assistants soon required one day per week to keep up with transcription demands. With a team of eight assistants, this time budget eventually led to the time equivalent of multiple months spent on transcribing data books to paper databases. More time still was subsequently spent by management staff entering those paper data onto a computer spreadsheet.

In 2013, the project replaced paper data collection with a digital data collection protocol (Figure 9.3). Assistants and researchers were provided with an Android tablet each and trained on its use. At the time, smartphones were not widely used in Tanzania, yet assistants still only needed a matter of days to familiarize themselves with the tablets. The ability to enlarge the font, the predictability of the questions, and the restricted choices all made for ease of use. Interobserver training lasted only a few weeks before assistants were comfortably and consistently recording digital data.

Open Data Kit (Table 9.1), using the free ODK Collect app and an online ODK Aggregate cloud server hosted on Google Cloud Platform, was introduced as the primary platform for data collection. While data are stored locally on tablets and can be exported, weak cellular internet service is available in some areas of the study site and means staff can upload data to the server on a ca. weekly basis. For most of the project's data collection, form file sizes are extremely small (≤ 2 kilobytes) because protocols do not include recording media such as photos or videos. As such, uploading forms (usually <5 megabytes total per week) can be a fast and typically inexpensive process that can be instigated using smartphone mobile hotspots. The process was made faster still with the introduction of a satellite internet connection in 2015. The server currently hosts >100,000 records from 50 data protocols at a monthly cost of ca. $\leq$\$2USD.

The transition to digital data collection has resulted in tangible benefits. By uploading data to a cloud server, project directors can view, download, and back up data remotely, indirectly increasing accessibility to data for collaborators and funders. Furthermore, ODK Aggregate allows permissions to be set for individual users. Management staff and researchers can be provided with limited access to the server to upload new protocols and verify data have uploaded successfully, reducing the need to contact and work through a single administrator (in this case, the project directors) each time.

Digital entry has eliminated most legibility issues and freed up time previously spent transcribing handwritten data. Limited, custom selections have

Figure 9.3 GMERC Project staff members Mlela Juma (top) and Sadiki Abeid, Mashaka Alimas, Shedrack Lukas (bottom) use Google Nexus 7 Android tablets running ODK Collect to record observations of chimpanzee vocalizations and phenological data, respectively, in miombo woodland (credits: EM; Christian Howell/GMERC).

greatly improved standardization of observations, particularly through multiple-choice questions and requirements that avoid skipping questions accidentally. Similarly, automatic background data collection, such as date and time stamps and GPS coordinates, have also reduced mistakes or gaps in data sets. Technological issues have also been relatively straightforward to identify and resolve because project directors and collaborators can view or be sent large numbers of data files easily by email or other cloud services, either from the study site itself or using stronger internet connections available in nearby villages and towns. Researchers and students can create, test, and familiarize themselves with data collection protocols in advance of fieldwork using the variety of XLSForms tools available online (Table 9.1).

At the Issa Valley, most costs of shifting to digital data collection have related to hardware. Even budget tablets are relatively expensive compared to pen and paper, and they are not as water-resistant as waterproof stationery. Tablets do get dropped and damaged accidentally, as can happen with any handheld device. Some issues such as incorrect time zones in device settings have led to later problems with behaviour or chronology reconstruction during data cleaning, although these have been relatively straightforward to rectify.

The other primary cost–benefit ratio relates to time; specifically, the time spent training staff in a new technology compared to time spent transcribing handwritten data. While all field sites vary in individual circumstances, it can be a worthwhile transition if time is more valuable than money. In the case of the Issa Valley, assistants now spend more time in the field, where their skills of behavioural observation, plant identification, and threat detection are used daily, and they no longer spend many hours each month transforming those data for later, remotely conducted analyses. That such a transition adds to a project fiscal budget was an acceptable cost to improving data standardization, streamlining the path from observation to analysis, and applying people's skills where they are most appropriate.

9.5 Limitations/Constraints

Although there is a focus on customization in many data collection apps, including those reviewed earlier, app choice can still be a limiting factor if compatibility and/or functionality do not meet a researcher's use case. While developing an app from scratch is the most effective way to ensure that an app functions exactly as required, this may not be an option for a majority of users who do not have the required technical knowledge to do so themselves, or time to find a collaborator who does (see Teacher et al., 2013, who provide a detailed walkthrough of the app-building process for researchers). As such, users may find themselves with a potentially convoluted workflow that relies on multiple apps or platforms, limiting the degree to which switching to a digital workflow will streamline data collection. Similarly, complex workflows may be more time-consuming to train staff or team members in, particularly those who do not already have experience with the hardware being used.

Logistically, as with any electronic device, power requirements can be a limitation. First, device battery life will always be a limiting factor for day-to-day use and will vary depending on battery capacity and usage intensity. Second, and more fundamentally, is the need for a power source to recharge devices. While the range of battery capacities and charging methods available (e.g. USB through AC, DC, or external battery/power bank) can alleviate this requirement to a certain degree, in remote field sites power may be limited or inconsistently available to the extent that electronic data collection is restricted to a small number of devices or not simply not feasible. Similarly, a lack of internet connection can restrict opportunities to send and store data remotely from a field site, although there is almost always an offline alternative for saving data locally (Table 9.1).

Hardware or software failure can result in data loss on any device. While backups are faster to make electronically, storage for digitized data on multiple external drives can become expensive depending on the storage capacity required and the type of drive used (e.g. solid-state drives are more expensive than hard drives, but also more reliable due to the absence of moving parts). For online cloud-based storage, expenses can be incurred either through subscriptions for the cloud service, or through the cost of obtaining an internet connection in remote locations (e.g. via satellite internet). In both cases, however, these issues are equally intrinsic to data collected by hand. Paper records can also be accidentally misplaced or destroyed and are arguably more difficult to back up and store without using an electronic device. Moreover, analogue data will need to be transcribed to a digital format at some point in the workflow in order to be analysed, which means any such benefit to collecting and keeping data in an analogue format is likely to be cancelled out by the time and/or financial costs of digitizing data (see 9.1).

9.6 Social impact/privacy

Digitizing data collection usually requires third-party software and almost always the use of third-party hardware. As such, maintaining data confidentiality will always depend on the extent to which hardware, software, and cloud storage providers can or will protect researcher data.

Regulations governing how manufacturers are obliged to protect user data differ between countries or region. Personal data protection in the European Union is mandated by the General Data Protection Regulation in the European Union, which differs from the United States, for example, where data protection laws vary between states. Researchers using third-party cloud services (e.g. Google Cloud Platform or Amazon Web Services) to store or transfer data should remain informed of where their data will be located throughout the process and how its storage will be regulated. For example, the Google Cloud Platform service for hosting cloud storage (among other uses) offers users a choice of country for server location. Alternatively, researchers in need of the strictest encryption should avoid outsourcing storage to third parties and instead use a personal home server, particularly one with an open-source operating system (e.g. see Ubuntu and Linux Mint as useful introductions to Debian Linux, and Manjaro as an introduction to Arch Linux). Depending on the researcher's needs, a low-power device may suffice to minimize running costs for home servers that run continuously (e.g. see Raspberry Pi and Intel NUC devices as starting points).

Researchers that collate data collected by other users with apps (e.g. citizen science projects) also have a responsibility to protect the data collectors' confidentiality (Bowser et al., 2014). Data that may be used for research purposes can also reflect personal information (e.g. GNSS locations; photographs that may inadvertently include individuals). As such, researchers should make clear to data collectors how and where the data will be stored, and which steps have been taken in line with appropriate legal frameworks to mitigate unintended uses (e.g. allowing data collectors to delete any data after submission; establishing strict schedules for how long data will be stored; deleting personally identifiable details from stored data—reviewed in Bowser et al., 2014).

9.7 Future directions

The distinction between mobile and desktop hardware and software is quickly becoming narrower. Each annual iteration of manufacturer flagship devices raises the bar for computing and graphics power, storage capacity, and connectivity to peripherals that mobile devices can support. As such, for many users, smartphones and certainly tablets now facilitate a level of productivity that would previously have been confined to desktop or laptop computers. Desktop software can be more effectively duplicated on mobile devices and several new form factors, including devices with foldable or dual screens (see the Samsung Galaxy Fold and Microsoft Surface Neo devices, for example) should further expand the opportunities for porting desktop workflows to mobile devices when they become available in lower-end devices with less exclusive prices. For conservation researchers, this means that stages in a workflow can be carried out on mobile devices in the field, allowing faster analysis and communication of data and, ideally, action based on those results.

The fast pace at which software and hardware are constantly evolving, and the wide range of applications needed by conservation researchers present a problem for standardizing software. More generalized apps can be improved and updated for new hardware faster by larger development teams, but inevitably will not be suitable for every protocol or study species/system. While building an app from the ground up and disseminating it are cornerstones of both Android and iOS, data collection software built by a small team designed to support a single project is less likely to see user uptake outside of the authors and is more likely to quickly depreciate (Teacher et al., 2013). We clearly observed this effect while reviewing currently available apps for this chapter, during which we identified many apps that did not fit our relatively broad criteria for inclusion in Table 9.1. These instances highlighted the degree to which apps are frequently developed for a very specific study species or system, released publicly—sometimes with media exposure or a dedicated journal article—and then no longer updated outside of a very short timeframe, or even at all. It could be argued that an app that suits at least one user's protocols for a single project can still be considered a success. Given the initial time costs of disseminating equipment and training users when switching to a digital data collection platform—especially for large teams—it can be inefficient for researchers to change data collection software at frequent intervals. Alongside the need to update and maintain apps in line with continuous advances in hardware and software, there is, therefore, a need for potential app developers to balance specific functionality against how much time can be committed to supporting a relatively small number of end-users in the future. As an alternative, researchers may consider adopting a platform that uses an open-source standard (e.g. XLSForms—see Table 9.1; 9.2.2), given that these apps benefit from larger user-bases that can provide technical support, greater flexibility for complex workflows, and less reliance on proprietary technology that may become deprecated at short notice.

9.8 Acknowledgements

We thank Fiona Stewart, Ineke Knot, and the Faculty of Science at Liverpool John Moores University for providing hardware and software tested in this review.

References

Aanensen, D. M., Huntley, D. M., Feil, E. J., Al-own, F., & Spratt, B. G. (2009). EpiCollect: linking smartphones to web applications for epidemiology, ecology and community data collection. *PLoS One*, 4(9), e6968.

Ahmed, Z., Zeeshan, S., Fleischmann, P., Rossler, W., & Dandekar, T. (2014). Ant-App-DB: a smart solution for monitoring arthropods activities, experimental data management and solar calculations without GPS in behavioral field studies. *F1000Research*, 3, 311.

Aitkenhead, M. J., Donnelly, D., Coull, M. C., & Hastings, E. (2014). Innovations in environmental monitoring using mobile phone technology—a review. *International Journal of Interactive Mobile Technologies*, 8(2), 42–50.

Andrachuk, M., Marschke, M., Hings, C., & Armitage, D. (2019). Smartphone technologies supporting community-based environmental monitoring and implementation: a systematic scoping review. *Biological Conservation*, 237, 430–442.

Ansell, S., & Koenig, J. (2011). CyberTracker: an integral management tool used by rangers in the Djelk Indigenous Protected Area, central Arnhem Land, Australia. *Ecological Management & Restoration*, 12(1), 13–25.

Bateman, H. L., Lindquist, T. E., Whitehouse, R., & Gonzalez, M. M. (2013). Mobile application for wildlife capture-mark-recapture data collection and query. *Wildlife Society Bulletin*, 37(4), 838–845.

Berger-Tal, O., & Lahoz-Monfort, J. J. (2018). Conservation technology: the next generation. *Conservation Letters*, 11(6), e12458.

Bowser, A., Hansen, D., He, Y., et al. (2013). Using gamification to inspire new citizen science volunteers. In Proceedings of the First International Conference on Gameful Design, Research, and Applications (pp. 18–25).

Bowser, A., Wiggins, A., Shanley, L., Preece, J., & Henderson, S. (2014). Sharing data while protecting privacy in citizen science. *Interactions*, 21(1), 70–73.

Delgado-Aguilar, M. J., Konold, W., & Schmitt, C. B. (2017). Community mapping of ecosystem services in tropical rainforest of Ecuador. *Ecological Indicators*, 73, 460–71.

Diefenbach, D. R., McQuaide, J. T., & Mattice, J. A. (2002). Using PDAs to collect geo-referenced data. *Bulletin of the Ecological Society of America*, 83(4), 256–9.

Graham, E. A., Henderson, S., & Schloss, A. (2011). Using mobile phones to engage citizen scientists in research. *Eos, Transactions of the American Geophysical Union*, 92(38), 313–5.

Johnson, C., Piel, A. K., Forman, D., Stewart, F. A., & King, A. J. (2015). The ecological determinants of baboon troop movements at local and continental scales. *Movement Ecology*, 3(1), 14.

Kim, S., Mankoff, J., & Paulos, E. (2013). Sensr: evaluating a flexible framework for authoring mobile data-collection tools for citizen science. In Proceedings of the Conference on Computer Supported Cooperative Work (pp. 1453–1462).

King, J. D., Buolamwini, J., Cromwell, E. A., et al. (2013). A novel electronic data collection system for large-scale surveys of neglected tropical diseases. *PLoS One*, 8 (9), e74570.

Le Bel, S., Chavernac, D., Mapuvire, G., & Cornu, G. (2014). FrontlineSMS as an early warning network for human–wildlife mitigation: lessons learned from tests conducted in Mozambique and Zimbabwe. *Electronic Journal of Information Systems in Developing Countries*, 60(6), 1–13.

Le Bel, S., Chavernac, D., & Stansfield, F. (2016). Promoting A mobile data collection system to improve HWC incident recording: a simple and handy solution for controlling problem animals in southern Africa. In Angelici, F. M. (ed.). *Problematic Wildlife*. Springer, Cham, Switzerland (pp. 395–411).

Leisher, C. (2014). A comparison of tablet-based and paper-based survey data collection in conservation projects. *Social Sciences*, 3(2), 264–271.

Liebenberg, L., Steventon, J., Brahman, N., et al. (2017). Smartphone Icon User Interface design for non-literate trackers and its implications for an inclusive citizen science. *Biological Conservation*, 208, 155–162.

Marshall, H. H., Griffiths, D. J., Mwanguhya, F., et al. (2018). Data collection and storage in long-term ecological and evolutionary studies: the Mongoose 2000 system. *PLoS One*, 13(1), e0190740.

Marvin, D. C., Koh, L. P., Lynam, A. J., et al. (2016). Integrating technologies for scalable ecology and conservation. *Global Ecology and Conservation*, 7, 262–75.

McCall, M. K., Chutz, N., & Skutsch, M. (2016). Moving from measuring, reporting, verification (MRV) of forest carbon to community mapping, measuring, monitoring (MMM): perspectives from Mexico. *PLoS One*, 11(6), e0146038.

McDonald, M., & Johnson, S. (2014). 'There's an app for that': a new program for the collection of behavioural field data. *Animal Behaviour*, 95, 81–7.

McLester, E., Brown, M., Stewart, F. A., & Piel, A. K. (2019). Food abundance and weather influence habitat-specific ranging patterns in forest- and savanna mosaic-dwelling red-tailed monkeys (*Cercopithecus ascanius*). *American Journal of Physical Anthropology*, 170(2), 217–31.

Meynecke, J.-O. (2014). Whale Trails—a smart phone application for whale tracking. In Ames, D. & Quinn, N. (eds.). *7th International Congress on Environmental Modelling and Software.* International Environmental Modelling and Software Society, San Diego, USA (pp. 1–7).

Newman, G., Wiggins, A., Crall, A., Graham, E., Newman, S., & Crowston, K. (2012). The future of citizen science: emerging technologies and shifting paradigms. *Frontiers in Ecology and the Environment*, 10(6), 298–304.

Newton-Fisher, N. E. (2012). Animal Behaviour Pro. University of Kent Academic Repository; https://kar.kent.ac.uk/44969/

Oviedo, A. F. P., & Bursztyn, M. (2017). Community-based monitoring of small-scale fisheries with digital devices in Brazilian Amazon. *Fisheries Management and Ecology*, 24(4), 320–329.

Pacha, M. J. (2015). *Community-Based Monitoring, Reporting,* and *Verification Know-How: Sharing Knowledge* from *Practice.* Report by WWF Forest and Climate Programme.

Piel, A. K., Bonnin, N., Ramirez-Amaya, S., Wondra, E., & Stewart, F. A. (2018). Chimpanzees and their mammalian sympatriates in the Issa Valley, Tanzania. *African Journal of Ecology*, 57(1), 31–40.

Pimm, S. L., Alibhai, S., Bergl, R., et al. (2015). Emerging technologies to conserve biodiversity. *Trends in Ecology & Evolution*, 30(11), 685–696.

Pratihast, A. K., Herold, M., Avitabile, V., et al. (2012). Mobile devices for community-based REDD+ monitoring: a case study for Central Vietnam. *Sensors*, 13(1), 21–38.

Rafiq, K., Bryce, C. M., Rich, L. N., et al. (2019). Tourist photographs as a scalable framework for wildlife monitoring in protected areas. *Current Biology*, 29(14), R681–R682.

Rainforest Foundation UK (2019). Real-*Time Forest Monitoring: Empowering Communities, Preventing Illegalities, Protecting Forests.* Overview produced by The Rainforest Foundation UK. Available at: https://www.rainforestfoundationuk.org/media.ashx/real-time-monitoring-2019.pdf

Ramirez-Gomez, S. O. I., Brown, G., Verweij, P. A., & Boot, R. (2016). Participatory mapping to identify indigenous community use zones: implications for conservation planning in southern Suriname. *Journal for Nature Conservation*, 29, 69–78.

Ross, M. R., Niemann, T., Wark, J. D., et al. (2016). *ZooMonitor.* Available at: http://www.zoomonitor.org

Schwartz, M. W., Cook, C. N., Pressey, R. L., et al. (2018). Decision support frameworks and tools for conservation. *Conservation Letters*, 11(2), e12385.

Snaddon, J., Petrokofsky, G., Jepson, P., & Willis, K. J. (2013). Biodiversity technologies: tools as change agents. *Biology Letters*, 9(1), 20121029.

Stevens, M., Vitos, M., Altenbuchner, J., Conquest, G., Lewis, J., & Haklay, M. (2013). Introducing Sapelli: a mobile data collection platform for non-literate users. In Proceedings of the 4th Annual Symposium on Computing for Development.

Sutherland, W. J. (1998). The importance of behavioural studies in conservation biology. *Animal Behaviour*, 56, 801–809.

Sutherland, W. J., Pullin, A. S., Dolman, P. M., & Knight, T. M. (2004). The need for evidence-based conservation. *Trends in Ecology & Evolution*, 19(6), 305–308.

Teacher, A. G., Griffiths, D. J., Hodgson, D. J., & Inger, R. (2013). Smartphones in ecology and evolution: a guide for the app-rehensive. *Ecology and Evolution*, 3(16), 5268–5278.

Verma, A., Van Der Wal, R., & Fischer, A. (2016). Imagining wildlife: new technologies and animal censuses, maps and museums. *Geoforum*, 75, 75–86.

Vitos, M., Lewis, J., Stevens, M., & Haklay, M. (2013a). *Making local knowledge matter: supporting non-literate people to monitor poaching in Congo.* Paper presented at the Proceedings of the 3rd ACM Symposium on Computing for Development.

Vitos, M., Stevens, M., Lewis, J., & Haklay, M. (2013b). Community mapping by non-literate citizen scientists in the rainforest. *SoC Bulletin*, 46, 3–11.

Waddle, J. H., Rice, K. G., & Percival, H. F. (2003). Using personal digital assistants to collect wildlife field data. *Wildlife Society Bulletin*, 31(1), 306–308.

Weisse, M. J., Petersen, R., Sargent, S., & Gibbes, S. (2017). *Places to Watch: Identifying High-Priority Forest Disturbance from Near-Real Time Satellite Data.* Technical note by World Resources Institute, Washington, DC, USA.

Zhang, S., Wu, Q., Van Velthoven, M. H., et al. (2012). Smartphone versus pen-and-paper data collection of infant feeding practices in rural China. *Journal of Medical Internet Research*, 14(5), e119.

Zimmerman, P. H., Bolhuis, J. E., Willemsen, A., Meyer, E. S., & Noldus, L. P. (2009). The Observer XT: a tool for the integration and synchronization of multimodal signals. *Behavior Research Methods*, 41(3), 731–735.

CHAPTER 10

Application of SMART software for conservation area management

Drew T. Cronin, Anthony Dancer, Barney Long,
Antony J. Lynam, Jeff Muntifering, Jonathan Palmer,
and Richard A. Bergl

10.1 Introduction

Biodiversity is in steep decline worldwide, with analyses indicating an average decline of 60% in the population sizes of vertebrates since 1970 (WWF, 2018) and around 1 million species already facing extinction largely as a result of human actions (Díaz et al., 2019). Larger vertebrates are especially vulnerable to these anthropogenic threats (Ripple et al., 2019), with declines primarily caused by overexploitation and habitat loss associated with an increasing human population and per capita resource use (Hoffmann et al., 2010; Maxwell et al., 2016; Ceballos et al., 2020). Exploitation is driven by numerous factors, but is often the result of illegal activities, such as hunting, logging, and wildlife trade. The global demand for wildlife products, fuelled by poverty, economic growth, and political instability have created an extensive global trade in wildlife rivaling that of illegal arms (Lawson & Vines, 2014).

Stopping the trade in wildlife will require many different solutions, including reducing demand for wildlife products, disrupting wildlife supply routes, and limiting the off take of wildlife. However, unlike the trade in drugs or arms, where most of the harm occurs at the user end, once the fauna or flora have been hunted/extracted, the harm to the species has already been done, even if interdiction disrupts further trade. Therefore, a critical component of efforts to curb exploitation

is prevention at the source, i.e. reducing illegal harvest of wild populations (Nurse, 2015; Dinerstein et al., 2019).

Protected areas, designed to safeguard threatened species and their habitats, are the foundation of biodiversity conservation (Watson et al., 2014). When well-managed, protected areas have shown to be essential and effective tools for conserving biodiversity which provide a broad range of ecosystem services (Laurance et al., 2012; Watson et al., 2014; Gill et al., 2017; Geldmann et al., 2018; Sala & Giakoumi, 2018). Metrics, such as species richness (terrestrial—10.6%; Gray et al., 2016; marine—21%; Sala & Rechberger, 2018), species abundance (terrestrial, 14.5%; Gray), and overall biomass (whole fish assemblages—670%; Sala & Giakoumi, 2018), all tend to be higher within protected areas as compared to adjacent unprotected areas. Research also suggests that lands managed by indigenous people—which often include 'rangers' appointed by and accountable to their local community (e.g., see Muntifering, 2019)—may support even greater levels of biodiversity than formally protected areas (Schuster et al., 2019). The recognition of conservation areas' effectiveness is reflected in the Aichi Biodiversity Targets (CBD, 2010), in which parties committed to an expansion of protected areas globally to cover 17% of terrestrial areas and 10% of marine areas by 2020. Current assessments of progress suggest that the coverage target

Drew T. Cronin et al., *Application of SMART software for conservation area managements*.
In: *Conservation Technology*. Edited by Serge A. Wich and Alex K. Piel, Oxford University Press.
© Oxford University Press 2021. DOI: 10.1093/oso/9780198850243.003.0010

of 17% will be met (with marine coverage (16.8%) increasing more rapidly than terrestrial coverage (14.9%); UNEP-WCNC et al., 2018), and that protected area representativeness and management are generally improving (Tittensor et al., 2014).

However, despite major expansions over recent decades, the existing global network of protected areas is inadequate to meet the global thresholds set to avoid significant climate change, to secure essential ecosystem services, or to prevent continuing biodiversity declines (Watson et al., 2014; UNEP-WCNC et al., 2018; Dinerstein et al., 2019; Geldmann et al., 2019). Many protected areas still have declining wildlife populations (Geldmann et al., 2013) and face significant human pressures (Laurance et al., 2012; Jones et al., 2018). In the IUCN Green List, effective management is considered foundational to achieving conservation outcomes (along with sound design and planning and good governance; IUCN & WCPA, 2017), but fewer than 25% of protected areas globally are effectively managed (Leverington et al., 2010) or have adequate staffing and financial resources (Coad et al., 2019). Rangers play a critical role in managing threats, effectively enforcing laws, and monitoring access and use (three of the seven criteria for effective management under the Green List (IUCN & WCPA, 2017), but rangers worldwide are undertrained, underequipped, and often undervalued (Belecky et al., 2019). The ability of those managing conservation areas to respond effectively and efficiently to an expanding array of threats (e.g., poaching), requires information on where the threats are occurring and ensuring there is adequate capacity in place to combat them. In reality, management and enforcement efforts are often poorly coordinated and operate in an ad-hoc manner (Nurse, 2015). Even protected areas with adequate staff numbers may not meet their conservation targets if they are poorly managed or lacking capacity or other resources (Venter et al., 2014; Barichievy et al., 2017; Coad et al., 2019). Poor conservation outcomes and/or ineffective management are not typically the result of a lack of will to make protected areas work, but rather a consequence of the combined effects of inadequate staff and capacity, limited resources, insufficient data upon which to make decisions, and a lack of decision-support systems that can improve overall effectiveness.

In recent years, a wide range of technologies has been proposed as solutions to these challenges. As conservation area management capacity and effectiveness are outpaced by increasing human pressures, 'technological solutionism' (Morozov, 2013) has become pervasive—an abundance of technology-based solutions seek to quickly and easily solve complex problems. However, these new technologies are often developed outside of conservation and local contexts, have limited ability to deploy at scale, or are provided with limited support. Despite these limitations, the allure of an exciting and dynamic technology solution can be hard to resist. Such approaches can overlook the site-specific impracticality of solutions, and often focus on the implementation of the technology, rather than the desired conservation outcomes. It is critically important that technology-based conservation tools be proven, field-tested solutions that can deliver cost-effective enforcement and management impact for the benefit of biodiversity.

The primary form of field-based monitoring and enforcement in protected areas around the world is patrols. These 'boots on the ground' can have various mandates, including research and monitoring, community engagement, and wildlife law enforcement (Marvin et al., 2016). Ranger patrols are the primary deterrent against illegal activities at many sites (Critchlow et al., 2017) and are fundamental for the success of most protected areas. However, law enforcement is often the largest expenditure in many protected areas (Critchlow et al., 2017) and, if poorly managed, can be ineffective at preventing biodiversity declines and may have consequences for thousands of species (Coad et al. 2019). As a result, the effectiveness and efficiency of ranger patrols (e.g., transparency of patrol effort; standardized data collection and analysis; rapid reporting and feedback; adaptive, data-driven patrol organization and deployment; etc.) must be maximized in order to meet conservation targets and ensure the protection of threatened species in protected areas.

The Spatial Monitoring and Report Tool (SMART)[1] platform, developed by the SMART Partnership,[2] was designed to address the challenges inherent to monitoring and evaluating the effectiveness of ranger patrols. The goal of SMART is to improve the effectiveness of conservation area management by providing the technology, services, and skills necessary to make better use of available resources. Since its initial release (2012), SMART has grown to become the most widely used technology solution for conservation area management globally. SMART is currently used in more than 1,000 sites (ranging from community conservancies smaller than 100 km^2 to national parks larger than 20,000 km^2) across more than 70 countries, providing a practical and freely available user-driven technology that meets the needs of frontline conservationists. This chapter demonstrates both how SMART has been used to improve conservation in protected areas and across national protected area systems through an adaptive management approach, and how complementary systems and emerging technologies can be integrated into a single unified platform for conservation area management. We also showcase how these integrated tools have empowered staff on the frontlines of conservation through a series of case studies that demonstrate their practical application.

10.2 New technology

10.2.1 Data collection: hardware

SMART is an integrated, multiplatform conservation area management system that uses data collected by rangers on patrol, along with data from a range of other sensors and sources, to inform protected area management. Rangers collect data in the field using a variety of methods (e.g., mobile devices, notebooks, Global Navigation Satellite System [GNSS] units). All data collected by ranger

must: (1) be georeferenced; (2) include information about observations or activities; and, (3) be brought or relayed back to a central location for analysis. While ranger data were, until recently, simply recorded in notebooks or on paper data sheets, technology is increasingly making these data easier to record, more accurate, and more readily available. UHF/VHF digital radios (e.g., Hytera[3]) allow for both voice communication and GNSS-based geolocation, field computers equipped with data collection software (e.g., SMART Mobile) provide more consistent and less error-prone data recording, and long-range, low-power communication networks (e.g., LoRaWAN,[4] Sigfox[5]) provide real-time tracking and data transmission.

One of the most consequential recent developments for ranger-based data collection has been the rapid proliferation of inexpensive, ruggedized smartphones. These devices have made mobile data collection in the field much easier by combining data collection, data storage, and multiple sensors in an easy to use, reliable package, while also providing access to technologies in resource-limited environments (Berger-Tal & Lahoz-Monfort, 2018). Given the broad range of technologies and devices currently available and the pace at which they are released, an exhaustive discussion of specific devices or functions is impractical. Rather, we discuss the general attributes of devices that should be considered when selecting devices for field use (Table 10.1).

The first factor to consider is the device's 'ruggedness' or durability. If devices will be treated roughly or the environment is challenging, a ruggedized device is a good investment. Most rugged devices are classified according to one or more international standards, the most common being the International Protection (IP) Rating (IEC, 2013). The IP rating is a measure of the device's degree of protection against intrusion of solid objects and water, and consists of two numerical ratings where the first digit indicates protection against solids (protection range of 1–6) and the second digit protection against

[1] https://smartconservationtools.org/
[2] Current SMART Partnership members include: Frankfurt Zoological Society, Re:wild (formerly Global Wildlife Conservation), North Carolina Zoo, Panthera, Peace Parks Foundation, Wildlife Conservation Society, Wildlife Protection Solutions, World Wildlife Fund, and Zoological Society of London.

[3] https://www.hytera.us/two-way-radios
[4] https://lora-alliance.org/about-lorawan
[5] www.sigfoxfoundation.org

Table 10.1 General considerations for selecting a mobile device for field-based data collection

Factor	Critical considerations
Ruggedness	Estimated abuse and everyday wear and tear on the device; Wetness/Dryness of the environment where device will be used
GNSS Sensitivity	Feasibility of GNSS acquisition (e.g., tree canopy cover, cloud cover, terrain, etc.)
Form factor	Overall device size and weight; size of screen; touchscreen vs. external buttons
Battery life	Duration of use between charges; ability to change battery in case of failure
Operating System	Variability/sustainability/functionality/flexibility and customizability between primary options (Android, Windows Mobile, iOS)

water (protection range of 1–8). Thus, an IP68 rating is the highest rating, signifying that no dust will penetrate the device (solids rating of 6) and that the device can be submerged continuously in water under manufacturer-specified conditions (liquids rating of 8). Higher IP ratings can help ensure the longevity of devices in environments that are typically challenging for electronics. The life of devices can also be extended by using separate cases and protective covers, if rugged devices are not available or are cost-prohibitive. There are numerous different brands and models of cases to choose from (e.g., Otterbox,[6] Lifeproof[7]), many of which are designed to fit specific smartphone models. Much like the devices themselves, some cases are better than others or are designed for specific purposes (e.g., wet environments), and users should carefully consider their needs and budget to ensure they select the best case for their devices, if necessary.

Related to the device's ruggedness, users should also consider its form factor, for example, device size, screen size, weight, etc. Larger screens make text entry easier and help prevent incorrect data entry, but increase power requirements and bulkiness, as well as the possibility of cracking screens. While there are many ruggedized tablets on the market, most users tend to favour devices similar in size to smartphones, as tablets are too bulky for most rangers to carry in the field. These trade-offs should be considered with inputs from the frontline staff who will use them in the field before investing in the purchase of the equipment.

The next factor to consider is the device's GNSS capability, specifically its ability to acquire location data in areas of limited satellite signal strength. Some older smartphone models do not have a built-in GNSS system—these models are not recommended for use in the field. GNSS receivers work using a line-of-sight method, and any physical objects or obstructions, such as tree canopy, cloud cover, or steep terrain, can block weak satellite signals. Unlike dedicated GNSS units, most smartphones also do not have a large antenna to amplify their signal. However, devices/hardware are rapidly improving, and newer phones can now communicate with more than one of the available GNSS (e.g., GPS + GLONASS). Users need to consider the challenges of the physical environment on both the device's GNSS sensitivity and the GNSS coverage where the unit will be used.

Another important consideration for field users is battery life, both in terms of how long can devices be used between charging and their overall life. Battery life is highly dependent on local conditions (e.g., heat can expand batteries and reduce battery life), device configurations (e.g., screen brightness, WiFi and/or Bluetooth scanning), and how the devices are used, i.e., the number of observations recorded each day, the frequency of tracklog point collection, and the level of detail recorded in each observation (lengthier entries require the screen to be on longer and using the screen is one of the major sources of power consumption). Users should note the number of milliamp hours (mAh)—a measure of electric power over time—a device's battery provides. The device's milliamp hours rating is a measure of the battery's overall energy capacity; that is, the greater

[6] www.otterbox.com
[7] www.lifeproof.com

the value, the longer the battery life. Also, especially at sites where power fluctuates, battery failure is one of the most common causes of device problems. Many mobile devices nowadays have built-in batteries that cannot be easily changed, meaning that when the battery dies, they must be returned to the manufacturer for service or replaced. If this a concern at a user's site, they might consider opting for a device with a removable battery that can be sourced locally, or investing in power conditioning hardware to mitigate charging current fluctuations.

Finally, we recommend that users consider the operating system of the device and whether it meets the needs of their site(s) and staff. The choice of operating system often comes down to a confluence of the factors listed earlier, cost, and availability; for example, opting to use an Android phone due to lower cost and its availability in local markets. With many mobile devices, the operating system is often tied to the device's specific manufacturer, with the two most common platforms being Google's Android, which is found on devices from many manufacturers, and Apple's iOS, found only on Apple devices. Some more specialized or legacy devices (e.g., Trimble Nomad) may also run Windows Mobile. The functionality provided by these operating systems for most ranger-based monitoring applications are comparable, but it is worth noting some specific use cases that may influence users' choice of device. Historically, CyberTracker[8]—the mobile data collection platform for SMART—has been compatible with Windows Mobile and later Android, but not with iOS. The SMART Partnership has now released SMART Mobile—a significantly enhanced mobile data collection app powered by CyberTracker—that supports Android and iOS, but not Windows Mobile, as the latter is now being deprecated (see next). For both Android and iOS, apps are installed from online app stores (Android— Google Play; Apple—App Store), but Android also allows for manual installation of 'apk' files via devices' standard file manager app. This provides for greater flexibility in the field by allowing offline transfer of files.

---

[8] https://www.cybertracker.org

10.2.2 Data collection and analysis: software

There are many data collection and analysis apps/software designed for specific applications, sites, and agencies—far more than can be considered in the applied scope of this chapter (but see Chapter 9 for further information). As this chapter focuses on the application of SMART for conservation area management, we will focus on both the SMART software platform and some optional third-party software applications with developed software integrations and/or long-term technology collaborations with the SMART Partnership. These third-party integrations have been developed carefully over time by the SMART Partnership, where the integrations met the following conditions: (1) the integration was pragmatic and met the needs of frontline practitioners; (2) the functionalities were additive to existing SMART capabilities; (3) where the third-party organization is focused on conservation outcomes; and (4) there is a long-term plan for sustainability (i.e., bug fixes, updates, maintenance, compatibility updates). We believe these conditions are critically important because they help users avoid 'tech for the sake of tech' and ensure that resources spent on hardware procurement and staff training are not lost if there is a loss of compatibility or apps are not updated.

10.2.2.1 SMART conservation area management platform

SMART encompasses three platforms: SMART desktop (Figure 10.1), the global-leading technology solution for conservation area management; SMART Connect, a cloud-based solution enabling centralized management and reporting and real-time alerts; and SMART Mobile, a mobile app for field-based data capture. These tools are aimed at gathering, analysing, and disseminating information critical to the management of protected areas. Those data are processed through user-defined iterations of an adaptive management approach that uses SMART (Figure 10.2) to inform management, decision-making, and patrol planning (Stokes, 2010; Marvin et al., 2016).

Though varying site by site, most SMART workflows follow the following approach. SMART

Figure 10.1 Rangers from the Bumi Hills Anti-Poaching Unit in Zimbabwe review patrol data on the configured conservation area map in the SMART desktop software application © James Slade/Re:wild.

desktop software is installed on a permanent computer at a site (e.g., park headquarters) and a SMART database is configured for the conservation area (e.g., national park or community conservancy). The configured database includes many important aspects about a site, including GIS layers (e.g., conservation area boundary, patrol sectors, key infrastructure, etc.), key patrol parameters (e.g., staff, patrol stations, transport types), and, most importantly, a site-specific data model that defines which data will be collected in the field and which will form the basis for all analysis and reporting. Patrol teams collect data on wildlife, threats, and ranger efforts with a mobile device configured with the SMART Mobile application. Those data are then uploaded to SMART via one of two pathways: (1) in real-time from the field via cellular/WiFi networks and SMART Connect (see 10.3.4); or, (2) after the completion of patrols into SMART desktop (e.g., back at headquarters). Once uploaded, they can be analysed, visualized, mapped, and reported on. Analyses and reports help managers understand what is happening at their site; for example, trends, where threats are occurring, or where wildlife are congregating. With those data, managers can better plan responses or future patrols for maximum impact. Patrol- teams then implement the adapted patrol plan and the

cycle begins again, with the aim of driving efficiency and increasing effectiveness. This final step is critical because, as noted earlier, technology can only ever be effective if used well, and vital to an effective application of SMART is adapting to changing conditions and ensuring rangers are motivated (e.g., through recognition of their efforts in the field), well-resourced, and supported.

SMART differs from other solutions in its integration of field data collection with robust analysis and reporting capabilities and in its facilitation of each step of the adaptive management cycle (see Figure 10.2). Users can create numerous types of queries (e.g., where were shotgun shells found or what was patrol effort in 2018; Figure 10.3) and summaries (e.g., individual ranger patrol effort; Figure 10.3) about all data at their site and create custom reports. Queries and reports can be configured to provide standardized analysis of trends (e.g., trends in snares per effort over time) or replicated at more than one site, allowing for 'cross-conservation area analyses'—analysis and reporting across multiple sites and entire protected area networks. Importantly, once analyses and reports are configured, they can then be largely automated, lowering the technical bar for users to interpret their data, as well as the time it takes for actionable information to be put in the hands of decision-makers.

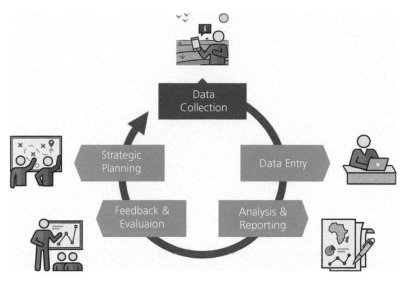

Figure 10.2 An adaptive management approach to ranger-based monitoring and conservation area management leveraging SMART. Point-based observations and tracklogs generated by ranger patrols are entered into a database and analysed, after which data summaries and reports can be produced, which are then in turn fed back to ranger teams as patrol plans. SMART was designed to facilitate each step in this cycle: (1) Data Collection—Patrol teams collect and record data on where they go and what they see while on patrol, such as threats (e.g., poaching signs), patrol results (e.g., arrests, confiscations of weapons), and wildlife observations. These data can be collected on smartphones equipped with SMART Mobile (or CyberTracker), standardizing and simplifying data collection, and adding increased efficiency and transparency. (2) Data Entry—patrols report their patrol activities, and patrol data and routes are checked and then stored in a SMART database. If using SMART Mobile, this process is automated, eliminating data entry errors. (3) Analysis & Reporting—data are processed semi-automatically into user-defined, highly visual tables, charts, and maps showing patrol effort, coverage, and results, forming the basis for patrol and trend analysis and rapid evaluation. (4) Feedback & Evaluation—regular meetings with rangers are held to discuss patrol effort and results to ensure all stakeholders are kept informed and to demonstrate the value of ranger efforts. (5) Strategic Planning—managers, rangers, and other stakeholders plan adaptive patrol strategies based on analysis of previous results, changing conditions, and/or temporal patterns, and set new patrol targets. Adapted from Marvin et al. (2016).

As mentioned earlier, users can also 'configure' their site-specific iteration of the adaptive management cycle by using different aspects ('modules') of the suite of SMART tools (Table 10.2). Most of these are part of the 'core' of SMART, including managing ranger-based monitoring data, generating maps, and analytic and reporting functions, but SMART has several other modules, such as 'Ecological Records' for biological surveys or 'Intelligence' for non-structured data collected outside of patrols, which can be turned on as needed at the site. In this way, the default SMART installation presents users with the foundation for a ranger-based conservation area monitoring system, enabling them to add further functions and complexity as needed and as capabilities evolve, thus limiting confusion.

10.2.2.2 SMART Mobile

SMART Mobile—the mobile data collection component of SMART—is powered by CyberTracker, a third-party platform created to simplify the collection of field data, that has been successfully deployed independently in hundreds of sites around the world. SMART Mobile (and CyberTracker) uses a GNSS-enabled mobile device to collect both observations (text or icon-based data entry and digital images) and GNSS data in a single unit (Figure 10.4). Observations and GNSS data can be transferred into SMART in real-time from the field to SMART Connect over local cellular or WiFi networks (see 10.3.4), or after completing patrols upon returning to the office in a semi-automated process. Released in mid-2020, SMART

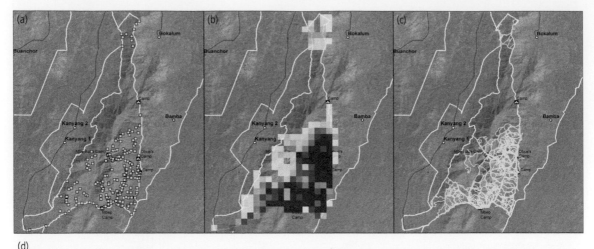

(d)

Name	Number of Patrols	Number of Patrol Days	Number of Patrol Hours	Total Patrol Distance(km)
Ranger 1	27	171	1145.7	1156.3
Ranger 2	27	174	1205	1168.6
Ranger 3	28	174	1150.3	1183.8
Ranger 4	28	170	1131.2	1146.1
Ranger 5	24	151	944.8	1045.3
Ranger 6	7	17	103.4	107.5
Ranger 7	27	174	1029.8	1142.2
Ranger 8	28	174	1214.5	1205.3
Ranger 9	2	10	50.3	54.3
Ranger 10	24	143	883.3	1023.2

Figure 10.3 Exported examples of general SMART query types displaying data from 1 January–31 December 2018 from Mbe Mountains Community Wildlife Sanctuary in Nigeria: (a) observation queries, showing used shotgun cartridges observed; (b) grid queries, showing patrol effort per 500 m^2; (c) patrol queries, showing tracks of all patrols, and (d) summary queries, showing patrol effort for individual rangers (Note: the table shown is a screenshot from a SMART report from which the names of rangers have been removed for confidentiality). The mapped visualizations of the three queries (excluding the summary query) can also be viewed in tabular format (i.e. a table of all observations and data relevant to the query). Data courtesy of Wildlife Conservation Society, Nigeria.

Mobile retains the core functions of Cybertracker; however, it is more tightly integrated with SMART, eliminating the need for CyberTracker software to be installed separately on users' devices or computers. The new software provides significant enhancements to the existing functionality of CyberTracker, including support for Android and iOS, touch/swipe controls, advanced mapping and navigation, support for all Unicode languages, and the ability to send data directly to SMART Connect. For the immediate future, SMART will support the use of both CyberTracker (now called 'CyberTracker Classic' in SMART) and SMART Mobile for field data collection, with the aim to migrate all users to the new system in the next few years.

10.2.2.3 SMART Connect

Since its public release in 2017, SMART Connect—the online extension to SMART desktop—has been gradually adopted across hundreds of sites around the world. It has now been implemented on multiple cloud infrastructures (e.g., Microsoft Azure[9], Google Cloud Platform[10], and Amazon Web Services[11]), numerous local servers, and even inexpensive Raspberry Pi[12] computers. The most widely reported benefit it has provided is centralized management of SMART deployments, which has

[9] https://azure.microsoft.com/en-us/
[10] https://cloud.google.com/
[12] https://www.raspberrypi.org/

Table 10.2 The primary functionalities and modules of the SMART conservation area management platform

SMART Modules		Functionality Summary
Core SMART Functionality	Database	Consolidate data streams into a centralized hub for holistic PA management
	Patrol	Monitor and optimize patrolling effort
	Maps	Generate maps
	Analyse	Conduct advanced analysis of SMART data, effort, and spatial coverage
	Report	Demonstrate impact via rapid, standardized, and semi-automated reporting
	Plan	Enhance decision-making and optimize deployment of conservation resources
Optional SMART Functionality	Survey	Rapid capture, collation, and analysis of ecological monitoring data
	Entities	Monitor and track entities (e.g., individually identifiable or collared animals) in time and space
	Intelligence	Advanced relationship and networks analyses to target and reduce wildlife crime
	Alerts	Manage and respond to real-time alerts
	Integrate	Complement SMART with additional data sources and systems
	Sensors	Manage, visualize, and analyse data from sensors together with patrol data
	Events	Internet of Things driven workflows
	Area Network	Centrally manage and deploy SMART at scale

Figure 10.4 A ranger from WCS Nigeria and the Conservation Association of the Mbe Mountains collects data using SMART Mobile while on patrol in the Mbe Mountains Community Wildlife Sanctuary in Cross River State, Nigeria. Using SMART on ruggedized devices allows standardized data to be easily captured by rangers with limited technology capacity and/or literacy in the field and streamlines data entry and management once patrols are completed.

allowed for information sharing of data, maps, and reports across entire protected area networks and enabled access to SMART analyses and reporting for non-SMART users. Connect has increased the security of SMART databases by not only providing users with the ability to back up data automatically and access data from anywhere (i.e. avoiding the risks associated with holding the database on a single computer), but also limiting the IT and security capacity needed by users onsite by providing enterprise-grade security and encryption of data. SMART Connect also facilitates data capture in as close to real-time as a site's infrastructure allows, making it possible for rangers to manage and respond to real-time threats. For example, SMART Connect now allows SMART Mobile users to attach alerts to specific branches or attributes in their configurable data model. When an observation with an associated alert is logged in the field, the application will send a notification via the mobile device's cellular or WiFi network (if available) to Connect and register the alert on the server. The Connect Alerts feature also works with data from other sources (e.g., Global Forest Watch; see Section 10.2.2.4 for explanation) and other commonly used field sensors, such as remote camera traps, which can transmit data to Connect via satellite, mobile, or long-range, low-power communication networks like LoRaWAN or Sigfox (see Section 10.2.2.6). Connect also offers exciting opportunities to use 'edge' devices, like TrailGuard AI[13] camera traps, which leverage machine learning algorithms on the device itself to filter false-positive photos (e.g., moving vegetation) and transmit only images of poachers to 'headquarters' over the existing communications network. Such use cases are most useful where data can be processed with simple artificial intelligence algorithms, significantly reducing effort and costs for data transfer (e.g., image, video, acoustics), and operations teams can be dispatched to respond to the alert.

10.2.2.4 Global Forest Watch

Global Forest Watch[14] (GFW) is a web platform that provides regularly updated satellite-based deforestation data for monitoring and action on the ground. The website provides hundreds of data sets, but the most relevant for SMART users are the GLAD deforestation alerts and VIIRS fire alerts (Hansen et al., 2016) from the University of Maryland. GLAD deforestation alerts are provided weekly, at a 30×30 m resolution, providing access to near real-time information about where and how forests are changing worldwide. These data are the most frequently updated and highest resolution freely available data of its type globally. VIIRS fire alerts show the number of fires, both human-ignited and naturally occurring, within a selected area in the last seven days. The fire data comes from NASA FIRMS (Fire Information for Resource Management

System),[15] which use data from the VIIRS satellite to map fires daily at 375-meter resolution. Through a partnership with Global Forest Watch, SMART users can incorporate these data into their SMART database as a map layer or as an alert sent to the Connect server (e.g., users can configure a VIIRS webhook trigger transmission of a Connect alert notifying of any new fires within the PA) in order to inform patrol planning and intelligence-led conservation area management. Users can also configure SMART Connect Alerts.

10.2.2.5 EarthRanger

Vulcan's EarthRanger[16] (formerly 'Domain Awareness System' or 'DAS') is an online software that collects, collates, and displays historical and real-time data available from a protected area. EarthRanger is an operational platform that can inform and help coordinate ranger responses by enabling users to visualize their protected area assets and make use of tracking capabilities in real-time. Alerts can also be triggered when critical threats are identified to facilitate an immediate ranger response. Users can leverage SMART Connect's integration with EarthRanger to share SMART data with EarthRanger, enabling a data-rich, real-time, and easy-to-understand view of the status of any activities in the areas that they seek to protect. At the time of writing, plans are being finalized to include EarthRanger as part of the SMART ecosystem, adding advanced real-time operations to the existing set of SMART tools and delivering a system that is fully adaptable to any protected area—from a single ranger in an occasionally connected environment to a 24/7-Internet of Things (IoT) powered operations room. Incorporating EarthRanger also provides integrations with more than 50 field sensors (i.e. camera traps, animal collars, radios, etc.) to existing SMART users, enhancing SMART Alerts, while also providing the suite of SMART applications to current EarthRanger users.

---

[13] https://www.resolve.ngo/trailguard.htm
[14] https://www.globalforestwatch.org/
[15] https://earthdata.nasa.gov/earth-observation-data/near-real-time/firms
[16] https://earthranger.com/

10.2.2.6 Sigfox

Sigfox Foundation, an arm of the French technology firm Sigfox that primarily builds Low Power Wide Area Networks (LPWAN) for IoT fleet management, has developed and piloted a field deployment of their infrastructure, data collection, and transmission system to be used alongside SMART in Bhutan. Updates from the deployed Sigfox devices are directly integrated with SMART Connect. The system, which uses inexpensive and long endurance tracking devices in communication with Sigfox antennas and base stations, can be deployed relatively inexpensively and can help sites overcome vast distances, limited resources, and a lack of connectivity in their sites. With an LPWAN in place, more IoT devices can be deployed in the future with a multitude of utilities, including but not limited to monitoring (e.g., park boundary, wildlife, patrol assets, tourists, weather), remote management, and basic communications).

10.2.2.7 R software environment

As mentioned earlier, SMART has a robust analytical and reporting framework built into its core platform. However, to facilitate more advanced analyses, SMART developed an integration with the R programming language and software environment for statistics and graphics[17] (R Core Team, 2015). The optional R Plugin allows users to run R scripts directly from within SMART. Results of SMART queries can be sent directly to an R script and the script results viewed within SMART. Although the plugin is installed in the same manner as all other SMART plugins, R software is a third-party software and must also be installed separately.

10.3 Wider applications: review of what has been done

SMART can be flexibly applied to myriad different use cases. As SMART adoption continues to grow, SMART users are implementing and adapting SMART to address many complex and diverse conservation needs at their sites. For example,

[17] https://www.r-project.org/

existing implementations have ranged from measuring the success of ranger patrols to the management of national forest monitoring programmes, or quantifying fisheries' offtake across marine protected area systems, to tracking energy consumption at field research facilities. This flexibility is a positive outcome of the fact that SMART is free, open-source, and fully customizable, although there are some trade-offs to a fully customizable approach. For example, sites can configure databases and data collection to their specific needs, but there may be issues collating data sets for larger-scale analysis. This can be addressed by scaling SMART implementation at the outset to accommodate broader analyses (e.g., landscape or protected area network), or at least planning appropriately at the outset to ensure the foundational elements are in place to facilitate scaling or collating, when possible. A challenging by-product of this flexibility and customizability is classifying a 'typical' SMART implementation. SMART also has a strong track record of adding new functionality and features year-on-year, so no list will remain static. For example, SMART is developing new features for monitoring human–wildlife conflict, managing infrastructure, and case-tracking for release in 2020. Next, we briefly describe the most common, generalized applications of SMART at the time of writing.

10.3.1 Law enforcement monitoring

SMART has evolved to become the world's leading tool for conservation law enforcement monitoring (LEM) and conservation area management, with implementation at more than 1,000 sites globally (Cronin, 2019). SMART LEM enables the collection, storage, communication, and evaluation of data on patrol efforts, patrol results, and threat levels. Implementation of SMART LEM has enhanced law enforcement effectiveness, improved morale of protection teams, and reduced threats to wildlife and other natural resources at numerous sites worldwide (e.g., Hötte et al., 2016; Critchlow et al., 2017). Although widely recognized for implementation at terrestrial sites, SMART can also be readily applied to marine conservation and marine protected area (MPA) management. Interest in SMART marine applications has been growing in recent years,[18] as marine patrols often use similar workflows to terrestrial patrols, and some users have begun to use SMART to monitor fisheries, such as local and commercial fish landing data in ports (SMART Partnership, 2018a). As of December 2019, more than 50 marine sites were implementing SMART globally (Cronin, 2019).

10.3.2 Ecological monitoring

SMART Ecological Records (ER) is an optional functionality that allows SMART users to capture, manage, map, and analyse data from systematic surveys of species, habitats, and disturbances in a framework that closely resembles the SMART patrol framework. Typically, practitioners have used Distance[19] software (Thomas et al., 2010) to design the survey framework, as this cannot be done in SMART, and then collected survey data in the field with CyberTracker and SMART. Field data are then managed and analysed in SMART, ensuring systematic data collection and management. The integration of SMART and R statistical software (see Section 10.2.2.7) allows for in-depth statistical analysis of formal monitoring surveys in R and makes SMART ER a powerful tool for site-based ecological monitoring. When effectively employed with other routine-protected area activities (e.g., LEM), SMART ER enables integration of ecological and law enforcement monitoring data, providing protected area managers with an understanding of conservation effectiveness over time, and information to improve decision-making for the management of fauna and flora. For example, in a given protected area there are monthly surveys using SMART ER that routinely survey the area for a critically endangered primate that occurs in low densities. Regular monitoring allows for a statistically sound understanding of the primate's population status in the protected area. However, there are more regular patrols which use the default SMART law enforcement module, and whose mandate is anti-poaching. Although the law enforcement patrols do not adhere to systematic monitoring protocols, they occasionally encounter the primate as well—outside of the survey area. By combining ecological monitoring and law enforcement frameworks in the same SMART database, the managers can have a holistic understanding of the primate's population status (via ecological monitoring) and range (via the inclusion of opportunistic encounters from anti-poaching patrols). These capabilities make it easier for managers to develop summaries and reports of biological data, and as a result provide greater leverage for conservation.

10.3.3 Planning

Managers need to monitor staff performance and make constant adjustments to the allocation of human resources and equipment to maximize efficiency. In order to accomplish this, SMART has an easy to use Planning function to support protected area managers' understanding of whether ranger teams are achieving their targets periodically (per patrol, monthly, quarterly, etc.) and planning of regular patrols. The targets (numeric, spatial, or administrative) are inputted into the Planning function, and can be defined for the entire conservation area, stations, teams, or individual rangers. Plans can be associated with a patrol or series of patrols, which then uses GNSS track information to calculate the success or failure of defined targets. When implemented effectively, SMART Planning can help better understand staff performance, enhance decision-making, and optimize the deployment of limited resources for conservation.

10.3.4 Profiles

SMART Profiles provides a platform for the management and analysis of multisource data related to a specific entity or entities (e.g., a poacher or collared rhino) and events, and the spatiotemporal relationships that exist among

---

[18] https://medium.com/wcs-marine-conservation-program/marine-protected-areas-are-getting-smart-55d022c2f985

[19] http://distancesampling.org/

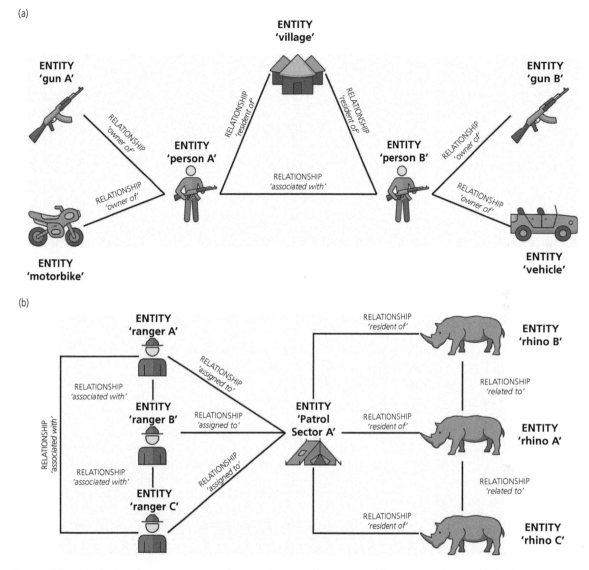

Figure 10.5 SMART Profiles allows you to capture and manage data on entities, events, and their spatiotemporal interrelationships in order to build up situational awareness over time. Shown here are two examples depicting different use cases: known poachers, their equipment, place of residence, and their associations (a), and monitoring of individual rhinos by rangers in a specific sector of a conservation area (b).

them (Figure 10.5). Profiles can be applied to any number of scenarios, such as recording sightings and developing a profile of individual animals, tracking traffic patterns through park entry points, recording the details of outreach projects in communities, maintaining records of staff participation in training events, tracking incidences of human–wildlife conflict around protected areas, or managing wildlife law-enforcement intelligence data. For management of enforcement data, Profiles provides conservation law enforcement officials with a system to manage offender data that can detail the actions, histories, and behaviours of individual wildlife offenders, and that can be readily used in conjunction with compatible third-party network analysis tools

Figure 10.6 Russian Amur tiger range (Goodrich et al. 2015), the location of Hötte et al. (2016)'s study sites, and the SMART sites added since 2016 in Primorskii Krai in the Russian Far East (left panel); and an Amur tiger. Map © Jonathan Slaght/WCS Russia; Photograph © Dale Miquelle/WCS Russia.

(e.g., Cytoscape,[20] Gephi,[21] IBM I2[22]) to provide actionable intelligence and inform response operations. By improving understanding of the offenders and their behaviour patterns, enforcement officials can craft more effective and targeted strategies aimed at preventing offences in their protected area.

10.4 Case studies

10.4.1 Primorskii Krai, Russia

Law enforcement is critical to the recovery of tiger numbers, as the primary threat for tigers in many of their key sites is poaching. SMART is now being implemented in 100+ tiger sites across Asia, and has become the standard tool for monitoring law enforcement work at the sites (SMART Partnership, 2017a). Hötte et al. (2016) implemented SMART as part of a holistic approach to LEM (Stokes, 2010) at four tiger sites in Primorskii Krai in the Russian Far East (Figure 10.6). They sought to define and evaluate specific indicators of success for the sites

by assessing change in outcomes following implementation, including: (1) patrol team effort and effectiveness; (2) catch per unit effort indicators (to measure reductions in threats); and (3) changes in target species numbers. Throughout their study, they observed a 3× increase in patrol effort, a partial reduction in threats, and tiger population numbers at each site remained stable or increased, leading to an overall increase in tiger numbers of 61% across all sites between 2011 and 2014. Since then, SMART has been introduced into nearly all protected areas with tigers in the Russian Far East (9 PAs), and is starting to be used by the wildlife enforcement agencies responsible for tiger protection outside PAs as well (Dale Miquelle, pers. comm.).

10.4.2 Cross River landscape, Nigeria

SMART has helped Wildlife Conservation Society (WCS) Nigeria improve its law enforcement patrols' effectiveness and to monitor illegal activities at all its sites more efficiently. Using CyberTracker together with SMART has reduced data collection and entry errors, improved the overall quality of field data collected by rangers by standardizing and streamlining data collection, quickened data uploads, and enabled rapid analyses and timely

[20] https://cytoscape.org/
[21] https://cytoscape.org/
[22] https://www.ibm.com/us-en/marketplace/enterprise-intelligence-analysis

reporting. Implementing SMART as part of an adaptive approach to patrol management has also contributed to improved patrol planning. For example, using data from previous patrols to develop intelligence-driven patrol plans that targeted hunting hotspots has improved patrol effectiveness in the community-managed Mbe Mountains. Rangers encountered an average of 40% fewer hunting signs on patrols between 2012 (3.08 signs/km) and 2016 (1.85 signs/km) (WCS Nigeria, unpublished data), meaning better protection overall for the Critically Endangered Cross River gorillas (*Gorilla gorilla diehli*) (IUCN, 2020).

These results align with those found more broadly in a comprehensive independent evaluation by the U.S. Fish & Wildlife Service on the use of SMART to strengthen LEM across its sponsored projects in the Congo Basin. Across several projects in Cameroon, Gabon, and Republic of Congo, the evaluation found that implementing the SMART approach was an effective method for improving wildlife LEM effectiveness, that LEM generally improved across the sites and countries as a result of SMART, and that, overall, LEM has improved in central Africa and that these improvements were often attributable to SMART (O'Neill & Honig, 2017).

10.4.3 Queen Elizabeth National Park, Uganda

Despite the importance of protected areas for wildlife conservation and the role that rangers play in safeguarding them, there have been few studies evaluating the efficiency of ranger deployments in the field. Where law enforcement is perceived to be more effective, the likelihood of illegal activity is lower (Fischer et al., 2014) and increased probability of detection improves enforcement effectiveness (Milner-Gulland & Leader-Williams 1992). However, as noted earlier, protection efforts around the world are limited by ineffective management (Leverington et al., 2010), inadequate staff and financial resources (Coad et al., 2019), and rangers that are unsupported, undertrained, and underequipped (Belecky et al., 2019). Efficient and effective deployment

of patrols is needed, but given the lack of empirical studies on the subject, few methods for improving ranger patrols exist. Critchlow et al. (2017) sought to address this by using a spatial crime mapping approach in Queen Elizabeth National Park, Uganda, as a potential method of improving ranger patrol allocation. The study focused on three primary aims: (1) identifying the efficiency of ranger patrol strategies relative to patterns of illegal activities; (2) determining an improved patrol strategy for any given set of conservation priorities using the existing patrol resources; and (3) assessing the impact of the improved strategies on the effectiveness of illegal activity detection. Rangers used SMART (and a precursor, Management Information System [MIST[23]]) and CyberTracker to collect data, which were then used to generate maps of detections of illegal activities. Statistical models (see Critchlow et al., 2015, 2017 for a detailed description of methods used) were then used to develop patrol strategies targeting different combinations of conservation priorities, and patrol effectiveness was calculated using catch per unit effort (CPUE; number of illegal activities detected per kilometre walked). Over 2 years (2014–2015), the new patrol allocation strategy improved patrol effectiveness and efficiency, leading to significant increases in the numbers of detections of illegal activities, typically by around 50%, but in some cases over 250%, without augmenting ranger resources. Despite the impressive results, the method described is relatively simple: using maps of ranger-based observations of illegal activity as the basis for adapting patrol strategies to improve their effectiveness over time. This also provides a good example of how using SMART as part of an adaptive management approach (Figure 10.2) can improve patrol effectiveness at a resource-limited site.

10.4.4 MPA system, Belize

Belize has been at the forefront of SMART marine implementation for years and remains the country

---

[23] http://www.ecostats.com/MIST

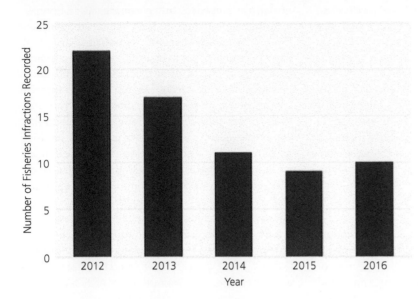

Figure 10.7 Observed fisheries infractions per year at Glover's Reef Marine Reserve. Adapted from Wildlife Conservation Society Belize and Belize Fisheries Department (2017).

with the most comprehensive marine implementation of SMART to date. Belize adopted SMART nationally across its MPA system in 2015 and was the first to pilot the use of SMART Connect in marine areas. SMART is implemented by the Belize Fisheries Department (BFD) throughout the country's MPA system and supported by nongovernmental organizations (NGO) co-managers in each MPA. The BFD also implements SMART out of its two central Conservation Compliance Units for all fisheries-related enforcement nationally. Wildlife Conservation Society Belize began SMART implementation in Belize when they deployed SMART at Glover's Reef Marine Reserve (GRMR) in 2012. The use of SMART at GRMR resulted in the identification of prioritized enforcement areas and improved management and deployment of patrols, which has reduced fuel costs—one of the major limitations of patrol deployments. Improved patrol management and its subsequent benefits have been replicated elsewhere throughout the MPA system (J. Maaz, pers. comm.), and, at GRMR specifically, have contributed to a 55% decline in the number of MPA fisheries infractions since 2012 (Figure 10.7; Wildlife Conservation Society Belize & Belize Fisheries Department, 2017). The success of BFD's SMART implementation in the MPA system also

led the Belizean Ministry of Agriculture, Fisheries, Forestry, Environment, Sustainable Development and Climate Change to adopt SMART nationally for the country's terrestrial protected area system in January 2018.[24]

In addition to monitoring human activity, commercial and recreational fishing, and wildlife, SMART is now also being implemented to monitor fisheries offtake. Using SMART Mobile and Connect, community researchers were provided with a simple and effective system to record catch in the field and easily transfer data to a centralized database for near real-time monitoring and analyses. Researchers monitored small-scale fisheries and collected data on more than 15,000 fishery products over 2 years, developing a much clearer understanding of the nature and availability of fisheries products in five of the largest coastal communities in Belize. Preliminary analyses indicated that several of the 65 species examined were being overexploited based on the presence of illegal size catch and low proportions of mature individuals (Tewfik et al., 2018). At the time of writing, efforts were underway to expand this system, still in its

[24] https://www.sanpedrosun.com/environment/2018/01/25/forest-fisheries-departments-get-smart/

pilot phase, and eventually transfer it to BFD to supplement or replace the current system hard-copy logbooks.

10.4.5 Effective community-based rhino monitoring in north-west Namibia

The communal lands of north-west Namibia support one of the last free-ranging black rhino populations, and the largest population to persist on formally unprotected land. Since the early 1990s, the rhino population has been monitored by local ranger teams mostly employed by a local NGO (Save the Rhino Trust—SRT) using basic paper-based tools to record individually recognized black rhino and managing these data with standard software. The upsurge in poaching pressure in 2009–10 motivated SRT to scale its rhino monitoring approach by expanding collaboration with numerous local institutions, which has led to dramatic increases in the number of rangers (~400%), patrol effort (1229%), and confirmed rhino sightings (458%) (Muntifering, 2019). The sheer amount of information 'overload' became un-manageable under the old system. SMART was explored as a solution to help better manage and make use of rhino monitoring information, and ultimately ensure the achievements from scaling were sustained. SMART, with its many associated functionalities, free access, and strong technical support network, proved a well-suited technological fix for the local context. For example, the ease of data entry using both smartphone devices and GNSS units into the central database and a direct link to a spatial summary map enabled quick processing and feedback to rangers and managers across institutions. The Entity function enabled managers to quickly and easily track 'missing' individual rhinos and plot recent sightings to guide search patrols, as well as to contribute towards more regular population performance assessments. The custom query builder enabled quick and accurate reporting on ranger performance, catalysing a new rhino award board system and annual recognition ceremony that boosted morale. What previously took hours, or even days, of manual data entry and shifting between independent software programs, can be completed in a matter of minutes, and, more importantly, be conducted by local senior staff and managers. Crucially, the implementation of SMART has allowed SRT to continue delivering on its mandate of timely and comprehensive reports to government and donors that capture the increased effort and activities without compromising crucial resource allocations or other critical functions. Maintaining ranger patrols and rhino sightings at a scale that serves to both effectively monitor the population and deter poachers (no rhino has been poached in north-west Namibia for in over 2.5 years; J. Muntifering, pers. obs.) is achieved by finding innovative means to equip and motivate rangers in the field. SRT is confident that SMART is, and will continue to be, fundamental to this success.

10.5 Considerations and Constraints

As noted previously, 'technological solutionism' (Morozov, 2013) remains a complex and pervasive challenge to overcome for conservation practitioners; users are recommended to carefully assess their site-specific needs and pursue the simplest fit-for-purpose solution. Considerations should include a solution's adaptability to local conditions, its stability and sustainability, whether it avoids duplication and confusion with other technologies/practices at your site, whether it meets a clear and specific need, and whether it is cost-effective compared to other approaches. These principles should guide the selection of any technology at a site.

An important factor when considering any conservation technology is its sustainability. With SMART, an international partnership of organizations is committed to the continuing support, development, and maintenance of the tool, ensuring its sustainability for the long-term. This is an important factor because instituting wholesale changes to operational procedures or existing technology frameworks can require substantial effort and resources. It should be clear to users before implementing such a change (e.g., adopting SMART), that not only will the tool be beneficial, but that the near-term investment will be worth it because the platform will continue to exist, function, and be supported for the foreseeable future.

There are no requirements for pre-existing management practices at a site to implement SMART, but as with any technology, using SMART software and establishing a patrol database will not, on its own, improve protection in a conservation area. Therefore, some basic foundational elements should be in place for the SMART implementation to succeed and to ensure that conservation resources are not wasted or diverted (for detailed breakdowns of enabling conditions for SMART success, see SMART Partnership, 2017b or Stokes, 2010). A formal management structure should be in place at the site where an adaptive management approach using SMART can be applied. Without an accepted and mandated management authority to manage the use and maintenance of equipment, provide ongoing training to rangers, evaluate patrol results, and lead decision-making, it will be difficult to affect measurable changes in patrol practices. The relevant management authority must also endorse an adaptive management approach using SMART—or it will, again, be difficult to affect a change in practices and outcomes at the site. There must be some level of active field patrolling at the site and a commitment to put resources into protection and monitoring activities, as well as to adapting protection strategies, management practices, or budget allocations based on results over time. Adequate resources must also be allocated to secure and maintain the necessary field equipment for collecting and managing patrol data (e.g., computers, GNSS units, batteries, battery chargers, power source) and for ensuring rangers are properly resourced, supported, and motivated in the field (e.g., field equipment, provisions). Ultimately, SMART is a spatial monitoring tool and requires, at minimum, a computer (with Windows OS and a min 4GB RAM, though 8 or higher is recommended) and rangers in the field using a GNSS or mobile device to collect spatial data.

There may also be further technical constraints associated with specific functions and integrations. Each new feature or technology (e.g., addition of digital radios to patrols) will add a layer of complexity and may increase the chance that workflows will breakdown, so the adoption of any new technology should be weighed against this and a stepwise approach relative to site capacity or over time as site capacity matures may be the best approach. SMART Connect requires basic, but reliable internet connectivity (i.e. a minimum speed of approximately 1 Mbps up and down) and either cloud-based server space or a single, secure, and reliable local server. Each server option comes with associated costs. A cloud-based subscription service will incur a monthly charge based on usage, the database size, and site-specific configurations, with basic usage and default configurations averaging roughly $100 per month (SMART Partnership, 2017b). More comprehensive usage (e.g., hundreds of data collectors regularly transmitting data from many sites) will be more expensive (i.e. likely more than $500 per month), but a cloud-based subscription service is still likely to be the most cost-effective long-term option for most sites. An onsite server can be very costly, as costs vary dramatically due to access to hardware, IT configuration and maintenance capacity, and local market costs variability. Either way, sites are recommended to have sufficient IT and information security expertise for at least basic server set-up, troubleshooting, and management. There are also numerous limitations/constraints associated with mobile devices, such as durability and battery life (as detailed in Section 10.2.1 earlier).

For SMART implementation to be successful, sites must also consider issues of staff capacity. It is often important for sites to identify specific SMART (or technology) focal points or 'champions'. These focal points are individuals who demonstrate effective leadership, understand patrol issues, and have the analytical skills for evaluating patrol data and providing feedback to rangers. Staff with strong technological proficiency are also recommended for database design, patrol data storage, and data management, and IT experts should be available on standby, either at the conservation area or remotely, to solve technical problems when they occur. As noted earlier, sites/users should commit to investing resources in ensuring these staff have enough capacity to manage the systems and workflows at the site. Investing in building and strengthening capacity at sites will not only benefit SMART implementation, as increased opportunities for learning has shown to lead to greater likelihoods of technology adoption in rangers (Sintov et al., 2019), but it will also contribute to greater ranger job satisfaction and engagement (Moreto et al., 2016). For sites considering implementing SMART, but lacking sufficient capacity and seeking a way to train staff, we recommend contacting the SMART Partnership[25] to discuss potential options, such as collaboration with (or contracting a trainer from) a nearby site/organization already implementing SMART. Both the SMART Partnership and its member organizations also regularly organize site-based and regional training workshops. Training opportunities are often made available for external participants (training workshops are typically offered at no cost, with participants covering their own travel, room, and board). For information on training and to connect with the global network of SMART users, we also recommend joining the SMART Community Forum.[26]

10.6 Social impact/privacy

As detailed earlier, implementing SMART is beneficial for area-based conservation management effectiveness, but, as with any software/hardware, users must take precautions to ensure data security and protect personal identifying information. Data collected in the field will be stored on mobile devices and uploaded to desktop software and potentially cloud-based infrastructure. While the SMART file store is encrypted and regularly and automatically backed up, maintenance of data security in typical SMART workflows will be dependent on the extent to which hardware and software providers protect data, which is governed by multiple regulations based on location, and the information security policies implemented for the site. A common example of basic risk factors is a single device or laptop which houses all site data. If the data are not backed up and securely stored elsewhere, any accident, corruption,

or theft that befell the laptop would represent significant lost investment, and, if the database contained sensitive data, could lead to potential harm or threat to humans and/or wildlife.

The SMART Partnership recommends numerous procedures and best practices for the maintenance of data security, summarized next (Table 10.3) and detailed at length in numerous publicly available manuals (SMART Partnership, 2017c, 2018b, 2019). These practices are important because developing and maintaining a SMART database represents a significant investment of time and resources and any disruption to the database, either from loss, corruption, or compromise, may result in loss and potential harm to an organization (SMART Partnership, 2018b). Implementing robust security policies represents insurance on the investment of time and resources and protects against any physical or reputational damage resulting from the improper use of data. With all this being said, the biggest observed risk to security at sites is the use of a generic password, often the one associated with the training data set from the initial SMART download, being used across staff.

10.7 Future directions

The SMART Partnership is working with a wide range of collaborators to integrate SMART with other technologies and approaches to provide a comprehensive and integrated suite of tools to those working on the front lines of conservation. The vision is one where rather than implementing independent technologies, complementary systems and emerging technologies can be integrated into a unified system for holistic protection of biodiversity and conservation areas. To that end, the SMART Partnership is developing numerous new features and collaborating with researchers and technologies to provide access to cutting-edge technology in all protected areas. SMART Collect will facilitate decentralized data collection (e.g., citizen-science projects or community-based human–wildlife conflict monitoring). SMART Connect as a service will provide the benefits of SMART Connect without the need for IT capacity at sites and, although at the time of writing the costs of this service are still being

[25] Email: info@smartconservationsoftware.org
[26] https://smartconservationtools.org/smart-forum/

Table 10.3 SMART best practice guidance for data protection and security

Data Protection	Security
Follow laws and regulations	Use a strong password
Follow organizational policies	Install and maintain antivirus software
Ensure data are managed ethically	Update your operating systems
Train staff in data protection	Password-protect mobile devices
Implement supporting processes	Automatically lock computers
Document and manage roles	Secure reports that are shared
Create a data management plan	Know what to do in event of a theft
Complete a risk assessment	Backup your data securely
Audit data protection efforts	
Align management structures	

SMART Connect
Maintain a firewall
Configure security and passwords
Encrypt transmission of data across open, public networks
Protect against malware, including updating antivirus and other programs
Restrict access to named users with secure passwords on a need-to know basis
Maintain security for all related systems and integrations
Restrict physical access to the SMART Connect server
Track and monitor all access to network resources and data
Regularly test security systems and processes
Maintain a policy that addresses information security for all personnel

evaluated, the SMART Partnership aims to provide the service at no cost to most users. SMART is developing several integrations that will enhance the data collection and analytical power of SMART. Integration with the Protection Assistant for Wildlife Security (PAWS), an artificial intelligence application, uses machine learning to predict poachers' behaviour based on SMART data, terrain features, and road networks and then produces heat maps of where patrols should focus on increasing chances of detecting and/or deterring wildlife crime and maximize impact. The SMART Partnership aims to significantly enhance the quality of SMART implementation and capacity of SMART users globally. With SMART now used in more than 1,000 sites and adopted nationally in more than 20 countries, it has become a widely recognized and powerful conservation tool. However, this broad adoption has led to a spectrum of implementation qualities and a range of apparent gaps in knowledge, with no formal way

of evaluating the areas needing improvement. To address this, the Partnership is developing a set of SMART-specific competence standards paired with the relevant protected area management competencies (Appleton, 2016). These SMART competencies will establish a clear professional structure for users, trainers, assessors, and managers and serve as a basis for developing universal, comprehensive training courses, assessments, and standards for different SMART users and establishing consistent global approaches to use, implementation, and adaptive management.

10.8 Acknowledgements

We want to thank the SMART community, especially the rangers and practitioners on the frontlines of conservation, for their commitment and service to securing a future for wildlife. We also thank Dale Miquelle, Michiel Hötte, Jonathan Slaght, Andrew

Dunn, Inaoyom Imong, Julio Maaz, and the Russia, Nigeria, Uganda, and Belize programmes of the Wildlife Conservation Society for their inputs for the case studies. We are grateful to an anonymous reviewer whose comments helped improve and clarify this chapter. Finally, thanks to the SMART Partnership leadership, all of our partners who have helped SMART grow from humble beginnings to the global standard for protected area monitoring, and the many funders, without whom SMART would not have been possible.

References

Appleton, M. R. (2016). *A Global Register of Competences for Protected Area Practitioners*. : IUCN, Gland, Switzerland.

Barichievy, C., Munro, L., Clinning, G., Whittington-Jones, B., & Masterson, G. (2017). Do armed field-rangers deter rhino poachers? An empirical analysis. *Biological Conservation*, 209, 554–560.

Belecky, M., Singh, R., & Moreto, W. (2019). *Life on the Frontline 2019: A Global Survey of the Working Conditions of Rangers*. Gland, Switzerland, WWF.

Berger-tal, O., & Lahoz-Monfort, J. J. (2018). Conservation technology: the next generation. *Conservation Letters*, 11(6), e12458.

CBD (2010). Decision X/2, The *Strategic Plan for Biodiversity* 2011–2020 and the Aichi Biodiversity Targets. Nagoya, Japan. Available at: https://www.cbd .int/decision/cop/?id=12268

Ceballos, G., Ehrlich, P. R., & Raven, P. H. (2020). Vertebrates on the brink as indicators of biological annihilation and the sixth mass extinction. *Proceedings of the National Academy of Sciences*, 117, 201922686.

Coad, L., Watson, J. E., Geldmann, J., et al. (2019). Widespread shortfalls in protected area resourcing undermine efforts to conserve biodiversity. *Frontiers in Ecology and the Environment*, 17(5), 259–264.

Critchlow, R., Plumptre, A. J., Alidria, B., et al. (2017). Improving law-enforcement effectiveness and efficiency in protected areas using ranger-collected monitoring data. *Conservation Letters*, 10(5), 572–580.

Critchlow, R., Plumptre, A. J., Driciru, M., et al. (2015). Spatiotemporal trends of illegal activities from ranger-collected data in a Ugandan national park. *Conservation Biology*, 29(5), 1458–1470.

Cronin, D. T. (2019). Leveraging SMART to support, motivate, and empower rangers globally in their efforts to protect wildlife. In 9th World Ranger Congress, Sauraha, Nepal, 16 November 2019.

Díaz, S., Settele, J., Brondízio, E., et al. (eds.) (2019). IPBES (2019): Summary for *Policymakers* of the Global *Assessment Report* on *Biodiversity* and *Ecosystem Services* of the Intergovernmental Science-Policy Platform on Biodiversity and Ecosystem Services. IPBES Secretariat, Bonn, Germany.

Dinerstein, E., Vynne, C., Sala, E., et al. (2019). A global deal for nature: guiding principles, milestones, and targets. *Science Advances*, 5(4), eaaw2869.

Fischer, A., Naiman, L. C., Lowassa, A., Randall, D., & Rentsch, D. (2014). Explanatory factors for household involvement in illegal bushmeat hunting around Serengeti, Tanzania. *Journal for Nature Conservation*, 22(6), 491–496.

Geldmann, J., Barnes, M., Coad, L., Craigie, I. D., Hockings, M., & Burgess, N. D. (2013). Effectiveness of terrestrial protected areas in reducing habitat loss and population declines. *Biological Conservation*, 161, 230–238.

Geldmann, J., Coad, L., Barnes, M. D., et al. (2018). A global analysis of management capacity and ecological outcomes in terrestrial protected areas. *Conservation Letters*, 11(3), UNSP e12434.

Geldmann, J., Manica, A., Burgess, N. D., Coad, L., & Balmford, A. (2019). A global-level assessment of the effectiveness of protected areas at resisting anthropogenic pressures. *Proceedings of the National Academy of Sciences*, 116(46), 23209.

Gill, D. A., Mascia, M. B., Ahmadia, G. N., et al. (2017). Capacity shortfalls hinder the performance of marine protected areas globally. *Nature*, 543(7647), 665–669.

Goodrich, J., Lynam, A., Miquelle, D., et al. (2015). *Panthera tigris. The IUCN Red List of Threatened Species*, e.T15955A50659951. Available at: http://www.uwice .gov.bt/admin_uwice/publications/publication_files /Reports/2015/IUCNTIGER.pdf

Gray, C. L., Hill, S. L. L., Newbold, T., et al. (2016). Local biodiversity is higher inside than outside terrestrial protected areas worldwide. *Nature Communications*, 7(1), 12306.

Hansen, M. C., Krylov, A., Tyukavina, A., et al. (2016). Humid tropical forest disturbance alerts using Landsat data. *Environmental Research Letters*, 11(3), 034008.

Hoffmann, M., Hilton-Taylor, C., Angulo, A., et al. (2010). The impact of conservation on the status of the world's vertebrates. *Science (New York, N.Y.)*, 330, 1503–1509.

Hötte, M. H. H., Kolodin, I. A., Bereznuk, S. L., et al. (2016). Indicators of success for smart law enforcement in protected areas: a case study for Russian Amur tiger (*Panthera tigris altaica*) reserves. *Integrative Zoology*, 11(1), 2–15.

IEC (2013). IEC Standard 60529: Degrees of Protection Provided by Enclosures (IP Code). 2.2. International Electrotechnical Commission, Geneva, Switzerland.

IUCN (2020). *IUCN Red List of Threatened Species. Version 2020–1*. IUCN 2020. IUCN, Gland, Switzerland.

IUCN and WCPA (2017). IUCN Green List of Protected and Conserved Areas: Standard, Version 1.1. IUCN, Gland, Switzerland.

Jones, K. R., Venter, O., Fuller, R. A., et al. (2018). One-third of global protected land is under intense human pressure. *Science*, 360(6390), 788.

Laurance, W. F., Useche, D. C., Rendeiro, J., et al. (2012). Averting biodiversity collapse in tropical forest protected areas. *Nature*, 489(7415), 290–294.

Lawson, K., & Vines, A. (2014). *Global Impacts of the Illegal Wildlife Trade: The Costs of Crime, Insecurity and Institutional Erosion*. Chatham House (The Royal Institute of International Affairs), London, UK.

Leverington, F., Costa, K. L., Pavese, H., Lisle, A., & Hockings, M. (2010). A global analysis of protected area management effectiveness. *Environmental Management*, 46(5), 685–698.

Marvin, D. C., Koh, L. P., Lynam, A. J., et al. (2016). Integrating technologies for scalable ecology and conservation. *Global Ecology and Conservation*, 7, 262–275.

Maxwell, S., Fuller, R., Brooks, T., & Watson, J. (2016). Biodiversity: the ravages of guns, nets and bulldozers. *Nature*, 536, 143–145.

Milner-Gulland, E. J., & Leader-Williams, N. (1992). A model of incentives for the illegal exploitation of black rhinos and elephants: poaching Pays in Luangwa Valley, Zambia. *Journal of Applied Ecology*, 29(2), 388–401.

Moreto, W. D., Lemieux, A. M., & Nobles, M. R. (2016). 'It's in my blood now': the satisfaction of rangers working in Queen Elizabeth National Park, Uganda. *Oryx*, 50(4), 655–663.

Morozov, E. (2013). *To Save Everything, Click Here: The Folly of Technological Solutionism*. Allen Lane, London, UK.

Muntifering, J. R. (2019). *Large-Scale Rhino Conservation in North-West Namibia*. Venture Publications, Namibia, South Africa.

Nurse, A. (2015). *Policing Wildlife: Perspectives on the Enforcement of Wildlife Legislation*. Palgrave Macmillan, London, UK.

O'Neill, E., & Honig, N. (2017). *Independent Evaluation of USFWS Support to Utilization of the Spatial Monitoring and Reporting Tool (SMART) in Central Africa (2009 – present)*. U.S. Fish & Wildlife Service, Arlington, VA.

R Core Team (2015). *R: A Language and Environment for Statistical Computing*. R Foundation for Statistical Computing, Vienna, Austria.

Ripple, W. J., Wolf, C., Newsome, T. M., et al. (2019). Are we eating the world's megafauna to extinction? *Conservation Letters*, 12(3), e12627.

Sala, E., & Giakoumi, S. (2018). No-take marine reserves are the most effective protected areas in the ocean. *Ices Journal of Marine Science*, 75(3), 1166–1168.

Sala, E., & Rechberger, K. (2018). Protecting half the ocean? In Desai, R. M., Kato, H., Kharas, H., & Mcarthur, J. W. (eds.). *From Summits to Solutions*. Brookings Institution Press, New York, NY (pp. 239–262).

Schuster, R., Germain, R. R., Bennett, J. R., Reo, N. J., & Arcese, P. (2019). Vertebrate biodiversity on indigenous-managed lands in Australia, Brazil, and Canada equals that in protected areas. *Environmental Science & Policy*, 101, 1–6.

Sintov, N., Seyranian, V., & Lyet, A. (2019). Fostering adoption of conservation technologies: a case study with wildlife law enforcement rangers. *Oryx*, 53(3), 479–483.

SMART Partnership (2017a). SMART Partnership 2017 Annual Report. SMART Partnership, New York, NY.

SMART Partnership (2017b). SMART: A Guide to Getting Started. 2.0. SMART Partnership, New York, NY.

SMART Partnership (2017c). Best Practice Security Advice for SMART Desktop & Connect Users. 1.0. SMART Partnership, New York, NY.

SMART Partnership (2018a). SMART Partnership 2018 Annual Report. SMART Partnership, New York, NY.

SMART Partnership (2018b). SMART Profiles Standards and Procedures Guide. *1.0*. SMART Partnership, New York, NY.

SMART Partnership (2019). Best Practice Data Protection Advice for SMART Desktop & Connect Users. *1.0*. SMART Partnership, New York, NY.

Stokes, E. J. (2010). Improving effectiveness of protection efforts in tiger source sites: developing a framework for law enforcement monitoring using MIST. *Integrative Zoology*, 5(4), 363–377.

Tewfik, A., Maaz, J., Alamina, V., & Ramnarace, J. (2018). A SMARTer approach to collection of catch data for conservation and sustainability. In 71st Gulf and Caribbean Fisheries Institute Conference, San Andres, Colombia, 2018.

Thomas, L., Buckland, S. T., Rexstad, E. A., et al. (2010). Distance software: design and analysis of distance sampling surveys for estimating population size. *Journal of Applied Ecology*, 47(1), 5–14.

Tittensor, D. P., Walpole, M., Hill, S. L. L., et al. (2014). A mid-term analysis of progress toward international biodiversity targets. *Science*, 346(6206), 241.

UNEP-WCNC, IUCN, & NGS (2018). Protected Planet Report 2018. UNEP-WCMC, IUCN, and National Geographic Society, Cambridge, UK and Washington, DC.

Venter, O., Fuller, R. A., Segan, D. B., et al. (2014). Targeting global protected area expansion for imperiled biodiversity. Edited by Craig Moritz. *PLoS Biology*, 12(6), e1001891.

Watson, J. E. M., Dudley, N., Segan, D. B., & Hockings, M. (2014). The performance and potential of protected areas. *Nature*, 515(7525), 67–73.

Wildlife Conservation Society Belize and Belize Fisheries Department (2017). *Spatial Monitoring and Reporting Tool (SMART) Implementation in Belize*. Wildlife Conservation Society Belize, Belize City, Belize.

WWF (2018). Living Planet Report—2018: Aiming Higher. Grooten, M., & Almond, R. E. A. (eds.) WWF, Gland, Switzerland.

CHAPTER 11

Challenges for the computer vision community

Dan Morris and Lucas Joppa

11.1 Introduction

The preceding chapters have addressed the extensive contributions technology is already making to conservation and the promise that emerging technology holds for conservation in the coming years. In this chapter, we will take a complementary approach, looking at the field of conservation technology from a computer scientist's perspective. Specifically, we will ask the question: *What challenging problems in computer vision (CV) do conservation problems require us to solve?*

Though 'computer vision' broadly refers to the use of *software* to extract high-level information from digital images, in this chapter we will define it more specifically as the use of *machine learning* to extract high-level information from digital images. We will give a 'tour' of several existing applications of CV in conservation, but this chapter primarily aims to highlight techniques we *have not* developed yet and serves as a call to action to the computer science community, to build these tools in partnership with ecologists.

We will look at several case studies in CV where tremendous progress has been made since the emergence of deep learning. However, this progress has been biased towards high-quality natural images from canonical data sets. Problems in conservation, on the other hand, often involve low-quality images (e.g. from camera traps) and extremely imbalanced data sets (e.g. from aerial images), which still represent fundamental CV challenges. Furthermore, inference is often required to run in remote regions with limited connectivity and unreliable power, requiring a rethinking of inference hardware and the model architectures we run at the edge.

This chapter will be structured as a series of CV problems. In each section, we will (1) introduce a technical problem in machine learning; (2) provide examples of conservation scenarios that solving this problem would accelerate; (3) discuss the obstacles that make this problem difficult; and (4) present the progress made on the problem, using examples from the conservation space. We will conclude with some practical 'lessons learned' from our work in applying machine learning to problems in conservation.

We argue that working on these problems will advance the state of the art in machine learning, CV, and numerous other computer science subfields and, more importantly, will advance the ways that we address conservation challenges.

11.2 Terminology: what's 'computer vision'?

Before we dive into specific CV applications in conservation, we will provide a brief overview of important terminology that we will use throughout the chapter.

11.2.1 Computer vision (CV)

CV is the field of computer science focused on using computer algorithms to derive useful information

Dan Morris and Lucas Joppa, *Challenges for the computer vision community*. In: *Conservation Technology*.
Edited by Serge A. Wich and Alex K. Piel, Oxford University Press.
© Oxford University Press 2021. DOI: 10.1093/oso/9780198850243.003.0011

from still images or video. This often includes *semantic understanding* of images—for example, 'this image contains a cat'. However, CV also includes the *transformation* of images to a new form that makes them easier for humans or other computer algorithms to work with—for example, extraction of depth and 3D structure information from images and video. The conservation-relevant examples we use in this chapter focus on the former type of CV: *semantic understanding*.

11.2.2 Machine learning (ML) and supervised ML

In fact, in this chapter we will focus on an even more specific subfield within CV, namely the use of *supervised ML* for semantic understanding of images. *Machine learning* is any process by which an algorithm learns to solve a problem using data, and *supervised ML* is the process by which a system learns to transform some input to the desired output based on examples of input/output pairs provided by the 'teacher' (i.e. the person developing the system). For example, if we are using supervised ML to train a system to recognize images of cats, we might show the system thousands of examples of images with cats and express that 'these are cats', and thousands of examples of non-cat images and express that 'these are not cats'.

As a counterexample, we could solve this problem with no ML at all: we might have a huge collection of images, some of which have cats and some of which do not, and when we receive a new image that our system needs to process, we could find the most similar image in our collection, and if that is a cat, we could assume that our new image has a cat. We might call this a *search-based* approach to understanding images, rather than an ML approach. This chapter will focus specifically on ML, in fact specifically on supervised ML, because this is where we have identified the broadest set of conservation-relevant examples, and because this represents the substantial majority of active CV research as of the time of this writing.

11.2.3 Image classification and object detection

Two problems within supervised ML for CV are sufficiently common in the conservation space that

it is important to understand them before laying out applications in conservation: *image classification* and *object detection*.

Image classification is the process of assigning one or more labels to an image: given an image it is never seen before, an image classifier might say 'this image contains a cat', or 'this image contains a cat and a dog'. It would not, however, tell us *where* in the image those objects are, or how many are present. Creating an image classifier generally requires examples of images and their corresponding labels.

Object detection is the process of determining *where in an image* specific objects of interest are. For example, given a new image, an object detector trained to detect cats would be able to identify *where* in the image each cat is. Object detectors can learn to recognize just one type of object, or they may learn to recognize many types of objects. Creating an object detector generally requires examples of images, along with boxes indicating where objects of interest are in those images.

11.2.4 Other terminology

An ML **model** is the result of teaching a system to solve a particular problem, for example, a particular classification or detection problem. One can think of a *model* as the actual file you get back at the end of all the work it takes to teach an ML algorithm to do something useful, that is, the file where all the learned information is stored.

Training is the process of teaching an ML algorithm to learn a task, typically involving the presentation of numerous examples of inputs and their corresponding outputs (e.g. images and their labels). ML practitioners will generally use 'training a model' and 'building a model' interchangeably.

Inference is the process of taking a model that has already been trained and running it on new images (i.e. *inferring* the meaning of those images from the image pixels).

Deep learning is a specific set of algorithms used in ML, using structures loosely inspired by human neurobiology. For the purposes of this chapter, what makes deep learning particularly relevant is that if we have sufficient data and sufficient compute resources to train them, deep learning algorithms generally provide more accurate results than other algorithms.

Transfer learning is the process of taking an ML model trained for some problem and using that as a starting point for a new problem, to reduce the training data and/or training time required for our new problem. For example, if we have an object detection model trained to detect cats using a training data set of one million labelled images, and we want to train a new model to recognize dogs but we only have ten thousand labelled images of dogs, we might proceed by adapting our cat model, rather than training a new model. Transfer learning is not necessarily specific to deep learning, but it is colloquially used to refer to adapting a deep learning model to a new problem.

A loss function is a metric an ML algorithm is trying to improve as it learns. This may seem obvious—for example, in a classification problem, we are usually trying to maximize accuracy—but loss functions can be subtle. Should an ML algorithm learning to classify species treat an error that confuses a cheetah with a leopard the same as an error that confuses a cheetah with a dolphin? If not, exactly how much 'worse' is the latter error? Designing loss functions is a key part of designing ML algorithms.

11.3 Fine-grained classification

11.3.1 Problem overview

Despite the success of image classification in a wide variety of applications, most successful image classification applications have benefitted from relatively coarse-grained classes: telling a human from a car, for example. In contrast, *fine-grained classification*—where classes are subjectively similar—remains challenging. While there is no universal definition of 'fine-grained', a reasonable intuition is that in a fine-grained problem, there are many pairs of classes in the problem space where a novice viewer would not immediately identify objects in those classes as even belonging to distinct classes. For example, if we show a novice an apple and an orange, even if that viewer has never seen apples or oranges, she will likely identify them as distinct. However, if we show a novice images of Eurasian lynxes (*Lynx lynx*) and Canadian

lynxes (*Lynx Canadensis*), she may not recognize them as distinct classes, and would certainly require training before being able to differentiate them reliably. Even experts struggle with this type of task, especially when image quality is poor, and when framing can obfuscate animal size.

11.3.2 Why fine-grained classification matters for conservation

Automated classification of animal and plant species from images has immense promise for accelerating conservation workflows, based on images from drones (Chapter 3), camera traps (Chapter 5), or handheld cameras. In many conservation scenarios, though, it is critical to identify entities to the species level, and in some cases it is critical to differentiate subtle sex- or age-related morphology. While tremendous contributions can be made at the coarse-grained level (e.g. differentiating just the megafauna in a single region), differentiating all entities to the species level—even within a region—is key to unlocking the full potential of ML-based workflows.

Some problems—for example, species classification for camera trap images—can rely on the fact that only a finite number of species exist in a region large enough to trigger a capture, leading to a coarse-grained problem. However, some problems require fine-grained classification even within a region (e.g. classification of arthropods), and some problems cannot leverage knowledge about species' typical range (e.g. classification of species at border crossings to support anti-trafficking efforts).

11.3.3 Why fine-grained classification is hard

The progress of image classification as a field has largely been evaluated on the ImageNet data set (Deng et al., 2009), which contains a huge diversity of objects, from vehicles to animals to household items. On a commonly used 1000-class subset of ImageNet, the state of the art as of 2019 is approximately 86% accuracy (Touvron et al., 2019). While there are many animals among these 1000 classes, including some fairly similar pairs (e.g. eight species of monkey, and 120 domestic dog breeds), these are mixed with a much larger number of non-animal classes (such as 'ice cream' and 'school bus'), so the challenge is a mix of coarse- and fine-grained classification. Even with this mixed granularity, the state of the art on just 1000 classes is only 86% accuracy, which—while an incredible technical achievement—leaves a long way to go in granularity, class set size, and accuracy before even the state of the art in image classification could meaningfully accelerate conservation workflows that depend on fine-grained classification.

Further exacerbating the problem is that nearly every success case in image classification has benefitted from *transfer learning* using models derived from ImageNet. That is, many open-source image classification models are available that were trained on ImageNet, and for real-world problems that do not have tens of millions of training images, standard practice is to start from a publicly available ImageNet-trained model and adapt it to the task at hand. Because the fine-grained classification task is so much more difficult than the coarse-grained ImageNet task, fine-grained problems in conservation may hit the limits of effective transferability to tasks that differ from the original task (Yosinski et al., 2014).

11.3.4 Progress on fine-grained classification

The most critical ingredient for accelerating most of the problems we cover in this chapter is not technical innovation *per se*, it is *data* that let us train models and rigorously evaluate competing solutions. This is true for much of ML: ImageNet was a key component that enabled recent massive advances in deep learning. Fortunately, the last several years have seen a dramatic increase in curated data sets for fine-grained classification related to wildlife conservation. The iNaturalist 2018 data set (Van Horn et al., 2018), for example, includes over 5000 species from multiple kingdoms; this data set includes over 850,000 images and continues to grow. The Caltech-UCSD Birds (Wah et al., 2011) and NABirds (Van Horn et al., 2015) data sets specifically focus on bird classification, while the Animals with Attributes data set (Xian et al., 2018) covers a more coarse-grained class distribution, but includes semantic part labels that may enable animal-specific fine-grained recognition techniques.

As a consequence of these data sets' availability, general-purpose image classification approaches are now frequently benchmarked for their applicability to conservation problems, and novel fine-grained classification methodology specific to conservation applications has emerged. Berg et al. (2014), for example, focus specifically on bird classification; they present a training paradigm that prevents excessively penalizing misclassifications when those examples may be visually indistinguishable from the true class (as often occurs in fine-grained data sets). Leveraging the fact that expert *humans* do not attend to all parts of an image equally when performing species classification, but rather have a model of relevant *parts* of an animal, Zheng et al. (2017) learn an attentional model for *parts* while training their classifier. Zhang et al. (2014) not only learn visual parts but learn geometric relationships among those parts. Meanwhile, Simon et al. (2017) introduce a method for *visualizing* the semantic parts being attended to by a classifier, focusing on bird classification. Visualizing regions of attention is critical to allowing human users to trust ML algorithms: if the parts of an image being used by an algorithm to differentiate species correlate with domain knowledge of visual differences, humans are more likely to trust that the algorithm has learned meaningful information, and the model is far more likely to generalize to new scenarios.

11.4 Generalizable classification and detection for low-quality images

11.4.1 Problem overview

Much of the recent success in image classification has benefitted from well-lit imagery, often focusing on images where a human photographer followed conventional rules of lighting, focus, and framing. While there is no universal definition of 'low quality', a reasonable intuition is that we are referring to images that do not have this property, that is, images where lighting, focus, and framing are subject to uncontrollable variability based on environmental conditions and unpredictable subject behaviour. This section will focus on the challenge of detecting and/or classifying objects in such images, and particularly on the development of models for challenging images that generalize to new regions.

11.4.2 Why CV for low-quality images matters for conservation

Monitoring wildlife populations is critical to planning the locations of protected areas, allocating anti-poaching resources, and advising governments and industry on the impact of human development on local biodiversity. For many scenarios—particularly for terrestrial megafauna surveys—camera traps (see Chapter 4) are among the best tools we have right now for objective, continuous monitoring of wildlife populations. However, a typical camera trap survey can produce millions of images, and labelling those images is intensely laborious. Even weeding out the ~75% of images that are typically empty strains the resources of most organizations who are tasked with measuring wildlife populations. CV provides a path to accelerating this process: first by developing algorithms that *detect* wildlife in images (reducing the number of images requiring labelling), later algorithms that *identify and count* wildlife in images, and finally algorithms that completely automate the estimation of wildlife populations from remote cameras, minimizing human intervention throughout the pipeline.

However, realizing this vision is subject to the key obstacle that *camera trap images are often extremely low-quality*, particularly night-time images. Unpredictable wake-up times and unpredictable animal movement mean that animals of interest are poorly framed, often at the margin of images or excessively close to the camera. For the same reasons, infrared (IR) flashes (which are concentrated near the centre of the image) rarely provide uniform illumination over the animal. Long exposure times in low light conditions lead to extreme motion blur, and even in daylight, animals of interest are often partially obscured behind rocks and vegetation. And finally, even in the best of conditions, there is no way to ensure an animal's consistent orientation relative to the camera.

While there are other conservation scenarios where image quality is a huge constraint (e.g.

marine video analysis in murky water), processing camera trap images represents such a ubiquitous problem—and has captured so much research attention—that we will focus this section specifically on camera traps.

11.4.3 Why CV for low-quality images is challenging

In the previous section, we highlighted that much of the recent progress on image classification tasks can be attributed to the ImageNet data set. Models trained on ImageNet are then used for *pretraining* for new problems: if we have a new classification problem to solve, rather than starting our training process from scratch, we would typically start from a model trained on ImageNet, and provide examples to adapt that model to our new problem. This lets us benefit from the huge volume of labelled images present in ImageNet, which we would rarely have for new problems.

The COCO data set (Lin et al., 2014) has played a similar role in advancing the state of the art in object detection and localization. In addition to differing from some conservation problems in the granularity of their classes (as per previous section), ImageNet and COCO also differ critically from camera trap data sets because they are almost entirely high-quality images, that is, the images are well-lit, free of blur, and generally well-framed.

Conservation problems rarely have the millions of labelled images that are typically required to train CV models from scratch, so this approach—*transfer learning* from ImageNet or COCO—is appealing. However, using this approach to leverage COCO- or ImageNet-trained models to build models, for example, for camera traps, presents the compound problem that (a) pretrained models based on ImageNet and COCO may have learned low-level features that are hard to 'un-learn' when fine-tuning, but may not be relevant to camera trap images, and (b) architecture and training methodology suitable for ImageNet-style images may not generalize to low-quality images, requiring novel methodology.

Furthermore, camera trap images present several challenges in the fundamental problem structure that are unique or unusual in CV.

The right loss function for an object detection problem, for example, may be highly asymmetric, that is, for many scenarios, *it is much more expensive to miss an animal than to ask a biologist to look at an empty image.* We suggest that the right error function for camera trap scenarios is very survey-specific, but often looks something like 'precision at 95% recall' or 'mean precision from 90% to 100% recall'. That is, we likely want to minimize the number of images an ecologist has to review while still guaranteeing that we find 95% of the animals. This is somewhat contradictory to the most common loss functions used to both train and benchmark object detection systems, which typically focus on mean average precision (mAP). mAP factors in the precision at very low recall values, which is essentially irrelevant for conservation scenarios. As we diverge from conventional error functions, we become less able to leverage existing training approaches and less able to simply select the 'best' object detector based on state-of-the-art results on COCO data. Furthermore, researchers are interested in certain species more than others and have varying tolerances for false negatives. We need to not only express those constraints in training but also *build tools that allow biologists to express loss functions directly.*

The *temporal* structure of camera trap images is also unique. Camera trap images are typically captured in 'bursts' of three to ten images, typically at a rate of around 1 Hz. While several approaches have been proposed for utilizing frame-to-frame differencing or movement tracking to highlight objects of interest (i.e. moving animals), this has proven difficult for three reasons: (1) animals are often stationary by the time the camera wakes up; (2) extensive movement of background vegetation is common; and (3) the low frame rate and lack of textured features in night-time images prevent the application of traditional optical flow approaches. That said, it is common for a human annotator to see an animal moving by paging through a sequence of images that would be completely undetectable in any single image, often because the primary visual feature is just a reflective eye glint. Consequently, we know that by not utilizing this sequence data, we are failing to take advantage of critical information. Furthermore, camera trap images are unique in a larger temporal sense: a typical survey might contain several million images, but those images are typically collected from several hundred cameras, each of which has a relatively stationary background that offers the opportunity to learn per-camera background information. *Fully exploiting this unusual temporal structure is an open problem for CV.*

Finally, recent work has shown that while detectors are robust within a set of cameras, they are highly sensitive to background texture; consequently, *generalization to new locations remains very difficult* (Beery et al., 2018). The real use case, of course, requires high accuracy on *new* camera trap deployments, where we do not have millions of annotations, or—often—any annotations at all. *Building generalizable models, intelligently selecting appropriate pretrained models for a new image set,* and *using interactive learning to bootstrap classifiers at new sites* remain critical open challenges for camera traps. Recent work on domain confusion (Tzeng et al., 2017)—a method for training deep networks to be less dependent on, for example, camera locations—may substantially improve the generalizability of models.

11.4.4 Progress in CV for low-quality images

As was the case with the fine-grained classification problem, progress in the camera trap space is as much about *data* as it is about *algorithmic innovation.* Consequently, progress in this area has been 'unblocked' by the recent emergence of large, labelled data sets of camera trap images. This trend was initiated by the release of the original, 1.3M-sequence Snapshot Serengeti data set (Swanson et al., 2015), and since then other authors have followed this as best practice, to allow reproducibility. Norouzzadeh et al. (2018), Tabak et al. (2019), Willi et al. (2019), Beery et al. (2018), and Yousif et al. (2019), for example, all either worked with publicly available data sets or released their data publicly along with their publications. Most of those data sets, and several others, are now available at the Labeled Information Library of Information: Biology and Conservation (LILA BC), an open data repository for conservation imagery.[1]

---

[1] http://lila.science

Similarly, a critical trend in this area has been towards the release of open-source code and models, allowing reproducibility, avoiding 'reinvention of the wheel', and allowing more systematic efforts to understand the limitations of published results (particularly for generalizability). Norouzzadeh et al. (2018), Tabak et al. (2019), and Beery et al. (2019) all released code associated with their publications, for example, and several other unpublished projects have contributed to the open-source community. The authors of this chapter curate a fairly exhaustive list of code repositories related to camera traps,[2] and the camera traps topic is used relatively consistently on GitHub to associate these repositories.

In addition to progress related to the release of data and code, a substantial amount of recent algorithmic work has focused specifically on addressing CV challenges for camera trap images. Some early success cases have emerged where ML has been deployed for camera trap image analysis.

Norouzzadeh et al. (2018) propose a multistage architecture, training separate deep networks to (a) identify empty vs. non-empty images, (b) identify species, and (c) count animals. They present state-of-the-art results on the publicly available Snapshot Serengeti data set (Swanson et al., 2015), showing 97% and 95% accuracy for presence/absence classification and species classification, respectively. Yousif et al. (2019) focus specifically on separating camera trap data sets into humans, animals, and empty images; they combine a classical-vision approach to region-of-interest identification (leveraging frame-to-frame differencing) with a deep learning approach to ROI classification. They report 99.58% accuracy on image-level empty vs. non-empty classification on Snapshot Serengeti data, the current state of the art. Tabak et al. (2019) achieve 98% top-1 accuracy on a data set of 3M from the United States, but confirm the work of Beery et al. (2018) in demonstrating substantial falloff in species classification accuracy when generalizing to a new region (82% top-1 accuracy on a data set from Canada). They demonstrate, however, that empty/non-empty classification generalizes substantially better,

achieving 94% empty/non-empty accuracy on a data set from Tanzania.

Willi et al. (2019) take the critical step of integrating an image classification system into a live experiment with human annotators; in this case, citizen-science volunteers. The use of ML results in a 43% reduction in the annotation time required of citizen scientists. To our knowledge, this is the only published result demonstrating a reduction in human effort for camera trap image annotation.

While all of this work demonstrates significant progress, none of these efforts address the generalization challenges raised earlier; metrics for camera trap vision are generally reported on a train/validation split that separates images by sequence but not by location, so claims cannot be made about generalizing to new cameras, even within the same ecosystem.

11.5 Supervised detection and classification in imbalanced scenarios

11.5.1 Problem overview

ML models learn both properties of the classes and objects of interest, for example, the appearance of a car and the *distribution* of those classes and objects, that is, the probability of observing a car in a scene. This raises significant challenges when rare classes or rare objects are of particular importance to a problem; in object detection scenarios, this often corresponds to environments where the background (empty) class is the vast majority of the data.

We will discuss some challenges associated with class imbalance for classification problems in the 'Partial Model Transfer' section next; in this section, we will focus on imbalance problems in object detection scenarios. Oksuz et al. (2019) present an excellent overview of the challenges associated with imbalanced data in objection detection problems.

11.5.2 Why CV with imbalanced training data matters for conservation

Two detection problems in wildlife conservation suffer from extreme imbalance problems, specifically a vast over-representation of the negative class:

² http://aka.ms/cameratrapsurvey

(1) detection of wildlife in aerial survey images; and (2) detection of wildlife sounds in acoustic (microphone/hydrophone) data. Even in areas that we think of as densely populated by wildlife, the percentage of image pixels occupied by wildlife is vanishingly small, due to both spatial sparseness and—for manned-flight surveys—high altitude.

Camera traps suffer from an imbalance problem as well, in that often 90% of images are empty, but camera traps benefit tremendously in this respect from motion triggering, which dramatically increases the probability that an image will contain an animal. However, this also limits the robustness of population surveys, since no *absence* data is available in motion-triggered surveys. Consequently, many camera trap surveys elect to use time-triggered capture, which frequently elevates the percentage of empty images to over 99%.

Nonetheless, even time-triggered camera trap surveys suffer nowhere near the class imbalance that impacts many aerial surveys. So we will use aerial surveys (from crewed aircraft, UAVs, or satellites) as our running example in this section.

11.5.3 Why CV with imbalanced training data is hard

Image sets in the aerial survey space are not large enough (in terms of positive examples) to train complex models from scratch, but provide very challenging images whose statistics are not like those of COCO, for example, which makes standard transfer learning approaches difficult. Images are of arbitrary orientation, the vast majority of images are empty (typically 99%), loss functions are highly asymmetric (false negatives are far more costly than false positives), and objects are blurry and very small relative to image resolution, such that even *positive* images are almost entirely *negative* pixels. The lack of suitability for transfer learning is exacerbated because many aerial data sets are multispectral (e.g. RGB+IR), complicating transfer learning from models trained on RGB images.

In addition to the core CV challenges, these data sets—though small in terms of positive examples—present significant challenges for CV *infrastructure*, requiring coordinated training over tens of TB of very-high-resolution images to ensure that the underlying distribution (i.e. the fraction of empty images) is preserved while ensuring adequate *positive* examples to train models effectively.

11.5.4 Progress in CV with imbalanced training data

The large size of data sets in this space has made open data sharing less practical than in the areas discussed in the previous sections. This is a substantial impediment to progress in this space at present. Naude and Joubert (2019) released a data set including over 2000 aerial images containing over 15,000 elephants, which is an excellent opportunity for standard benchmarking on a species of great interest for aerial surveys, but this is still small compared to a typical survey, making it difficult to assess imbalance challenges. A large labelled data set of RGB/IR images from a seal survey was recently released;[3] this is the largest publicly available aerial wildlife survey data set of which we are aware.

Despite the lack of publicly available data and therefore the challenges in reproducibility and reuse, tremendous progress has been made on addressing these challenges for aerial surveys. Torney et al. (2019), for example, modify a YoloV3 detector to accommodate specific features of aerial survey images, namely the relatively consistent scale (compared to natural images) and the huge bias towards negative space (in this case addressed with a modified loss function). Kellenberger et al. (2018) comprehensively address the challenges unique to UAV images, leveraging class weighting and overrepresentation of positive examples early in training (adding hard negatives later on). They also introduce evaluation metrics that represent a more appropriate estimate of time-saving for wildlife surveys than traditional CV metrics, particularly evaluating the number of images requiring human review, rather than applying the object-level accuracy statistics that are common in CV.

Corcoran et al. (2019), Hong et al. (2019), Rivas et al. (2018), and Kellenberger et al. (2017) all demonstrate progress using deep learning to detect

---

[3] http://lila.science/datasets/arcticseals

animals (koalas, birds, cattle, and savanna mammals, respectively) in UAV images. Guirado et al. (2019) and Borowicz et al. (2019) extend this trend to detecting whales in satellite images. To allow reasonable performance over the very large areas that become addressable with satellite surveys, Guirado et al. (2019) leverage a low-cost binary classifier as a screening mechanism, then run a more expensive object detector on high-probability images.

11.6 Emerging topics

The challenges we looked at in previous sections remain open problems, but they have been addressed to a sufficient extent that translational work has begun in conservation. In this section, we will survey—in slightly less detail—several areas of critical importance to conservation problems that have received little or no explicit attention in the conservation technology literature, that is, where translational work has barely begun.

11.6.1 Partial model reuse

In conservation scenarios, ecologists frequently want to take a model that has been trained to recognize plant or animal species in one region and use that to bootstrap a model in a region that shares most—but not at all—of the species. For example, anecdotally, we have seen high interest in taking pretrained models for camera trap data in East Africa and applying them to camera trap data in Southern Africa, removing classes unique to East Africa and adding new data only for classes not observed in the training region.

While this is sometimes casually referred to as transfer learning, this is a common misunderstanding, and this is not a conventional transfer learning scenario. Transfer learning reuses much of the information stored in a model, but discards the original output classes, and requires retraining on a new, *complete* set of the target classes. The purpose of transfer learning is to *reduce* the number of examples required compared to what would be required to train a network from scratch, but it does not transfer the *classes* from the original model. *Deep learning does not yet have a robust way to reuse some classes from*

a trained model without having access to the original training data.

There is emerging literature around this problem in the CV space, but to our knowledge this has not yet matured to the point where it has been consumed by conservation applications. In CV, this is a subproblem within *incremental learning*, where new classes can be added without disrupting a model's memory of existing classes, and without access to all of the original training data. The incremental learning literature typically addresses the typically addresses the dynamic *addition* of classes; less attention has been given to allowing models to selectively *forget* classes, for example, species that are not present in the target region.

Shmelkov et al. (2017) address this challenge for object detectors specifically: they introduce a novel loss function that penalizes a *change* in the probabilities of existing classes when training on new data to add new classes, that is, they discourage 'forgetting' information about existing classes. Tasar et al. (2019) take a similar approach specifically for remote sensing data. Neither of these approaches requires any of the original training data to be present at the time of class expansion. Castro et al. (2018) use a similar loss to penalize forgetting information on existing classes in classification problems, requiring only a few examples from the original classes to be present at the time of class-expansion training.

11.6.2 Human-in-the-loop annotation

Nearly all of the work in the space of CV for conservation has focused on automating classification and detection using statically trained models. This is appropriate for scenarios where (a) projects have enough ecosystem-specific data to train ML models and (b) accuracy is high enough that human review is no longer necessary. In practice, neither of these things are true for most problems. Consequently, there is a need for incorporating human expertise directly into the model-training process, to bootstrap models from limited data, and to interactively refine models to improve accuracy for new domains, regions, etc.

The incorporation of a human labeller into the ML process, typically by leveraging machine intelligence to select useful images for a human to annotate, is referred to as *active learning*. To our knowledge, the only work to apply active learning to problems in wildlife conservation is that of Norouzzadeh et al. (2019), who take an active learning approach to adapt species classification models for camera traps to new species distributions; describe an architecture that uses a large embedding model (trained infrequently) and a small classification model (trained frequently), along with active learning selection strategies. Though only in a simulated active learning environment, they show the same results on Snapshot Serengeti with 14.1k labels as was previously achieved with 3.2M labels. Though not directly related to wildlife conservation, Robinson et al. (2019) and Tuia et al. (2011) take an active learning approach to the related space of land use segmentation in aerial images.

Deeper integration of the human expert into the training process is a growing opportunity for this field, potentially addressing many of the data scarcity and generalization challenges raised earlier in this chapter.

11.6.3 Simulation for training data augmentation

The cost of data collection in the wild—particularly for rare species or in remote areas—leaves CV for conservation scenarios in a chronically 'data-poor' state. The use of simulation to augment training data sets is a promising way to 'fill gaps', or add balance to data sets biased towards common classes. Tremblay et al. (2018) and Shah et al. (2018) have started to address this for CV in general and for autonomous vehicle navigation specifically, but simulated training data is still in its infancy, even in the broad sense. That said, Bondi et al. (2018) present early results in applying simulation approaches to aerial imaging scenarios, while Beery and Morris (2019) explore the application of data simulation for adding missing classes in camera trap data.

11.7 Practical lessons learned and challenges to the community

We will conclude with some anecdotal remarks on the main practical challenges we have encountered as technologists working to accelerate conservation workflows; we believe that these are not research problems *per se*, but rather are primarily non-technical problems we can address as a community.

First and foremost, we have observed that ecologists share our enthusiasm regarding the ability of deep learning methods to accelerate their workflows, but are often disappointed when their high expectations are not met regarding the 'readiness' of this technology. We have seen this most frequently as an overestimation of existing models' ability to generalize to new problems—or to nearly identical problems—even in related ecosystems. This is partially a failure on the part of technologists to communicate the fundamental limitations of transfer learning and domain adaptation, particularly the issues raised previously, in the 'partial model transfer' section of this chapter; for example, adding and removing classes is not a straightforward process. This is also a consequence of some laxity on our part, as technologists, to explicitly match the claims we make to the validation approaches we use in the literature. In particular, much of the literature in this space trains models on a subset of images, then evaluates those models on a non-overlapping subset that *includes images from the same regions and even the same stationary cameras*. This is not 'wrong', but it limits the claims we can make about generalizability, and when CV literature does not explicitly *restrict* the claims of generalizability, the target audience frequently overestimates the applicability of approaches to new problems. **We challenge the conservation technology community to develop standards and consistent terminology around validation approaches, favouring cross-location validation whenever possible. We explicitly (rather than implicitly) restrict the claims we make about generalizability based on the validation approaches we use.**

Another major practical challenge has been the communication of *required accuracy* between ecologists and technologists, particularly the mapping of traditional CV accuracy metrics (classification accuracy, precision, recall, etc.) to the *amount of time saved* for real workflows. Literature reports of 90% or 95% accuracy may sound excellent, and for some workflows, that level of error is minor compared to other sources of noise. However, some workflows need higher accuracy, and for those situations the gap may make machine learning—even at 95% accurate—not a significant workflow accelerator. For example, if we are using ML to identify invasive rodents in camera trap images, it may be essential to find nearly *every* image containing a rodent, to allow rapid response before the invasive population grows. Furthermore, we have been lax as a technology community in reporting meaningful metrics around efficiency gains. For the most part, we relied on traditional metrics that allow comparison to existing CV papers, but are not necessarily indicative of what the benefits are for the ecology community in adapting ML-based workflows. **We challenge the conservation technology community to begin reporting results in estimated workflow efficiency gains, and whenever possible, to perform human-subjects experiments that allow us to *measure* workflow efficiency gains.**

Perhaps most importantly, the literature in this area has focused primarily on algorithmic innovation, which often leaves a gap between 'good ML work' and 'a tool that ecologists can use'. Only very limited development of tools and workflows has occurred in this area, and to our knowledge this work has not typically been published in academic venues. **We challenge the conservation technology community to prioritize the rigorous design, study, and publication of tools, workflows, and efficiency gains, engaging with the human factors and human–computer interaction (HCI) communities for collaboration and joint publication.**

Finally, the lack of publicly available labelled data sets has presented a consistent challenge in developing new techniques, engaging the ML community in conservation problems, applying existing techniques to new species distributions, and consistently benchmarking new approaches. Some open data repositories have emerged recently (e.g. LILA BC[4]), but these sample a tiny slice of the world's species. These repositories are heavily biased towards handheld images and camera traps, with only a very limited amount of labelled bioacoustic data available, and virtually no open labelled data available from aerial wildlife surveys. **We challenge the conservation community to encourage the release**

of labelled data and the use of standardized data sets, particularly through the peer-review process. Some reticence to release data has been based on the sensitivity of location information about at-risk species; we stress that in the vast majority of cases, precise location information is not important to advancing the state of the art.

11.8 Other reviews

Finally, we will highlight several other review papers that overlap with the present survey. Humphries et al. (2018) survey machine learning applications in ecology; this survey is not primarily focused on CV or deep learning, rather on classical ML applications in ecology and conservation. Weinstein (2018) surveys CV applications specifically, and Lamba et al. (2019) focus specifically on deep learning. Several online surveys also aggregate materials relevant to this space[5,6,7]. Two of these are maintained by the authors of this chapter and include many of the materials presented here.

References

Beery, S., Liu, Y., Morris, D., et al. (2019). Synthetic examples improve generalization for rare classes. arXiv preprint arXiv:1904.05916.

Beery, S., & Morris, D. (2019). Efficient pipeline for automating species ID in new camera trap projects. *Biodiversity Information Science and Standards*, 3, e37222.

Beery, S., Van Horn, G., & Perona, P. (2018). Recognition in terra incognita. In Proceedings of the European Conference on Computer Vision (ECCV) (pp. 456–473).

Berg, T., Liu, J., Woo Lee, S., Alexander, M. L., Jacobs, D. W., & Belhumeur, P. N. (2014). Birdsnap: large-scale fine-grained visual categorization of birds. In Proceedings of the IEEE Conference on Computer Vision and Pattern Recognition (pp. 2011–2018).

Bondi, E., Dey, D., Kapoor, A., et al. (2018, June). Airsim-w: a simulation environment for wildlife conservation with UAVS. In Proceedings of the 1st ACM SIGCAS Conference on Computing and Sustainable Societies (p. 40).

[4] http://lila.science

[5] http://aka.ms/cameratrapsurvey
[6] http://aka.ms/aerialwildlifesurvey
[7] https://github.com/patrickcgray/awesome-deep-ecology

Borowicz, A., Le, H., Humphries, G., Nehls, G., Höschle, C., Kosarev, V., & Lynch, H. J. (2019). Aerial-trained deep learning networks for surveying cetaceans from satellite imagery. *PloS One*, 14(10), e0212532.

Castro, F. M., Marín-Jiménez, M. J., Guil, N., Schmid, C., & Alahari, K. (2018). End-to-end incremental learning. In Proceedings of the European Conference on Computer Vision (ECCV) (pp. 233–248).

Corcoran, E., Denman, S., Hanger, J., Wilson, B., & Hamilton, G. (2019). Automated detection of koalas using low-level aerial surveillance and machine learning. *Scientific Reports*, 9(1), 3208.

Deng, J., Dong, W., Socher, R., Li, L. J., Li, K., & Fei-Fei, L. (2009, June). Imagenet: a large-scale hierarchical image database. In 2009 IEEE Conference on Computer Vision and Pattern Recognition (pp. 248–255).

Guirado, E., Tabik, S., Rivas, M. L., Alcaraz-Segura, D., & Herrera, F. (2019). Whale counting in satellite and aerial images with deep learning. *Scientific Reports*, 9(1), 1–12.

Hong, S.-J., Han, Y, Kim, S.-Y., Lee, A.-Y., & Ghiseok, K. (2019). Application of deep-learning methods to bird detection using unmanned aerial vehicle imagery. *Sensors*, 19(7), 1651.

Humphries, G. R., Magness, D. R., & Huettmann, F. (Eds.). (2018). *Machine Learning for Ecology and Sustainable Natural Resource Management*. Springer, Cham, Switzerland.

Kellenberger, B., Marcos, D., & Tuia, D. (2018). Detecting mammals in UAV images: best practices to address a substantially imbalanced dataset with deep learning. *Remote Sensing of Environment*, 216, 139–153.

Kellenberger, B., Volpi, M., & Tuia, D. (2017, July). Fast animal detection in UAV images using convolutional neural networks. In 2017 IEEE International Geoscience and Remote Sensing Symposium (IGARSS) (pp. 866–869).

Lamba, A., Cassey, P., Segaran, R. R., & Koh, L. P. (2019). Deep learning for environmental conservation. *Current Biology*, 29(19), R977–R982.

Lin, T. Y., Maire, M., Belongie, S., et al. (2014, September). Microsoft COCO: common objects in context. In European Conference on Computer Vision (pp. 740–755).

Naude, J., & Joubert, D. (2019). The aerial elephant dataset: a new public benchmark for aerial object detection. In Proceedings of the IEEE Conference on Computer Vision and Pattern Recognition Workshops (pp. 48–55).

Norouzzadeh, M. S., Morris, D., Beery, S., Joshi, N., Jojic, N., & Clune, J. (2019). A deep active learning system for species identification and counting in camera trap images. arXiv preprint arXiv:1910.09716.

Norouzzadeh, M. S., Nguyen, A., Kosmala, M., et al. (2018). Automatically identifying, counting, and describing wild animals in camera-trap images with

deep learning. *Proceedings of the National Academy of Sciences*, 115(25), E5716–E5725.

Oksuz, K., Cam, B. C., Kalkan, S., & Akbas, E. (2019). Imbalance problems in object detection: a review. arXiv preprint arXiv:1909.00169.

Rivas, A., Chamoso, P., González-Briones, A., & Corchado, J. (2018). Detection of cattle using drones and convolutional neural networks. *Sensors*, 18(7), 2048.

Robinson, C., Ortiz, A., Malkin, K., et al. (2019). Human-machine collaboration for fast land cover mapping. arXiv preprint arXiv:1906.04176.

Shah, S., Dey, D., Lovett, C., & Kapoor, A. (2018). Airsim: high-fidelity visual and physical simulation for autonomous vehicles. In Field and Service Robotics (pp. 621–635).

Shmelkov, K., Schmid, C., & Alahari, K. (2017). Incremental learning of object detectors without catastrophic forgetting. In Proceedings of the IEEE International Conference on Computer Vision (pp. 3400–3409).

Simon, M., Gao, Y., Darrell, T., Denzler, J., & Rodner, E. (2017). Generalized orderless pooling performs implicit salient matching. In Proceedings of the IEEE international conference on computer vision (pp. 4960–4969).

Swanson, A., Kosmala, M., Lintott, C., Simpson, R., Smith, A., & Packer, C. (2015). Snapshot Serengeti, high-frequency annotated camera trap images of 40 mammalian species in an African savanna. *Scientific Data*, 2, 150026.

Tabak, M. A., Norouzzadeh, M. S., Wolfson, D. W., et al. (2019). Machine learning to classify animal species in camera trap images: applications in ecology. Methods in Ecology and Evolution, 10(4), 585–590.

Tasar, O., Tarabalka, Y., & Alliez, P. (2019). Incremental learning for semantic segmentation of large-scale remote sensing data. *IEEE Journal of Selected Topics in Applied Earth Observations and Remote Sensing*, 12(9), 3524–3537.

Torney, C. J., Lloyd-Jones, D. J., Chevallier, M., et al. (2019). A comparison of deep learning and citizen science techniques for counting wildlife in aerial survey images. *Methods in Ecology and Evolution*, 10(6), 779–787.

Touvron, H., Vedaldi, A., Douze, M., & Jégou, H. (2019). Fixing the train-test resolution discrepancy. arXiv preprint arXiv:1906.06423.

Tremblay, J., Prakash, A., Acuna, D., et al. (2018). Training deep networks with synthetic data: bridging the reality gap by domain randomization. In Proceedings of the IEEE Conference on Computer Vision and Pattern Recognition Workshops (pp. 969–977).

Tuia, D., Pasolli, E., & Emery, W. J. (2011). Using active learning to adapt remote sensing image classifiers. *Remote Sensing of Environment*, 115(9), 2232–2242.

Tzeng, E., Hoffman, J., Saenko, K., & Darrell, T. (2017). Adversarial discriminative domain adaptation. In Proceedings of the IEEE Conference on Computer Vision and Pattern Recognition (pp. 7167–7176).

Van Horn, G., Branson, S., Farrell, R., et al. (2015). Building a bird recognition app and large scale dataset with citizen scientists: the fine print in fine-grained dataset collection. In Proceedings of the IEEE Conference on Computer Vision and Pattern Recognition (pp. 595–604).

Van Horn, G., Mac Aodha, O., Song, Y., et al. (2018). The iNaturalist species classification and detection dataset. In Proceedings of the IEEE Conference on Computer Vision and Pattern Recognition (pp. 8769–8778).

Wah, C., Branson, S., Welinder, P., Perona, P., & Belongie, S. (2011). The Caltech-UCSD Birds-200-2011 dataset. Computation & Neural Systems Technical Report, CNS-TR-2011-001.

Weinstein, B. G. (2018). A computer vision for animal ecology. *Journal of Animal Ecology*, 87(3), 533–545.

Willi, M., Pitman, R. T., Cardoso, A. W., et al. (2019). Identifying animal species in camera trap images using deep learning and citizen science. *Methods in Ecology and Evolution*, 10(1), 80–91.

Xian, Y., Lampert, C. H., Schiele, B., & Akata, Z. (2018). Zero-shot learning—a comprehensive evaluation of the good, the bad and the ugly. IEEE *Transactions* on *Pattern Analysis* and *Machine Intelligence*. PP. 10.1109/TPAMI.2018.2857768.

Yosinski, J., Clune, J., Bengio, Y., & Lipson, H. (2014). How transferable are features in deep neural networks?. In Advances in Neural Information Processing Systems (pp. 3320–3328).

Yousif, H., Yuan, J., Kays, R., & He, Z. (2019). Animal Scanner: software for classifying humans, animals, and empty frames in camera trap images. Ecology and Evolution, 9(4), 1578–1589.

Zhang, N., Donahue, J., Girshick, R., & Darrell, T. (2014, September). Part-based R-CNNs for fine-grained category detection. In European Conference on Computer Vision (pp. 834–849).

Zheng, H., Fu, J., Mei, T., & Luo, J. (2017). Learning multi-attention convolutional neural network for fine-grained image recognition. In Proceedings of the IEEE International Conference on Computer Vision (pp. 5209–5217).

CHAPTER 12

Digital surveillance technologies in conservation and their social implications

Trishant Simlai and Chris Sandbrook

12.1 Introduction

Digital technology is rapidly changing economies, societies, cultures, and lives throughout the world. Indeed, many commentators and scholars have argued that human society has entered the 'Information Age' (Castells, 2018). Large amounts of public and private information now flow through digital networks created by the internet and other communication technologies, resulting in new modes of governance, business, and communication. As the chapters in this volume make clear, this digital and technological revolution has had a considerable impact on the social practices of individuals and organizations involved in nature conservation. From tracking the movement of wild animals to detecting illegal wildlife trade online (Sonricker Hansen et al., 2012), digital technologies and applications are gaining increasing prominence in nature conservation and are rapidly reshaping the discourses of conservation science and practice (Newman et al., 2012; Joppa, 2015). These technologies are rapidly influencing how scientists, government officers, and members of the public think, perceive, and engage with nature (Kahn, 2011; Verma et al., 2015).

Researchers, conservationists, and policymakers often welcome the technologies of the information age as they promise large amounts of data, rapid processing speeds, better access to information, unique visual representations, and efficient decision-making capability (Arts et al., 2015). However, some have argued that there is a downside to this story, due to the practical challenges and the social consequences created by the use of new technology (Humle et al., 2014; Maffey et al., 2015; Wich et al., 2017). This chapter reviews some of these social implications, with a focus on camera traps and drones. We argue that digital technologies are not the panacea for all conservation-related problems, and interventions based on these technologies must be carefully reviewed before use. We also argue that it is important to consider and reflect on who controls, benefits from, is affected by and pays for these technologies. Finally, we stress the need for ethical guidelines for the use of these technologies and also frameworks of best practice and regulation that promote democratization of the use of technology from its outset. Before turning to the specific case of conservation technologies, we begin by introducing some of the broader issues raised by the social lives of digital surveillance technologies.

12.1.1 Privacy, civil liberties, and freedom

The growing intrusion of surveillance by state and corporate actors into citizens' daily lives all around the world is increasingly common knowledge. However, many people remain unaware of the extent of this intrusion, and the social, political, and ethical issues that surround it. Foremost among

Trishant Simlai and Chris Sandbrook, *Digital surveillance technologies in conservation and their social implications.*
In: *Conservation Technology.* Edited by Serge A. Wich and Alex K. Piel, Oxford University Press.
 DOI: 10.1093/oso/9780198850243.003.0012

such concerns is the possibility that surveillance may constitute an invasion of privacy. A widely accepted definition of privacy is still being debated. Some scholars have called it 'an unusually slippery concept' (Whitman, 2004: pp. 1151–1221), while some have referred to it as a 'concept in disarray' (Solove, 2008: p. 12). The question of what is and is not private varies culturally and geographically. For example, entering a room without knocking on the door might be considered a serious privacy violation in one culture yet permitted in another. Privacy hence may not be understood as a universal given but as a social construct, which changes according to cultures and geographies (Shapiro & Baker, 2001; Moore, 2008; Gomes de Souza, 2015). However, consensus exists on the fact that privacy is comprised of multiple dimensions that are specified as privacy of a person, personal behaviour, personal data, and personal communication (Clark, 2006).

Despite debates over its definition, privacy is recognized as a basic human right, just like freedom of speech and freedom of assembly. Article seven of the European Charter of Human Rights (The European Commission, 2000) and Article 12 of the United Nations Charter of Human Rights recognize the right to privacy as a basic human right (United Nations, 1948). The right of the state to protect itself and its community at large, as opposed to the right of an individual to privacy, has always been marred with tension throughout recorded history (Bannister, 2005). Today, digital technologies are changing at remarkable speed the scale and dimensions of how scientists, researchers, hobbyists, and the state inquire into the lives of individual citizens. For example, drones have been used to take pictures of high-profile individuals like television celebrities relaxing on beaches, sportspersons in training, and even politicians without their consent (Hyde, 2016; Berkowitz, 2017).

A key contemporary debate around technology and privacy relates to the use of digital surveillance technologies by the private sector. The case of Cambridge Analytica's theft of Facebook user data that was subsequently used to profile people and influence their political decisions brought to light the larger context of surveillance in modern day capitalism (Manokha, 2018). It has been argued that user data on digital platforms is fast becoming a 'fictitious commodity' that is being used in large amounts by commercial market entities as 'raw material' (Zuboff, 2019). Modern digital intelligence collection is increasingly relying on private companies (Joh, 2017). Data sets of interest to state enforcement agencies and state security agencies are being collected and generated in large amounts by private companies through a multitude of digital softwares that range from social networking applications to fitness and health applications (Manokha, 2018). These data sets are created and owned by private companies and can be purchased, legally compelled, or hacked, resulting in enormous consequences for privacy and civil liberties.

Surveillance technologies do not only raise privacy concerns—they also affect civil liberties (Lyon, 2001). The use of surveillance technologies is known to discourage or deter individual participation in social movements or activities of dissent, leading to the inhibition of an individual's freedom of assembly or freedom of expression (Cunningham & Noakes, 2008). Using surveillance technologies can also reinforce existing social inequalities, particularly marginalization along the lines of race, class, gender, age, and sexuality (Coleman & McCahill, 2011). Surveillance technologies can also adversely affect an individual's freedom of movement; it is argued that surveillance systems enable privileged mobility of some individuals over others, leading to what has been called social sorting (Lyon, 2010). For example, certain social groups are targeted over others through surveillance and subjected to random security checks, body scans, and paperwork at airports and immigration queues (Graham & Wood, 2003). These restrictions can disproportionately impact some groups of already marginalized travellers, such as Muslim women, for whom participation in body scanning systems impinges on their religious beliefs (Rohen, 2010).

Privacy, like freedom, is a valued good, which is traded voluntarily by citizens for other goods and benefits such as public services and security. In conservation practice, the use of surveillance technologies to monitor local communities living inside or next to protected areas (whether deliberately or inadvertently) may seem like a small thing to ask in relation to the benefits gained overall for the protection of certain species; however, it is precisely an accumulation of such small judgements that may cumulatively affect the perception of infringements of people's privacy, liberty, and freedom. In the following sections, we suggest that the use of surveillance technologies for conservation provides a clear example of this phenomenon, raising serious concerns about civil liberties and privacy.

12.2 Digital technologies and surveillance

A wide range of digital devices now readily available for nature conservation were originally developed for surveillance and law enforcement. This has opened up a scholarly discussion on the impacts that these devices can have on human well-being. Academic discourse on the impacts of surveillance exists as an entire branch in criminological research called surveillance studies (Lyon, 2001; Ball & Haggerty, 2005). However, these discussions have not looked into the impacts of surveillance or surveillance technologies when used in the field of environmental conservation.

Surveillance is a French word that literally means 'to watch from above' (Mann & Ferenbok, 2013). Surveillance has often been associated with a dystopian world where the Orwellian 'big brother' is always watching subversive behaviour (Lyon, 1994). However, the debate around surveillance often tends to overlook certain useful tendencies of surveillance in shaping and ordering modern citizens and societies (Lyon, 2001). In other words, not all surveillance is bad. It has been argued that like other digital technologies, surveillance technologies are value-neutral until they are applied towards specific uses (Ferenbok & Clement, 2012). However, we contest this notion and demonstrate in the latter part of this chapter that surveillance technologies may never truly be value-neutral, and that it largely depends on who is doing the surveillance and their objectives for doing so.

A key point of entry for understanding such surveillance is the notion of the 'panopticon' developed by the English philosopher Jeremy Bentham and conceptualized further by Michel Foucault. Bentham's panopticon is an imaginative work of architecture that comprises a central tower surrounded by cells. In the central tower is a watchman, and

in the cells are prisoners. Through the tower, the watchman sees all but people in the cells cannot see the watchman and hence have to assume that they are always under observation. The subject being observed is always at a disadvantage and is often oblivious to when s(he) is being watched, causing clear asymmetries in power distribution (Foucault, 1977; Ferenbok & Clement, 2012). Foucault further conceptualizes the panopticon and its watcher from an external entity to something that is more internalized and omnipresent (Foucault, 1977). The gaze of the watcher is internalized by the inmates of the prison to such an extent that each prisoner becomes his/her own guard, leading to what Foucault calls 'Panopticism' (Foucault, 1977).

Panopticism in governance creates docile and self-disciplined subjects (Lyon, 2006). New digital surveillance technologies are transforming the possibilities of how surveillance itself is practised due to reduced costs, flexibility, and increased efficiency, ushering in a new era of geographies of surveillance (Dobson & Fisher, 2010). Digital devices such as remotely triggered acoustic sensors, cameras, and drones open up new dimensions in how individual people are recorded. The potential to identify and track people through artificial intelligence and facial recognition algorithms increases the ability of states to enforce law and exercise power through a security apparatus—in effect, creating a new kind of conservation panopticon (Sandbrook et al., 2018). It is to the conservation surveillance regime that we now turn.

12.3 Conservation surveillance

Surveillance technologies are used in nature conservation primarily as a means to monitor wildlife populations. These technologies are shifting the paradigm of biodiversity conservation and management by helping conservationists and managers around the world to monitor wildlife, vegetation, and environmental degradation with precision and efficiency. The monitoring of wildlife includes studying and keeping track of animal movements, studying, and estimating populations, habitat condition and loss, and the identification of potential threats to wildlife such as poaching and human–wildlife conflict. The development of devices such as camera traps, radio collars, and drones for conservation has ushered in a new era for wildlife monitoring and management (Pimm et al., 2015). The use of these devices is rapidly changing the way conservation is perceived by citizens, particularly in protected areas (Shrestha & Lapeyre, 2018) and even more so outside them (Simlai, 2015). Modern technologies are known to mediate human–wildlife relations and in turn shape the way conservation policies are chosen and implemented (Büscher, 2016).

Surveillance technologies not only contribute to the intensification of conservation territorialization by demarcating spaces for people and nature, but also intensify management of the resulting boundaries (Adams, 2017). For example, tracking tigers using GPS collars or remote cameras can result in a change in management strategies when the tiger crosses protected area boundaries (DeFries et al., 2010; Carter et al., 2012). The public can see animals that are monitored by digital tracking technologies as the responsibility of those who monitor their movements (Cooke et al., 2017). For example, considerable outrage is generated when individual tigers that are collared or monitored using remote cameras raid livestock—or worse, kill a human—around the protected areas of India (Lenin, 2014).

Another implication of surveillance technologies in conservation is their contribution towards coercive conservation strategies (Peluso, 1993). Surveillance and control measures can be used on people as well as animals; in fact, digital tracking devices are currently being widely used to monitor anthropogenic activities inside protected areas, and there is a growing call for them to be extensively used to monitor poaching, illegal logging, and collect evidence to catch offenders (Hossain et al., 2016). Moreover, the use of surveillance technologies is central to the militarization of conservation (Duffy, 2014; Lunstrum, 2014; Büscher & Ramutsindela, 2015; Duffy et al., 2019). The Kenyan Wildlife Service regularly uses surveillance technologies such as drones, camera traps, and helicopter-mounted infrared cameras to combat (increasingly sophisticated and organized) poaching (Haslam, 2016). The ease with which surveillance technologies can shift objectives between tracking animals

and monitoring people or between warfare and securitized conservation is crucial in understanding how conservation is being increasingly drawn into narratives of global security (Duffy, 2016). There has been very limited work on the ethics and social risks of using these technologies for conservation (Sandbrook, 2015). In fact, intensive surveillance regimes have now been identified as an integral component to conservation governance in tandem with other intelligence gathering techniques based on classic military-styled counter-insurgency methods (Duffy, 2016). The perception of threat of physical enforcement exacerbated through surveillance has been argued to be as important as actual violence (Lombard, 2016).

It can be argued that conservation actors use surveillance technologies as part of a wider strategy to create disciplined conservation actors (Sandbrook et al., 2018) and establish new forms of governance (Agrawal, 2005). However, these processes and the resultant responses towards them may occur differently according to varying cultural contexts. It is important to highlight here that a large proportion of work on surveillance theory and the social impacts of surveillance is from the global north, while conservation surveillance technologies are most widely used in the global south. This is an important avenue for further research.

Conservation interventions exercise rules that constrain and restrict people's movement and rights, and enforce rules that are aided by the use of surveillance technologies. These practices exercise power over people and attempt to turn them into subjects that support conservation objectives decided by the state or private organizations. In some cases, this might be justified by the local context—for example, as a way to discourage people from entering strictly protected areas from which they are forbidden. In others, any social harms associated with surveillance may not be a price worth paying for any real or perceived conservation benefit. In the following sections, we will discuss the social and political implications of specific conservation surveillance technologies in detail and according to certain relevant themes they fall under. Although there are many kinds of surveillance technologies used in conservation that may have social and political implications, such as high-resolution satellite imagery, live feed video surveillance cameras and mobile applications to monitor ranger patrolling, we focus on the use of camera traps and drones as their use is widespread, and we have studied them most closely in our own work.

12.4 The social and political implications of camera traps and drones

Camera traps or motion-activated cameras are perhaps the most widely used surveillance technology in conservation and ecological research (Karanth & Nichols, 1998; Rovero & Zimmerman, 2016; Adams, 2017; Caravaggi et al., 2017). See Chapter 5 for further details. Although camera traps are normally deployed with the objective of taking images of wildlife, they are often triggered by humans (Sandbrook et al., 2018). The collection of human images in camera traps is often accidental or unplanned (Butler & Meek, 2014)—a phenomenon that has been termed 'human bycatch' (Sandbrook et al., 2018). However, camera traps have also been specifically deployed to monitor human activity for conservation goals such as quantifying anthropogenic pressure on ecosystems (Betts, 2015), monitoring human–wildlife interactions (Pusparini et al., 2018) and for anti-poaching (Hossain et al., 2016) and there are increased calls for their use in these sectors. Photographs of people taken by camera traps have the potential to change the nature of conservation law enforcement and give rise to new regimes of surveillance, 'conservation surveillance', 'that may be defined as close watch kept over someone or something for conservation purposes' (Sandbrook et al., 2018: p. 494).

The use of Drones or Unmanned Aerial Vehicles (UAVs) has rapidly evolved and developed in the past decade after being primarily designed for military use. Drones are now used for a range of purposes, from disaster management and relief to biodiversity conservation and management. See Chapter 3 for further details. In the context of direct conservation applications, drone use has been advocated for law enforcement and the monitoring of illegal activities (Sandbrook, 2015). Examples include patrolling protected area boundaries (Mulero-Pazmany et al., 2014), gathering evidence of illegal logging (Koh & Wich, 2012),

photographing ongoing illegal activity with a high-resolution camera as evidence to secure prosecution (Snitch, 2014) and functioning as a deterrent (Schiffman, 2014). Although the use of drones in the conservation sector has a wide range of potentially relevant applications, their use has largely been experimental to date (Sandbrook, 2015). However, it seems highly probable that drones will be used for the above-mentioned applications regularly as there are several examples of conservation agencies and research institutions actively engaged in developing their use (Vidal, 2013; Gorman, 2014; Wilkie & Rose, 2014). In the following subsections, and by using a thematic approach, we discuss the various social and political implications that can arise with the use of camera traps and drones. We also discuss some of the potential positives such as community-led resistance and countermapping arising out of the use of these technologies.

12.4.1 Infringement of privacy and consent

The use of conservation surveillance technologies for processing data on human activities raise concerns about civil liberties, freedom, and infringement of privacy. For example, camera traps set up in the woods of the Austrian state of Carinthia captured images of a local politician engaging in sexually explicit behaviour (Day, 2012). Pebsworth and LaFleur (2014) identify ethical concerns with the use of surveillance technology related to how best to protect people's identities, their privacy, and what to do with images of people engaging in illegal behaviour. Using camera traps for surveilling people in the context of conservation also follows the rise of a larger surveillance discourse in wider society, particularly in global security discourse and policing by the state (Sandbrook et al., 2018; Duffy et al., 2019). Camera traps may not seem as pervasive as drones in terms of surveillance, but they replicate the same intensification of conservation enforcement and governance regimes (Sandbrook et al., 2018; Shrestha & Lapeyre, 2018). When deployed in and around protected areas or even outside protected areas in human-dominated landscapes (Karanth & Defries, 2010), camera traps capture people's images and allow for the monitoring of human activities in the area. Images captured in the process are often used to inform research, law enforcement, and management activities that may adversely affect people who may not have consented to be photographed. For example, in the Rajaji and Corbett National Parks in India, a network of camera traps monitored the presence of tigers, other wild animals, and also people belonging to a particular indigenous and pastoral community called van gujjars (Simlai, 2015). The data gathered were used to estimate daily activity patterns and population of wild tigers and their prey species, and also quantify human presence and their activity as 'anthropogenic pressure' or 'illegal human presence'. Furthermore, processed results from the data went on to inform conservation policy that subsequently led to displacement of the van Gujjar community from both national parks (Simlai, 2015).

The potential of conservation drones to support data collection and law enforcement is well established; however, this itself should not be regarded as adequate justification for their extensive adoption (Sandbrook, 2015). Civil and military applications of drones have attracted much discussion on their social and ethical implications in which they might lead to pervasive restriction of civil liberties and infringe upon privacy (Sparrow, 2009). Concerns have been raised about whether monitoring people from the air, without their consent or knowledge, is ethically acceptable and at what point might this cross boundaries and become an infringement to privacy (Finn & Wright, 2012). The small and subtle nature of drones can make them access areas and spaces that might otherwise be considered private (Luo et al., 2014). Conservation drones and UAVs used for natural resource management are very likely to collect information about human activities and about human presence, leading to identification at an individual level. In the case of law enforcement, this is done deliberately and with the intention to prosecute. It has been suggested that drones should be used to monitor vehicular activity on public roads near protected areas to deter and detect illegal activities (Snitch, 2014). It has also been suggested that drones should be used covertly to observe potential illegal hunters (Mulero-Pazmany et al., 2014). These practices are ethically questionable when taking place on public land particularly when certain groups, often marginalized and

vulnerable, are targeted (Sandbrook, 2015; Duffy et al., 2019).

12.4.2 Psychological well-being and fear

Conservation surveillance technologies like drones and camera traps can also lead to considerable fear and confusion, generating hostility among people that are being monitored (Campbell & Veríssimo, 2015; Sandbrook, 2015). Many areas of conservation importance are located in remote regions of developing countries where communities of people have little exposure to modern technologies. This may generate suspicions and conspiracy theories about these technologies and the reasons for which they are being operated. While doing fieldwork, one author of this chapter has personally witnessed panic-stricken women running for cover when a drone appeared in an area where they regularly and legally collected firewood (Simlai, unpublished). Drones can carry an image of warfare and destruction, which could mean people having misconceptions about their purpose. Indeed, their use in areas of conflict or in areas with histories of violence could lead to people believing they have been sent by militaries, terrorist groups, or private companies, further fuelling conflicts or creating new ones (Simlai, 2015; Sandbrook, 2015, Duffy et al., 2019). Many such areas in the world have long-standing difficult relationships with state interventions and could very well transfer such suspicions to any new technology introduced to the area.

The extent and determination with which camera traps are vandalized or damaged in theft attempts, even when they are code-locked, has been well documented in Australia (Meek, 2017). Similarly, a survey of camera trap users in Australia and the United States revealed that a large proportion of them experienced some or the other form of damage to their equipment (Butler & Meek, 2014). In a more recent survey, over 75% of respondents reported objections to or direct interference with deployed camera traps confirming resistance or opposition to them (Sandbrook et al., 2018). These events on camera traps demonstrate that people on the ground are sufficiently concerned about them to attack them, although research on motivations is lacking.

Camera traps and drones increase the power of conservation governance by providing evidence of illegal activities and in turn help enforce the traditional ideas of distinction between spaces for nature and spaces for people (Bluwstein & Lund, 2018). When camera traps are deployed by non-state actors such as non-governmental organizations they extend power and authority and governance structures beyond the state. Discussion on the possible social implications of camera traps for conservation is very limited in academic literature in contrast to the number of scientific papers that base their research on camera trap surveys (Sandbrook et al., 2018). Some studies that use camera traps have mentioned negative effects such as theft and vandalism of people on camera traps (Bernard et al., 2014; Clare et al., 2015); however, very few mention the negative effects of camera traps on people (Rupf et al., 2011; Villaseñor et al., 2014).

The use of surveillance technology in most spaces without publicly agreed rules and collective transparency could provoke conflicts among community members (Paneque-Gálvez et al., 2014). Conflicts arising out of such use can affect partner organizations and in turn conservation in the long term. In some places, conservation organizations have a history of conflict with local communities resulting from repetitive evictions and exclusionary policies (West et al., 2006). It seems that negative perceptions and resistance towards drone deployment in such places will be highly likely. In some cases, the use of drones can be used deliberately to create a climate of fear and as a deterrent against illegal activities (Simlai, 2015; Sandbrook, 2015, Duffy et al., 2019). Demonstrations of drone use by local authorities and conservation agencies coupled with a media awareness campaign are done particularly to serve these ends as seen in South Africa and India (Mulero-Pazmany et al., 2014; Simlai, 2015). For example, the use of drones in the Balule nature reserve seems to have created a rumour among 'poachers' that machines in the sky are observing them continuously and even at night, reducing poaching incidents (Snitch, 2014). Such interventions that create widespread fear resulting in a particular kind of behaviour desired by the interveners has been described as 'disciplinary governmentality' (Foucault, 1977; Fletcher, 2010). Such approaches

may work in the short term, but the use of 'fear' as a tool for conservation raises ethical questions,[1] does not always work and may shift illegal activities to an alternate location that is not under surveillance.

12.4.3 Wider issues in conservation practice

Legislation specific to each country may lead to researchers who use conservation surveillance technologies themselves being prosecuted for disseminating images of humans, distributing them or for even deploying them in the first place (Butler & Meek, 2013; Meek & Zimmerman, 2016). For example, a group of Iranian biologists were prosecuted, charged, and face the death penalty under suspicion of espionage after they deployed camera traps to study Asiatic cheetahs but got caught in a power struggle between Iran's revolutionary guards and the relatively moderate administration under Iranian President Hassan Rouhani (Stone, 2018). In some cases, government authorities specifically request images taken by camera traps to pursue leads on wildlife crimes; moreover, in some countries like India, handing over images to authorities is compulsory. Although this may seem to be in accordance with the law and help solve cases of wildlife crimes, it still raises serious ethical questions for researchers about using camera traps when there is a risk that government authorities might use the images inappropriately.

As well as having various impacts on resident human populations who are surveilled by conservation technologies, the same devices could have significant impacts on labour practices within the conservation movement. For example, an unintended effect of digital technologies that automate surveying and replace human effort could be 'de-skilling' of natural history and loss of traditional ecological knowledge (Arts et al., 2015, Shrestha & Lapeyre, 2018). Where technologies replace activities that were previously labour intensive (such as collecting ecological data or conducting foot patrols in

---

[1] The use of surveillance technologies might be considered legitimate and a moral imperative to detect illegal activities, but lying about the capabilities of the technology and device in order to generate fear crosses ethical boundaries.

and around PAs), they could result in job losses on the ground (Adams, 2017). This would reduce local opportunities to benefit economically from conservation, and might also take away an important point of contact between local communities and conservation staff, which can promote good relations and be a valuable source of information. Conservation practitioners often maintain that most low-cost conservation drones cannot replace boots on the ground due to their limited flight times and inability to penetrate forest canopy (Simlai, 2015). Finally, surveillance devices used by park rangers can be used to surveil themselves as well as other people and wildlife, particularly if the device has a GPS unit. This kind of 'workplace monitoring' has been shown to be detrimental to staff well-being elsewhere (Ball, 2010; Rodríguez et al., 2012), but has not yet been investigated in a conservation context.

The use of drones is glamorous, fashionable, and may help connect the wider public to conservation issues by providing high-resolution images and other data. However, there is a risk that the use of drones may support misinformed, simplistic, and often counterproductive narratives about how conservation is practised among the general public (Sandbrook, 2015). Public comments made to an editorial in the *Guardian* newspaper (2014) titled 'In praise of drones' advocated for drones to be armed 'ideally with hellfire missiles' and sent to 'Africa' to deal with poachers. Such comments reveal that many people associate drones with military applications and warfare, an association actively promoted by some users of conservation drones (Sandbrook, 2015). There is ample evidence suggesting that conservation problems such as illegal wildlife trade are highly complex (Duffy et al., 2015; Duffy et al., 2016), and the use of surveillance technologies such as conservation drones to address this problem may give rise to simplistic narratives that undermine the understanding of this complex issue (Humle et al., 2014).

12.4.4 Data security

There are also concerns about how data collected by camera traps and drones may be protected and the risks of it being leaked. An argument can be made by a law enforcement agency for the use of conservation surveillance technologies to prevent poaching. However, what if the data being gathered is then sold on to a political or commercial entity such as a political party, business interest or an advertiser (Sandbrook, 2015)? A state security apparatus may also use such data to curb civil liberties in sensitive areas and quell public protests (Guardian, 2013). Data from drones can also be stolen, as they are particularly susceptible to be shot down, recovered, and dismantled by those wishing to get access (Hartmann & Steup, 2013). Indeed, in 2012, a drone was shot down in Malta as it was being used to enforce unpopular conservation legislation among local communities (Times of Malta 2012). Protection of digital data is becoming an increasing concern throughout the world following the recent revelations and cases of Cambridge Analytica, Facebook, and WhatsApp (Cadwalladr & Graham-Harrison, 2018; Isaak & Hanna, 2018). With such concerns about data use and security, it seems appropriate to ask whether people 'on the ground' whose data are captured by drones should be given the opportunity to consent to be surveyed from the air (Sandbrook 2015)? Consent processes are a standard requirement of social research ethics review committees in most academic institutions when considering traditional methods such as interviews, but are rarely considered for remote surveillance.

12.5 Counter-mapping and social justice

The use of digital surveillance technologies in conservation science and practice is focused largely on ecological monitoring of biodiversity and on law enforcement measures, mostly by trained researchers and scientists from academic institutions, NGOs, or the state. There has been limited attention on the potential use of these technologies to promote participatory management of community forests, participatory governance, and counter-mapping. The practice by which indigenous communities map their customary rights or ancestral land to lay legal claims to the land or territories is usually referred to as counter-mapping. Since the 1970s, indigenous communities in North America started using counter-mapping and participatory mapping to legally assert authority on their ancestral lands (Willow, 2013). Since then, the counter-mapping movement has become widespread across the world, with many nation states such as Canada, Nicaragua, Panama, Indonesia and Suriname recognizing counter maps created by local indigenous communities (Radjawali & Pye, 2017).

In recent years there has been some work on the use of drones for community-based counter-mapping that suggests that digital surveillance technologies like drones may be appropriated and used as tools for social and environmental justice (Radjawali & Pye, 2017; Millner, 2020). Indigenous communities may be more empowered by the use of drones to map and accurately identify illegal activities on indigenous lands such as logging, water pollution, mining, and attempts to grab land through encroachment. It is also argued that in addition to new environmental and geographic knowledge created by the use of drones, gaining digital technology-related skills could bridge digital divides and gaps that exist between communities and wider societies, thus reducing exploitation and marginalization (Paneque-Gálvez et al., 2017). The relatively easy availability and affordability of drones, cameras and open-source software have given grassroots environmental activists access to tools that were previously thought to be privileges of extremely powerful actors. High-resolution imagery produced by drones in the hands of indigenous communities and grassroots activists can challenge spatial planning and cartographies created by corporate interests and the state. Research in the Kapuas River in Indonesia has shown that indigenous communities can also use drones for mapping socioecological dynamics such as draining of wetlands, customary forest concessions and pollutants in rivers (Radjawali & Pye, 2017). Data and evidence from such imagery can be crucial to environmental justice movements worldwide.

Though digital surveillance technologies like drones have the potential to empower indigenous communities, support environmental justice movements and aid community-based conservation, it is argued that such technologies and the GIS skills needed to make use of them are still very complex and technical for them to be fully appropriated by indigenous communities (Radjawali & Pye, 2017). Technical expertise will still be needed to interpret maps and data so that information may be effectively communicated to the state, policymakers, or the courts. Such technical expertise inadvertently will have to come from international donors or financially powerful NGOs. Hence, it is likely that participatory mapping and counter-mapping will never become 'a science of the masses' (Peluso, 1993). Moreover, the source of the funding and expertise required for interpreting technical information would introduce new kinds of power relations around the use and control of such technologies, defeating the primary objective of community empowerment (Radjawali & Pye, 2017).

In the use of digital surveillance technologies, it is particularly important to consider who controls the technology and to what end. Placing drones and cameras in the hands of indigenous communities may seem politically neutral and benevolent but as argued by Radjawali and Pye (2017) this largely depends on the political ecology of the context. For example, a counter-mapping project for community-based natural resource management by the indigenous San group called the Khwe resulted in them asserting ethnic authority in the region by excluding other ethnic communities (Taylor, 2008). Similarly, placing digital technologies in the hands of government forestry departments is likely to cement and exacerbate existing power structures. Hence, it is important to consider that the use of digital surveillance technologies may never be truly neutral and will always depend on who is surveilling and towards what end.

12.6 Conclusions and recommendations

Conservation researchers and practitioners are increasingly advocating the use of digital technologies as a central element in the way they collect data, make management decisions, and do communications. It is highly likely that this will only increase in the future. This enthusiasm is enhanced by media platforms that present digital technologies as a panacea to all kinds of conservation problems (Arts et al., 2015). Such fervour may be long-lived as digital technologies get entrenched in the structures by which organizations function (Arts et al., 2013). However, they can also be short-lived if they come to be seen as a short-term fix that does not address the underlying root causes of conservation problems (Huesemann & Huessemann, 2011). They may also fall victim to hype, which either ends with the object of the hype becoming ubiquitous or when realization dawns that it is not as useful as made out to be (Meijer et al., 2009). Hypes and fads have been common occurrences in the field of nature conservation (Redford et al., 2013), and an emphasis on short-term 'fixes' echoes well with the crisis-oriented and mission-driven character of conservation science (Meine et al., 2006). Such a narrative may find itself at odds with a growing paradigm of evidence-informed conservation in which technology-related interventions are not taken for granted, but tested (Sutherland et al., 2013).

Conservation as a discipline tends to quickly embrace narratives propagating 'good news' (Arts et al., 2015). This predisposition is present in scientific literature and even more so in popular media stories discussing conservation and digital technology, wherein challenges, backlashes, failures, and setbacks that the projects face are hardly discussed (Arts et al., 2013). Many projects primarily involving digital technologies die a slow death due to the lack of long-term funding and a paucity of trained, skilled staff (Joppa, 2015). Many organizations and even governments use digital technologies as a way to garner attention and look glamorous in a bid to attract funding and gain media attention (Arts et al., 2015). The dominant narratives advocating the use of digital technologies in nature conservation paint a simplistic and sometimes naïve logic of it being a magic wand to solve conservation problems. Such narratives can negatively affect conservation outcomes by prematurely closing useful discussion on how conservation should be practised (Arts et al., 2015). The use of digital technology in conservation should be considered a force that can be guided and directed for certain purposes (Castells, 2018). Currently, this force of digital technologies needs to feed less into hypes and techno-fixes and more into long-term carefully implemented interventions.

Questions of political economy such as who controls, benefits from, pays for or is affected by digital technologies are of utmost importance to nature conservation (Arts et al., 2015; Duffy et al., 2019). Conservation has a mixed historical track record

on how it is practised with regard to its social impacts and exercising of powers (Adams, 2004, Brockington et al., 2008). Critical examination with this as a backdrop is hence necessary in the application of digital technologies. Frameworks that promote the democratic use of technology from its outset through integration in its development and design and use by underrepresented or marginalized societal groups may be of help (Haklay, 2013).

The use of digital technologies in conservation also needs a framework of best practice and regulation (Maffey et al., 2015; Sandbrook, 2015; Duffy et al., 2019). The absence of such regulation potentially hampers the long-term sustainability of a promising field as it increases the risk of missteps that could have been avoided (Arts et al., 2015). Questions need to be asked about who should be given permission to deploy surveillance devices, where they should be allowed to do so, and under what regulations and frameworks. For example, these technologies may be used to exercise power by state or non-state authorities to serve institutional goals and objectives at the cost of civil liberties (Arts et al., 2015). Therefore, it is imperative to direct research towards coming up with best-practice methods, taking into account ethical considerations, legislation, and international regulations (Sandbrook et al., 2021).

It is also important to consider the ethics of how the knowledge, data, and visualizations created by these devices under missing regulations and frameworks get incorporated into environmental activism and fundraising by NGOs and the state. Images from drones and camera traps shared across media have as much chance of being misused (e.g. for wildlife trafficking) as being beneficial for conservation education and awareness. For example, overhead images of indigenous communities and their settlements in remote rainforests could be used for social campaigns and environmental activism without their consent. Camera trap images of a tigress suspected of killing more than 10 people in central India made it to multiple campaigns, including a Change.org petition opposing the capture and killing of the animal. It is a matter of concern when camera trap images of high-profile and endangered species can travel through different forms of media with ease, increasing the risk of misinformation and ultimately affecting conservation negatively.

As mentioned before, the conservation of nature has an inconsistent record in terms of its social impacts. It has led to the displacement and eviction of indigenous communities, propagated fortress conservation, and generally lacked diverse stakeholder involvement in conservation decision-making (Adams, 2004). Hence, it is imperative to pay attention to and critically examine who benefits from the use of technologies in conservation and who does not—or indeed, suffers from it. It is also important to critically examine who controls how information flows and is processed, and how these processes can be democratized (Arts et al., 2015). For example, researchers in the Canadian territories of Manitoba and Nunavut are involving indigenous communities in the camera trapping process and tapping into their traditional ecological knowledge of the region (Clark et al., 2018).

Finally, we wish to emphasize the importance of not seeing digital technologies in conservation from a good-bad binary perspective. It is a force with the potential to transform the work of protected area managers, conservation agencies, and scientists. In that light, we hope that more interdisciplinary engagement on the topic is galvanized in academic and non-academic sectors. Such an engagement will result in concerted thinking that ensures that the application of digital technologies in conservation remains ethical and democratic. Conservationists must capitalize on the opportunities provided by digital technologies while being mindful of their associated social impacts and willing to take steps to mitigate potential harm where necessary. Doing so will not only benefit people—it is also likely to benefit conservation itself in the long term.

References

Adams, W. (2004). *Against Extinction*. Earthscan, London, UK.

Adams, W. (2017). Geographies of conservation II. Technology, surveillance and conservation by algorithms. *Progress in Human Geography*, 43(2), 337–349.

Agrawal, A. (2005). *Environmentality: Technologies of Government and Political Subjects*. Duke University Press, Durham, NC.

Arts, K., Van Der, W. A. L., Adams, W. M. (2015). Digital technology and the conservation of nature. *Ambio*, 44, 661–673.

Arts, K., Webster, G., Sharma, N., Melero, Y., Mellish, C., Lambin, X., & Van Der Wal, R. (2013). Capturing mink and data: interacting with a small and dispersed environmental initiative over the introduction of digital innovation. Framework for Responsible Research and Innovation in ICT. Available at: https://aura.abdn.ac.uk/bitstream/handle/2164/3358/Arts_et_al._2013_Capturing_mink_and_data_interacting_with_a_small_and_dispersed_environmental_initiative_over_the_introduction_of_digital_innovation_0.pdf?sequence=1&isAllowed=y

Ball, K. (2010). Workplace surveillance: an overview. *Labor History*, 51(1), 87–106.

Ball, K., & Haggerty, K. D. (2005). Editorial: doing surveillance studies. *Surveillance and Society*, 3, 129–138.

Bannister, F. (2005). The panoptic state: privacy, surveillance and the balance of risk. *Information Polity*, 10, 65–78.

Berkowitz, D. (2017). Stop the 'nazzi': why the United States needs a full ban on paparazzi photographs of children and celebrities. *Loyola of Los Angeles Entertainment Law Review*, 37(2), 175–206.

Bernard, H., Baking, E. L., Giordano, A. J., Wearn, O. R., & Hamid, A. (2014). Terrestrial mammal species richness and composition in three small forest patches within in an oil palm landscape in Sabah, Malaysian Borneo. *Mammal Study*, 39(3), 141–154.

Betts, D. (2015). Panthera's PoacherCam: a closer look. Available at: https://www.panthera.org/panthera-poachercam-closer-look

Bluwstein, J., & Lund, J. F. (2018). Territoriality by conservation in the Selous–Niassa corridor in Tanzania. *World Development*, 101, 453–465.

Brockington, D., Duffy, R., & Igoe, J. (2008). *Nature Unbound: Conservation, Capitalism and the Future of Protected Areas*. Earthscan, London, UK.

Büscher, B. (2016). Reassesing fortress conservation? New media and the politics of distinction in Kruger national park. *Annals of the American Association of Geographers*, 106(1), 114–129.

Büscher, B., & Ramutsindela, M. (2015). Green violence: rhino poaching and the war to save southern Africa's peace parks. *African Affairs*, 115, 1–22.

Butler, D., & Meek, P. (2013). Camera trapping and invasions of privacy: an Australian legal perspective. *Torts Law Journal*, 3 (20), 235–264.

Butler, D., & Meek, P. (2014). Now we can 'see the forest and the trees too' but there are risks: camera trapping and privacy law in Australia. In Meek, P., &

Flemming, P. (eds.). *Camera Trapping: Wildlife Management and Research*. CSIRO Publishing, Collingwood, VIC (pp. 331–345).

Cadwalladr, C., & Graham-Harrison, E. (2018). Revealed: 50 million Facebook profiles harvested for Cambridge Analytica in major data breach. *The Guardian*. Available at: https://www.theguardian.com/news/2018/mar/17/cambridge-analytica-facebook-influence-us-election

Campbell, B., & Verissímo, D. (2015). Black stork down: military discourses in bird conservation in Malta. *Human Ecology*, 43, 79–92.

Caravaggi, A., Banks, P. B., Burton, C. A., et al. (2017). A review of camera trapping for conservation behaviour research. *Remote Sensing in Ecology and Conservation*, 3, 109–122.

Carter, N. H., Riley, S. J., & Liu, J. (2012). Utility of a psychological framework for carnivore conservation. *Oryx*, 46(4), 525–535.

Castells, M. (2018). Globalisation, networking, urbanisation: reflections on the spatial dynamics of the information age. *Urban Studies*, 47(13), 2737–2745.

Clare, J. D. J., Anderson, E. M., MacFarland, D. M., & Sloss, B. L. (2015). Comparing the costs and detectability of bobcat using scat-detecting dog and remote camera surveys in central Wisconsin. *Wildlife Society Bulletin*, 39(1), 210–217.

Clark, C. (2006). Against confidentiality? Privacy, safety and the public good in professional communications. *Journal of Social Work*, 6(2), 117–113.

Clark, D. A., Brook, R., Oliphant-Reskanksi, C., Laforge, M. P., Olson, K., & Rivet, D. (2018). Novel range overlap of three ursids in the Canadian sub-Arctic. *Arctic Science*, 5(1), 62–70.

Coleman, R., & McCahill, M. (2011). *Surveillance and Crime*. Sage, London, UK.

Cooke, S. J., Nguyen, V. M., Kessel, S. T., Hussey, N. E., Young, N., & Ford, A. T. (2017). Troubling and unanticipated issues at the frontier of animal tracking for conservation and management. *Conservation Biology*, 31(5), 1205–1207.

Cunningham, D., & Noakes, J. (2008). What if she's from the FBI? the effects of covert forms of social control on social movements. In Deflem, M. (ed.). *Surveillance and Governance: Crime Control and Beyond*. Emerald Group Publishing Limited, Bingley, UK (pp. 175–198).

Day, M. (2012). Wildlife camera catches Austrian politician having sex in forest. The Telegraph. Available at: https://www.telegraph.co.uk/news/worldnews/europe/austria/9334182/Wildlife-camera-catches-Austrian-politician-having-sex-in-forest.html

Defries, R., Karanth, K. K., & Pareeth, S. (2010). Interactions between protected areas and their surroundings in human dominated tropical landscapes. *Biological Conservation*, 143(12), 2870–2880.

Dobson, J. E., & Fisher, P. F. (2010). The panopticon's changing geography. *Geographical Review*, 97(3), 307–323.

Duffy, R. (2014). Waging a war to save biodiversity: the rise of militarised conservation. *International Affairs*, 90(4), 819–838.

Duffy, R. (2016). War by conservation. *Geoforum*, 69, 238–248.

Duffy, R., St John, F., Büscher, B., & Brockington, D. (2016). Towards a new understanding of the links between poverty and illegal wildlife hunting. *Conservation Biology*, 30(1), 14–22.

Duffy, R., St John, F., Büscher, B., & Brockington, D. (2015). The militarization of anti-poaching: undermining long-term goals? *Environmental Conservation*, 42(4), 345–348.

Duffy, R., Massé, F., Smidt, E., et al. (2019). Why we must question the militarisation of conservation. *Biological Conservation*, 232, 66–73.

Ferenbok, J., & Clement, A. (2012). Hidden changes: from CCTV to smart video surveillance. In Doyle, A., Lippert, R., & Lyon, D. (eds.). *Eyes Everywhere: The Global Growth of Camera Surveillance*. Routledge, London, UK (Chapter 13).

Finn, R. L., & Wright, D. (2012). Unmanned aircraft systems: surveillance, ethics and privacy in civil applications. *Computer Law & Security Review*, 28(2), 184–194.

Fletcher, R. (2010). Neoliberal environmentality: towards a poststructuralist political ecology of the conservation debate. *Conservation and Society*, 8, 171–181.

Foucault, M. (1977). *Discipline and Punish: The Birth of a Prison*. Translated by Sheridan, A. 2. Vintage Books, New York, NY.

Gomes De Souza, A. A. (2015). The moral relativism of privacy and the social construct of WhatsApp. *Management and Administrative Sciences Review*, 4(5), 801–812.

Gorman, J. (2014). Drones on a different mission. *The New York Times*, 21 July. Available at: https://www.nytimes.com/2014/07/22/science/drones-on-a-different-mission.html

Graham, S., & Wood, D. (2003). Digitizing surveillance: categorization, space, inequality. *Critical Social Policy*, 23(2), 227–248.

Guardian (2013). Edward Snowden section. *The Guardian*. Available at: http://www.theguardian.com/us-news/edward-snowden

Haklay, M. (2013). Neogeography and the delusion of democratisation. *Environment and Planning A*, 45(1), 55–69.

Hartmann, K., & Steup, C. (2013). The vulnerability of UAVs to cyber attacks—an approach to the risk assessment. In Podins, K., Stinissen, J., & Maybaum, M.

(eds.). In 5th International Conference on Cyber Conflict (CYCON).

Haslam, C. (2016). The snipers trained to protect rhinos. *BBC News Online*. Available at: https://www.bbc.co.uk/news/magazine-35503077

Hossain, A. M. N., Barlow, A., Barlow, C. G., Lynam, A. J., Chakma, S., & Savini, T. (2016). Assessing the efficacy of camera trapping as a tool for increasing detection rates of wildlife crime in tropical protected areas. *Biological Conservation*, 201, 314–319.

Huesemann, M., & Huesemann, J. (2011). *Techno-Fix: Why Technology Won't Save Us or The Environment*. New Society Publishers, Gabriola Island, BC.

Humle, T., Duffy, R., Roberts, D. L., Sandbrook, C., Fav, J., & Smith, R. J. (2014). Biology's drones: undermined by fear. *Science*, 344, 1351

Hyde, M. (2016). Drone-hacking: the scourge of sunbathing celebrities. The Guardian. Available at: https://www.theguardian.com/lifeandstyle/lostinshowbiz/2016/apr/15/drone-hacking-scourge-sunbathing-celebrities-richard-mabeley

Isaak, J., & Hanna, M. J. (2018). User data privacy: Facebook, Cambridge Analytica, and privacy protection. *Computer*, 51(8), 56–59.

Joh, E. E. (2017). The undue influence of surveillance technology companies on policing. Available at: https://www.nyulawreview.org/wp-content/uploads/2017/08/NYULawReviewOnline-92-Joh_0.pdf

Joppa, L. N. (2015). Technology for nature conservation: an industry perspective. *Ambio*, 44(Suppl 4), 522–526.

Kahn, P. H. Jr. (2011). Wild technology. *Ecopsychology*, 4(3), 237–243.

Karanth, K. K., & Defries, R. (2010). Conservation and management in human dominated landscapes: case studies from India. *Biological Conservation*, 143(12), 2865–2964.

Karanth, K. U., & Nichols, J. D. (1998). Estimation of tigers in India using photographic captures and recaptures. *Ecology*, 79, 2852–2862.

Koh, L. P., & Wich, S. A. (2012). Dawn of drone ecology: low-cost autonomous aerial vehicles for conservation. *Tropical Conservation Science*, 5, 121–132.

Lenin, J. (2014). Locals fearful of suspected killer tiger released near their village in India. The Guardian. Available at: https://www.theguardian.com/environment/india-untamed/2014/dec/04/locals-fearful-suspected-killer-tiger-released-near-village-india

Lombard, L. (2016). Threat economies and armed conservation in Northeastern Central African Republic. *Geoforum*, 69, 218–226.

Lunstrum, E. (2014). Green militarization: anti-poaching efforts and the spatial contours of Kruger national park. *Annals of the American Association of Geographers*, 104, 816–832.

Luo, C., Li, X., & Dai, Q. (2014). Biology's drones: new and improved. *Science*, 344(6190), 1351.

Lyon, D. (1994). *The Electronic Eye: The Rise of Surveillance Society*. Polity, Cambridge, UK.

Lyon, D. (2001). Facing the future: seeking ethics for everyday surveillance. *Ethics and Information Technology*, 3, 171–180.

Lyon, D. ed. (2006). *Theorizing Surveillance*. Willan, Devon, UK.

Lyon, D. (2010). Surveillance, power and everyday life. In Kalantzis-Cope, P., & Gherab-Martín, K. (eds) *Emerging Digital Spaces in Contemporary Society*. Palgrave Macmillan, London, UK (pp. 107–120).

Maffey, G., Homans, H., Banks, K., & Arts, K. (2015). Digital technology and human development: a charter for nature conservation. *Ambio*, 44, 527–537.

Mann, S., & Ferenbok, J. (2013). New media and the power politics of sousveillance in a surveillance dominated world. *Surveillance and Society*, 11, 18–34.

Manokha, I. (2018). Surveillance: the DNA of platform capital—the case of Cambridge Analytica put into perspective. *Theory and Event*, 21(4), 891–913.

Meek, P. (2017). How to stop the thieves when all you want to do is capture wildlife in action. The Conversation. Available at: https://theconversation.com/how-to-stop-the-thieves-when-all-we-want-to-capture-is-wildlife-in-action-73855

Meek, P., & Zimmerman, F. (2016). Camera traps and public engagement. In Meek, P., & Flemming, P. (eds.). *Camera Trapping: Wildlife Management and Research*. CSIRO Publishing, Collingwood, VIC (Chapter 11).

Meijer, A., Boersma, K., & Wagenaar, P. (2009). Hypes: love them or hate them. In Meijer, A., Boersma, K., & Wagenaar, P. (eds.). *ICTs, Citizens and Governance: After the Hype!* IOS Press, Amsterdam, the Netherlands (pp. 3–9).

Meine, C., Soulé, M., & Noss, R. F. (2006). 'A mission driven discipline': the growth of conservation biology. *Conservation Biology*, 20(3), 631–651.

Millner, N. (2020). As the drone flies: configuring a vertical politics of contestation within forest conservation. *Political Geography*, 20, 102163.

Moore, A. (2008). Defining privacy. *Journal of Social Philosophy*, 39(3), 411–428.

Mulero-Pazmany, M., Stolper, R., Van-Essen, L. D., Negro, J. J., & Sassen, T. (2014). Remotely piloted aircraft systems as a rhinoceros anti-poaching tool in Africa. *PLoS One*, 9 (1), e83873.

Newman, G., Wiggins, A., Crall, A., Graham, E., Newman, S., & Crowston, K. (2012). The future of citizen science: emerging technologies and shifting paradigms. *Frontiers in Ecology and the Environment*, 10(6), 298–304.

Paneque-Gálvez, J., Mccall, M. K., Napoletano, B. M., Wich, S. A., & Koh, L. P. (2014). Small drones for community-based forest monitoring: an assessment of their feasibility and potential in tropical areas. *Forests*, 5(6), 1481–1507.

Paneque-Gálvez, J., Vargas-Ramírez, N., Napoletano, B. M., & Cummings, A. (2017). Grassroots innovation using drones for indigenous mapping and monitoring. *Land*, 6(4), 86.

Pebsworth, P. A., & Lafleur, M. (2014). Advancing primate research and conservation through the use of camera traps: introduction to the special issue. *International Journal of Primatology* 35(5), 825–840.

Peluso, N. L. (1993). Coercing conservation? The politics of state resource control. *Global Environment Change*, 3(2), 199–217.

Pimm, L. S., Alibha, S., Bergl, R., et al. (2015). Emerging technologies to conserve biodiversity. *Trends in Ecology and Evolution*, 30(11), 685–696.

Pusparini, W., Batubara, T., Surahmat, F., et al. (2018). A pathway to recovery: the critically endangered Sumatran tiger *Panthera tigris sumatrae* in an 'in danger' UNESCO World Heritage Site. *Oryx*, 52(1), 25–34.

Radjawali, I., & Pye, O. (2017). Drones for justice: inclusive technology and river related action research along the Kapuas. *Geographica Helvetica*, 72, 17–27.

Redford, K. H., Padoch, C., & Sunderland, T. (2013). Fads, funding and forgetting in three decades of conservation. *Conservation Biology*,27(3), 437–438.

Rodríguez, A., Negro, J. J., Mulero, M., Rodríguez, C., Hernández-Pliego, J., & Bustamante, J. (2012). The eye in the sky: combined use of unmanned aerial systems and GPS data loggers for ecological research and conservation of small birds. *PLoS One*, 7(12), e50336

Rohen, P. (2010). The emperor's new scanner: Muslim women at the intersection of the first amendment and full body scanners. *Hastings Women's Law Journal*, 22(2), 339–360.

Rovero, F., & Zimmerman, F. (2016). Introduction. In Rovero, F., & Zimmerman, F. (eds.). *Camera Trapping for Wildlife Research*. Pelagic Publishing, Exeter, UK (p. 320).

Rupf, R., Wyttenbach, M., Köchli, D., et al. (2011). Assessing the spatio-temporal pattern of winter sports activities to minimize disturbance in capercaillie habitats. *Ecological Monitoring*, 3(2), 23–32.

Sandbrook, C. (2015). The social implications of using drones for biodiversity conservation. *Ambio*, 44, 636–647.

Sandbrook, C., Luque-Lora, R., & Adams, W. M. (2018). Human-bycatch: conservation surveillance and the social implications of camera traps. *Conservation and Society*, 16, 493–504.

Sandbrook, C. Clark, D. Toivonen, T. et al. (2021). Principles for the socially responsible use of conservation monitoring technology and data. *Conservation Science and Practice*, https://doi.org/10.1111/csp2.374

Schiffman, R. (2014). Wildlife conservation drones flying high as new tool for field biologists. *Science*, 344, 344–459.

Shapiro, B., & Richard Baker, C. (2001). Information technology and the social construction of information privacy. *Journal of Accounting and Public Policy*, 20(4), 295–322.

Shrestha, Y., & Lapeyre, R. (2018). Modern wildlife monitoring technologies: conservationists versus communities? A case study: the Terai-Arc landscape in Nepal. *Conservation and Society*, 16(1), 91–101.

Simlai, T. (2015). Conservation 'Wars': global rise of green militarisation. *Economic and Political Weekly*, 50, 39–44.

Snitch, T. (2014). Poachers kill three elephants an hour. Here's how to stop them. The Telegraph. Available at: http://www.telegraph.co.uk/news/earth/environment/conservation/10634747/Poachers-kill-three-elephants-an-hour.-Heres-how-to-stop-them.html

Solove, D. J. (2008). *Understanding Privacy*. Harvard University Press, Cambridge, MA.

Sonricker Hansen, A. L., Li, A., Joly, D., Mekaru, S., & Brownstein, J. S. (2012). Digital Surveillance: a novel approach to monitoring the illegal wildlife trade. *PLoS One*, 7(12), e51156

Sparrow, R. (2009). Building a better WarBot: ethical issues in the design of unmanned systems for military applications. *Science and Engineering Ethics*, 15(2), 169–187.

Stone, R. (2018). In letter, researchers call for fair and just treatment of Iranian researchers accused of espionage. *Science, People and Events, Scientific Community*. Available at: https://www.sciencemag.org/news/2018/11/letter-researchers-call-fair-and-just-treatment-iranian-researchers-accused-espionage

Sutherland, W. J., Bardsley, S., Clout, M., et al. (2013). A horizon scan of global conservation issues. *Trends in Ecology and Evolution*, 28, 16–22.

Taylor, J. J. (2008). Naming the land: san countermapping in Namibia's West Caprivi. *Geoforum*, 39(5), 1766–1775.

The European Commission (2000). Charter of Fundamental Human Rights, 2000/C 364/01. *Official Journal of the European Communities*. Available at: https://www.europarl.europa.eu/charter/pdf/text_en.pdf

Times of Malta (2012). Maltese hunters 'shooting protected birds' in Egypt. *Times of Malta*. Available at: http://www.timesofmalta.com/articles/view/20120215/local/Maltese-hunters-shooting-protected-birds-in-Egypt.406905

United Nations (1948). *Universal Declaration of Human Rights*. Available at: https://www.un.org/en/about-us/universal-declaration-of-human-rights

Verma, A., Van Der Wal, R., & Fischer, A. (2015). Microscope and spectacle: on the complexities of using new visual technologies to communicate about wildlife conservation. *Ambio*, 44, 648–660.

Vidal, J. (2013). Drones are changing the face of conservation. The Guardian. Available at: https://www.theguardian.com/environment/2013/may/28/drones-changing-face-conservation

Villaseñor, N. R., Blanchard, W., Driscoll, D. A., Gibbons, P., & Lindenmayer, D. B. (2014). Strong influence of local habitat structure on mammals reveals mismatch with edge effects models. *Landscape Ecology*, 30(2), 229–245.

West, P., Igoe, J., & Brockington, D. (2006). Parks and peoples: the social impact of protected areas. *Annual Review of Anthropology*, 35, 251–277.

Whitman, J. Q. (2004). The two western cultures of privacy: dignity versus Liberty. *Yale L.J*, 113, 1151–1221.

Wich, S., Scott, L., & Koh, L. P. (2017). Wings for wildlife: the use of conservation drones, challenges and opportunities. In Sandvik, K. B., & Jumbert, M. G. (eds.). *The Good Drone*. Routledge, London, UK (Chapter 7).

Wilkie, D., & Rose, R. (2014). A challenge to the world: build a better conservation drone. Available at: http://www.policyinnovations.org/ideas/innovations/data/000254

Willow, A. J. (2013). Doing sovereignty in native North America: Anishinaabe counter-mapping and the struggle for land-based self-determination. *Human Ecology*, 41, 871–884.

Zuboff, S. (2019). *The Age of Surveillance Capitalism: The Fight for a Human Future at the New Frontier of Power*. Profile Books, London, UK.

CHAPTER 13

The future of technology in conservation

Margarita Mulero-Pázmány

13.1 Current scope of conservation technology

Understanding and documenting biodiversity as well as the functioning of ecosystems constitute an ambitious scientific challenge that needs urgent attention (Dirzo & Raven, 2003; Barnosky et al., 2011). Habitat loss and fragmentation, invasive species, pollution, overexploitation of resources, and climate change are acting together to threaten species worldwide and the interactions among them (Arroyo et al., 2015).

As an interdisciplinary science—incorporating the physical, biological, and social sciences as well as engineering—conservation relies on a variety of techniques, instruments, and methods including technological advances that facilitate data collection and analysis. Conservation technology specifically refers to devices, software platforms, computing resources, algorithms and biotechnology methods that support the conservation community (Pearce, 2012; Greenville & Emery, 2016; Lahoz-Monfort et al., 2019). This book dedicates a chapter to each of several key conservation technologies, describing hardware and software characteristics, function and application for research and management, as well as the social implications of using these devices. This chapter will provide a general overview of conservation technology, summarizing current applications of these technologies, describing the temporal and spatial resolution they provide, and reviewing their implications for animal welfare and environmental impact (Table 13.1).

13.1.1 Research questions and conservation applications

Conservation science has a wide scope and addresses questions that range from individual behaviours to species distribution and ecosystem functions and their social context (see Sutherland et al., 2009 where 'a hundred questions of importance to the conservation of global biological diversity' are suggested). Within this extensive conceptual framework, conservation technology can be categorized into (1) data collection and (2) data analysis. Technology helps to determine the status and population trends of wildlife species, conduct habitat monitoring, identify ecosystem threats, and support management decision-makers (Berger-Tal & Lahoz-Monfort, 2018). It also contributes directly to conservation action in the field, helping to detect illegal (e.g. poaching) or habitat restoration activities (Berger-Tal & Lahoz-Monfort, 2018).

The data collected with conservation technology are extremely variable. Aerial images acquired by drones and satellites serve to obtain spatial data about flora, fauna, and their habitats, helping to estimating abundance and density, and contributing to studies about wildlife distribution, animal behaviour, and social organization (Chapters 2 and 3). The image and sound data captured by camera traps and acoustic sensors facilitate the detection and identification of animal species at specific locations providing critical data for analyses into their distribution, density, behaviour, and habitat selection patterns (Chapters 4 and 5).

Margarita Mulero-Pázmány, *The future of technology in conservation*. In: *Conservation Technology*.
Edited by Serge A. Wich and Alex K. Piel, Oxford University Press.
© Oxford University Press 2021. DOI: 10.1093/oso/9780198850243.003.0013

Table 13.1 Current applications of conservation technology

Technology	Type of data	Main current applications	Species	Spatial resolution	Temporal resolution	Animal welfare and environmental impact
Remote sensing and GIS	Georeferenced images: RGB, thermal, multi, or hyperspectral	Mapping conservation targets distribution, viability, health, and threats. Planning conservation actions	Plants Large terrestrial vertebrates Habitat studies of various species	cm-m	Depends on satellite. Potentially daily	Minimal (space debris)
Drones	Georeferenced aerial images and video: RGB, thermal, multi, or hyperspectral	Animal and plant detection, distribution, abundance Spatial ecology Ecosystems and land cover mapping Animal behaviour Management (illegal activities, infrastructures)	Plants Medium/large terrestrial and marine animals Habitat studies of various species	cm-m depending on GPS, RTK or the use of ground control points (GCPs)	Real or quasi-real-time	Varies depending on species, phenology, drone characteristics, and flight parameters Generally low (if following guidelines)
Camera traps	Images, video	Individual identification Species presence, population metrics Habitat preference Human–wildlife conflicts Threat assessments Assessment management actions Dissemination	Medium/large terrestrial animals	cm/m	Real-time if connected or upon download	Minimal (mainly associated with human presence)
Acoustic sensors	Sound records	Behaviour: social organization Community richness Population: occupancy, abundance, density Ecology: habitat use, animal movement Management anthropogenic disturbance, poaching	Marine and terrestrial animals that make characteristic sounds	m	Real-time if connected or upon download	Minimal (the associated with human presence)

Technology	Data records	Application	Taxa	Spatial resolution	Temporal resolution	Risk/impact
Animal-borne technology	Animal locations	Animal movement (e.g. migrations, wildlife corridor mapping) and behaviour, social interactions, energetics, human–wildlife conflict, foraging strategies	Insects, fish, birds, mammals	Light-level geolocators <1 km; Argos >150 m, GPS a few metres	Variable. VHF constant transmission; Argos 6–14 locations/day; GPS up to several locations/second	Generally low (if following guidelines) Can negatively impact species survival, reproduction, and parental care
Field labs	Molecular and physiological indicators	Nutritional physiology Behaviour Energetic and social stress Social and reproductive strategies, life history Individual and population health status Genetic biodiversity assessment Conservation: human–wildlife conflict, captive breeding, reintroductions	Vertebrates	Several m, depending on user's GPS	Real-time or upon further lab analysis	Minimal (the associated with human presence)
Data collection apps	Images, locations, or other data, as recorded by users or sensors embedded in cell phones	Community richness Species presence	Animal and plant species	Several m, depending on user's GPS	Real-time or upon user decision to upload	Minimal (the associated with human presence)
Computer vision	Identification or detection records for images processed	Species presence and derived parameters from drone or camera trap images	Mainly mammals and birds	Not applicable	After data processing	Not applicable
SMART conservation software	Rangers locations Animals/plant locations and data as provided by end-user	Protected areas management Law enforcement monitoring, ecological monitoring, planning Connect/data sharing, profiles for the management and analysis of multisource data	Animal, plant, and environment. Human (e.g. ranger) data	Several m, depending on user's GPS	Real-time if connectivity allows	Not applicable
Environmental DNA	DNA records	Species presence	Animals, plants, fungus, bacteria	Several m, depending on user GPS but requires additional research to determine DNA original location	Determination in quasi-real-time or upon further lab analysis. Assigning a time to the sample is complex	Minimal (the associated with human presence)

Animal-borne technologies such as global navigation satellite systems (GNSS) and very high frequency (VHF) units provide animal locations that serve to better understand animal movements and help to conduct more efficient wildlife management programmes (Chapter 6). Genetic information obtained with environmental DNA technology allows to detect and identify species presence, while molecular and physiological indicators are used in field labs to help determine the health status of specific individuals and wildlife populations. These data are useful for conservation research, contribute to nutritional physiology, behavioural and life history studies, and for the management of human–wildlife conflicts, captive breeding programmes, and reintroductions (Chapters 7 and 8). Finally, data collection apps facilitate increased data collection and faster processing, contributing to greater knowledge about biodiversity richness, and in estimating population parameters (Chapter 9).

Conservation technologies for data analysis have numerous applications. Geographic Information Systems (GIS, Chapter 2) software serves to process and visualize spatial information, which helps understanding ecosystem processes from local to global scales. Commercial software packages (e.g. Agisoft for photogrammetry with drone data), together with increasingly popular open-source platforms (e.g. camtrapR (Niedballa et al., 2016)) facilitate more rapid and sophisticated data classification and analysis. The recent advances in artificial intelligence and machine learning are helping to reduce manual effort and costs otherwise dedicated to data analysis by automatically processing large volumes of data (e.g. automatic recognition of animals in camera trap images (Chapter 11, Tabak et al., 2019)). Finally, recent advances such as SMART (Chapter 10) allow integrating information from different field devices (e.g. mobile devices, global position system (GPS) units) used by ranger patrols and to visualize data in a way that has improved management and enforcement, especially in rapid ranger responses to illegal activities (e.g. poaching).

13.1.2 Resolution

Broadly, conservation technology improves temporal and spatial resolution. Animal-borne technologies such as light-level geolocators provide errors that can range between a few and hundreds of km (higher in latitude than in longitude), which is considered sufficient to study broad-scale movements such as bird migrations (Fudickar et al., 2012; Lisovski et al., 2020). Argos satellites perform successive communications with animal transmitters and obtain locations ~150 m, generally considered acceptable for habitat selection, migration, and resource selection studies (Hebblewhite & Haydon, 2010). Similarly, triangulation using VHF transmitters can have up to 500 m errors (Millspaugh et al., 2012). The locations recorded with data collection apps that are installed on smartphones use GNSS, such as the GPS, and provide an accuracy of ~5 m and additional smartphone corrections can improve this to 1–2 m (Garmin, 2020; National Coordination Office for Space-Based Positioning Navigation and Timing, 2020). Low-cost drone images can be georeferenced with an error of a few meters, which is generally sufficient to perform wildlife censuses and plant or animal counts in, for example, spatial ecology studies (e.g. Mulero-Pázmány et al., 2015; Bushaw et al., 2019). The use of additional RTK (real-time kinematic) systems, differential GPS, ground control points, and postprocessing techniques can further reduce the spatial error of drone images to centimetres, allowing fine-scale photogrammetry products such as point clouds, digital surface models, and ortho-mosaics, where individual plants, animals and topographic features can be accurately detected and positioned (Casella et al., 2017; Forlani et al., 2018; Casella et al., 2020).

The immediacy at which field data are obtained is an important parameter in conservation, especially in management, because numerous threats such as wildlife poaching, pollution, or wildfires require rapid responses to be effective. In practice, immediacy depends on three steps: (1) data collection, (2) analysis, and (3) transmission to the end-user. For data collection, remote devices (e.g. camera traps, acoustic sensors) or those that include a mobility component (e.g. drones) now offer on-board processing, and when networked via wireless communications [e.g. mobile phone networks, WiFi, Bluetooth (Augustin et al., 2016)], can relay data to stakeholders in real-time. Wireless data transmission does require additional infrastructure, is typically more expensive than traditional systems and requires higher power consumption. The key advantage to onboard processing is the reduction in file size for transmission. Machine learning can now be implemented in edge computing to allow that, for example, only camera trap images that contain animals are transmitted to the end-user and those with no relevant targets are discarded (Elias et al., 2017).

13.1.3 Animal welfare and environmental impact

The impact of conservation technology on animal welfare varies with the technology in question. Animal-borne technologies require capturing and manipulating individual animals, and for animals to wear transmitters for a specified time. The tags can be embedded in collars (for mammals), attached as backpacks (for birds), or glued to different parts of the body (e.g. feathers, skin, turtle shells). Studies about transmitter effects have reported different results, from no significant effects to variable negative effects on the animals' behaviour, survival (injury, mortality), reproduction, and parental care as a result of the capturing, handling, sampling, and tagging processes (Peniche et al., 2011; Bodey et al., 2018; Omeyer et al., 2019; Dore et al., 2020; Puehringer-Sturmayr et al., 2020). Safer tag materials are under development to avoid undesirable effects on marked individuals (e.g. Kay et al., 2019). Recommendations about the most appropriate transmitter weights and volumes for different purposes are periodically updated (Kenward, 2001; Silvy et al., 2012).

Camera traps are static, small, and discrete, so they are generally accepted as a non-intrusive method of studying animals (Long et al., 2008). Nevertheless, cameras produce sounds within most mammals' hearing range, and some animals detect the noise or light that signals the camera is recording (Meek et al., 2014). Animals may respond to cameras with changes in their behaviour: staring at or fleeing from cameras or avoiding cameras altogether (Meek et al., 2014).

Drones have a variable impact on wildlife depending on aircraft type, size, engine type, and flight pattern—all of which influence animal behaviour. The animal response also depends on the animal themselves, for example, age and sex, time of year, and the phenological context in which flights are conducted (Mulero-Pázmány et al., 2017). Animal reactions range from no response to the drone (or at least not visually noticeable), to physiological responses, fleeing behaviours, or attacking the aircrafts (Ditmer et al., 2015; Ramos et al., 2018; Rebolo-Ifrán et al., 2019; Mesquita et al., 2020). Following the ethical guidelines and recommendations for using these systems helps reduce drone impact on wildlife (Vas et al., 2015; Lyons et al., 2018; Mustafa et al., 2018; Rebolo-Ifrán et al., 2019). These include reducing noise levels, maintaining the recommended distances for take-off and landing, manoeuvring smoothly and respecting minimum flight heights above the animals (Vas et al., 2015; Hodgson & Koh, 2016; Mulero-Pázmány et al., 2017; Mesquita et al., 2020).

Another type of environmental impact potentially derived from conservation technology is pollution, mainly caused by devices and material that for various reasons end up abandoned in the field, such as biohazard residues in field labs or eDNA, batteries, and drone rests (e.g. Fuchs, 2008). Additionally, like with all field research, human presence is itself a disturbance, as people are required to install, check, and maintain field devices, collect samples and data, or use field labs. There is a lack of literature comparing the impacts on animal behaviour of conservation technology with traditional methods. Such systematic comparisons would provide a more complete picture of the costs and balances associated with new conservation tools.

13.2 Current limitations and expected improvements in conservation technology

Despite the recent explosion of conservation technology, there are limitations to any new device. We review them in this section along with the advances that may help to overcome them in the near future, in the light of the latest developments (Table 13.2).

13.2.1 Power supply and data storage

Power supply is a general concern for conservation technology users. It constrains the autonomy

Table 13.2 Current limitations and expected improvements in conservation technology

Technology	Current hardware limitations	Desired hardware improvements	Current software limitations	Desired software improvements	Other limitations
Remote sensing and GIS	Limited resolution	Higher spatial and temporal resolution	Difficulties in handling large image files	Faster image-data processing	High cost of high-resolution images Out-of-date local, global, and regional data sets
Drones	Autonomy Range Lenses affected by low light Battery/cost for increasing range Unpredictable durability Unreliability for prototypes and new models	Long-lasting batteries at lower costs	Big data management and storage Target detection and classification is time-consuming and costly	User-friendlier open-source packages for image processing (detection and classification)	Bureaucracy High-cost hardware payloads and software (licences) Subjected to good weather (no rain, low wind, appropriate light conditions) Open vegetation areas Training to pilot Legislation constraints
Camera traps	Difficult to detect small or cold blood species Isolation and resistance problems	More power and storage Connectivity Faster triggers More sensitivity 360° wider area Improve video quality Adding other sensors (stereo cameras, LIDAR, polarize, thermal, sound sensors)	Big data management Data privacy Too specific software	Artificial intelligence solutions to automate data analysis Desirable edge computing Statistical models to interpret results	Stolen or damaged by the public High cost

Acoustic sensors	Power and storage Too large size and weight	More power Connectivity Tools integration	Non-standardization of monitoring procedures Time-consuming acoustic analyses Limited data curation Data sharing resources limit the benefits of PAM	Automated species detection (including with variation in acoustic structure), noise control Real-time monitoring. Edge computing Big data management	High-cost acoustic units, batteries, and software Privacy
Animal-borne technology	Power Cost	Reduce size and weight Long-lasting batteries at lower costs	Advanced analytical models require prohibitive computational power and timeframe Challenges in terms of storage and processing power for big data analysis	Machine learning for efficient data collection, maximize data resolution and deployment durations	Cost, that constrains sample size Privacy and security issues surrounding animal-borne data. Risks with open-data sharing (e.g. human–wildlife conflict, wildlife trade, potentially exposing the animals to risk) Development and support for a particular solution can wane over time
Field labs	Limited to electricity availability Connectivity Reliability of portable equipment Shelf-stable reagents or refrigerate options Heavy and non-portable equipment	Power supply solutions Less chemically intensive procedures	Non identified	Non identified	Training (protocols, equipment, safety) High costs portable equipment Lack of access to reagents Facilities biohazard waste Bureaucracy Funding

continued

Table 13.2 *Continued*

Technology	Current hardware limitations	Desired hardware improvements	Current software limitations	Desired software improvements	Other limitations
Data collection apps	Power supply	New form factors for exporting desktop workflows to mobile devices	Bloatware slows performance and battery and adds distractions Compatibility issues Difficulty in developing apps from scratch Too many specific apps	Standardize software Faster analysis and communication Less reliance on proprietary technology Open-source platforms for technical support and flexible workflows	Data confidentiality and protection Regulatory local differences Apps cost
Computer vision	Non-identified	Faster computers	Lack of publicly available data	More available/public data to train algorithms	Engineer-end user communication problems
SMART conservation software	Ruggedness; GPS sensitivity; form factor; battery life Operating system of the data collection devices New technology may increase the chance of workflow breakdown	Reliable internet connectivity Cloud-based server space or a single, secure, and reliable connected server onsite	Technical constraints associated with specific functionalities and integrations	Sufficient IT and information security expertise	Staff capacity: effective leadership, patrol understanding, analytical skills, and strong technological proficiency IT experts available to solve technical problems Data security and protection of personal identifying information
Environmental DNA	Limited to lab-validated target species Reliability of handheld eDNA devices Biases from environmental factors, reagents, lab procedures against reliable detection of rare species Some biosensors can produce false positives	Handheld field-deployable DNA sequencers	Complex data management and analysis	Improve data management and analysis Standardized protocols	Non-identified

of field-deployed systems such as camera traps or acoustic sensors, limiting the amount of data that can be obtained and forcing revisits to remote installations, or recaptures of individuals marked with animal-borne technologies in order to change batteries. Drone users would like batteries that last longer and are less expensive, because these restrict the number of flights that can be made per mission. In addition, stable electricity supply is necessary for long periods in order to recharge drone batteries (e.g. 1–2 h/6S 8A battery). The weight and volume of the batteries and multiple chargers frequently exceed that of the drone itself, which complicates transportation and logistics. Similarly, field lab and eDNA equipment need a power supply for cooling reagents, samples, and analyses. As the general need for improvements in fast, low-cost, and low-weight power supply is not exclusive to conservationists, but shared for other purposes by millions of other global users, the industry forecasts are for future batteries to last longer, be lighter, recharge faster, and produce less pollution (Baran et al., 2019; Bitenc et al., 2020; Langridge & Edwards, 2020; Liu et al., 2020).

In the meantime, solar panels constitute an alternative that has been integrated to provide power supply to camera traps (Voltaic, 2016), bioacoustics sensors (Rach et al., 2013), power freezers, charge drone batteries, and other field devices. Solar GPS tags with rechargeable batteries are pervasive in studies investigating bird movements because these are light, long-lasting, and capable of sending the data remotely (Silva et al., 2017). However, solar radiation is not constant and in low availability can lead to battery drains producing seasonal and circadian biases for solar GPS tags (Silva et al., 2017). Solar energy is also being exploited for drones to add power (which is translated into autonomy) without compromising the weight, size, or manoeuvrability of the aircrafts (Alta Devices, 2019). There are numerous examples of solar cells integrated on drone wings and fuselage, but their use in conservation studies is not generalized yet. Hydrogen fuel cells constitute a new option for powering drones, allowing them to operate longer than those equipped with batteries, and with lower thermal and noise signatures (Ballard Power, 2020).

Bio-batteries are a new generation of fuel cells that convert chemical energy into electricity using low-cost biocatalyst enzymes, potentially obtaining the energy from organic components abundant in nature (Allan et al., 2018). This technology is still in its infancy, but it seems promising for conservation research, potentially allowing scientists to gather large volumes of data from remotely deployed devices.

Data storage limitation is another constraint that forces conservationists to revisit remote installations when the internal hard drives or memory cards are full, and no new data can be recorded in the field devices. Sensors are also evolving and produce higher spatial or temporal resolution files of increasing sizes (e.g. 8K drone images). According to manufacturers, memory cards with more capacity (e.g. up to 128 TB maximum capacity) and faster registration protocols (e.g. 104–985 MB/s) are expected to be widely available soon at affordable prices (Tung, 2018). The safe and fast storage and exchange of large volumes of data are also critical issues for camera trap users. As an example, 60 trap cameras with 32 GB SD can produce around 2TB in a standard 3-month field deployment. Copying and emptying the memory cards to a computer to redeploy them involves considerable dedication that—unless transmitted via wireless—needs to be done in the field. Storing these data volumes can cost around €200/year in external hard drives (without considering backup copies). In the era of big data, and although there are accessible solutions for large companies, the standard academic user still experiences difficulties in transferring the material of a field campaign to a collaborator, since academic institutions do not usually have storage or transfer services for these large data volumes, and free solutions (Dropbox Inc., One Drive or Google Drive) only allow for very limited storage space. It is expected that more efficient and safer platforms, such as state-of-the-art Enterprise File Sync and Share (EFSS) (Harrison, 2017) will help addressing the problems of data storage and sharing at decreasing costs.

13.2.2 Image quality

The quality of the images and videos captured with drones, camera traps, or mobile phones determines the subsequent data processing options and the usefulness of the resulting data, especially if the objective is using automatic detection, classification, and individual identification with machine learning algorithms. Low-quality images may be enough to detect species presence, but higher resolution ones may be needed to identify individuals (for example, a specific fox according to its natural marks (Dorning & Harris, 2019)). Forecasts indicate that the quality of the lenses and sensors will improve, along with the complementary systems that affect the final quality of the images (e.g. flashes, stabilizers, gimbals). In addition, the advances in the field of computational photography in the frame of edge computing allow analysts to treat the image on the device itself, for example combining several frames to obtain a better quality one (HDR) or dynamically focusing the object that moves between the different ones that appear framed (Chen, 2019). While the increase in image quality is desirable, the potential of it is sometimes wasted because the image processing software (photogrammetry, GIS software, and machine learning) struggles to handle heavy files. Further, the increase in images quality or the treatments applied to these may have consequences for analysing the images with machine learning. Consequently, drone and camera trap users often end up working with low-quality versions of the images, splitting or downsampling the raw data obtained in the field.

13.2.3 Connectivity

Connectivity allows data transmission from deployed devices to the end-user. Data are generally transmitted via wireless connections and users access the data via the internet or through mobile phone networks. Where data transmission costs are low, wireless communications can reduce the costs and efforts associated with field visits and accelerate data access for research or management purposes. The vast majority of the current data collection systems, such as camera traps or bioacoustic sensors incorporate digitalization, which is the first step necessary for efficient data transmission.

Once the data are digital, the transmission depends on the networks' availability. Mobile (e.g. 3–4G), WiFi networks, satellite, or radio are the most common ones. For example, 868 MHz can be used for drone edge devices, and low-power wide-area network (LPWAN) such as LoRa (LoRa Alliance, 2020), that uses licence-free sub-gigahertz radio frequency bands like 433 MHz, 868 MHz (Europe), 915 MHz (Australia and North America), and 923 MHz (Asia) is gaining popularity because it enables long-range transmissions (>10 km) with low power consumption.

The field of wireless communications is rapidly improving in data transmission speed and spatial coverage. Several large companies have launched initiatives for global internet through satellite constellations (Crane, 2019; Oneweb, 2019; Sheetz, 2019) and stratospheric balloons such as Loon (X Development LLC, 2019), some of them with the promise of free service (chinadaily.com.cn, 2018). Others have suggested imaginative solutions, such as using solar drones at high altitudes that would act as repeaters to provide greater spatial coverage (Tafintsev et al., 2019). In parallel, there are significant advances in the development of 5G (GSMA, Intelligence, GSMA, & Intelligence, 2014) that will support wireless networks with much higher speeds than the current 4G (Deans, 2020). 5G availability depends on a variety of factors, including government regulation and technological improvements by mobile operators. Information on where 5G is available can be found online on different websites (e.g. Asay, 2019).

There is an increase in the number of protected areas that have created local networks or repeaters that provide mobile or WiFi coverage that can be used for conservation purposes. Several examples of these can be found in Europe [e.g. large-scale Singular Scientific-Technical Infrastructures such as Doñana National Park (RBD-CSIC, 2020), Africa (Serengeti, Ngorongoro Crater), and United States Parks (US National Park Services, 2020)]. Other initiatives, such as The Things Network (The Things Network, 2020) promote communities with low power devices to use long-range gateways to connect to an open-source, decentralized network to exchange data with applications in several parts of the world, including Asia, Australia, and South

America. Platforms such as LoRa (IRNAS, 2018; LoRa Alliance, 2020) also promote long-range connectivity in low power ways that are particularly interesting for research and management purposes. For example, Smart Parks, in cooperation with African Parks, recently equipped an elephant with a LoRaWAN®/GPS-sensor in Liwonde National Park (Malawi) that will send reliable and consistent locations every 15 minutes for 8 and a half years (more frequent than the current GPS collars that would allow one location/hour for 5 years). This technology makes a difference for the rangers because getting the animals' locations more frequently allows them to more rapidly react in their mission to protect elephants and avoid conflicts with surrounding communities (Smart Parks, 2020).

13.2.4 Sensor standardization and integration

Implementing technology over conventional methods typically allows for improved standardization, repeatability, and systematization of data collection (e.g. Hodgson et al., 2017). Data obtained by camera traps, drones, or acoustic sensors can be reviewed afterwards, research design is repeatable across sites and seasons, and the sources of bias derived from the individual variability in terms of level of detection (auditory or visual) of human observers can be avoided (Anderson & Gaston, 2013). While technology offers improved standardization, data collection devices, even with similar manufacturer-listed specifications, still may have different sensitivities. For example, Heiniger and Gillespie (2018) showed that similar camera trap models varied substantially in their performance, providing different detection probability of small mammals in northern Australia. Because of this variation, it has been suggested that researchers include in their methods the specifications and configurations of the technological devices used and incorporate them in the analyses (Heiniger & Gillespie, 2018). Automatic data analysis tools such as image recognition by machine learning (e.g. camera traps for wildlife detection or vegetation classification in drone images) offer high levels of data processing standardization. These tools are convenient for systematic comparisons, but it is important to account

for their variable detection and identification capabilities (see Chapter 11).

Specialized users of conservation technology require sensors integration, sometimes in the same device. For example, the use of camera traps with acoustic sensors is promising to monitor the ecological impact of human disturbance (Buxton et al., 2018), and some research groups have begun to use these techniques together at large scales (Alberta Biodiversity Monitoring Institute, 2020; Colorado State University, 2020). The integration of camera traps with stereo cameras or LIDAR has been suggested to contribute to estimating animal characteristics, such as the individuals' size (Chapter 5). Because of the specificity of these sensor combinations, that may not be of interest of other industrial or consumers markets, it seems that the user community would need to push for their development via research and development projects or in collaboration with manufacturers.

13.2.5 Regulations

Legal barriers often limit scientific work that otherwise is technically feasible. For example, drones can be ready to fly, parked in solar charging stations, use take-off/landing pads located anywhere, and perform flights without any human intervention at different ranges. Potentially, these autonomous missions could be performed in natural areas that are scarcely populated minimizing the risks over humans. However, drone legislation precludes fully autonomous operation in most countries, limiting the range of the flights (often to line of sight) and stating that operators should always be physically present and capable of controlling the aircraft in the field. This suggests there is a need to adapt legislation to the risks, addressing different drone uses (e.g. leisure, research) and areas (e.g. populated vs. unpopulated) and to seek consensus among countries (Cracknell, 2017).

13.2.6 Cost

Cost is an important limiting factor for the use of conservation technology. While being up-to-date using the last advances can be expensive, the access to technologies that have been on the market for some years is generally affordable and can complement or substitute traditional fieldwork (Koh & Wich, 2012; Welbourne et al., 2015; Welbourne et al., 2020). The price of technological devices tends to decrease as technology evolves once development costs are amortized and production is standardized (Belton, 2015). However, there are some exceptions, such as mobile phones market, where prices—so far—have continued to rise for the new models (Priestley, 2020) and where outdated devices that can be less expensive produce functionality issues precluding customers to use them. While the final price of technological devices is influenced by production costs and, in principle, tends to decrease with time, other pricing strategies that consider the amount the customer is willing to pay will influence price elasticity and are affected by competition and other factors.

Some studies compare the cost and efficacy of traditional methods with technological solutions, or different devices with each other. Welbourne et al. (2020) compare camera traps with traditional survey methods, concluding that camera traps are a more cost-effective technique for surveying terrestrial squamates. Vermeulen et al. (2013) provide information of the cost of manned aircrafts versus drones for aerial surveys of elephants, indicating that while the drone cost is lower, the cost per area covered is almost ten times higher than that of a manned aircraft, but they exclude the costs of humans for analyses. Koh and Wich 2012 highlight the advantages of using inexpensive drones (US $2000) as a low-cost alternative to satellite and airborne sensors for surveying and mapping forests and biodiversity. Mulero-Pázmány et al. (2015) compare the use of GPS collars (€33,000) with drones (€5700) to obtain cattle locations and discuss the complexity of deciding which is the most cost-efficient method, because these provide complementary information. Mulero-Pázmány et al. (2014) discuss the costs and benefits of using drones for rhinoceros anti-poaching, including estimations on the personnel training costs and the lifespan of the specific drone used, concluding that the cost of integrating drones in antipoaching tasks could be assumed by a medium-size security company or institutions dedicated to anti-poaching surveillance. Jiménez López and Mulero-Pázmány (2019) review

several studies that compare the cost of traditional methods with drones for conservation in protected areas in tasks such as water sampling, estimating birds' nest densities, and performing forest inventories. While these comparative studies provide useful information on methodological considerations, technology evolves so fast that the costs, technical specifications, training, quality of the data obtained, and lifespan of the technological devices should be constantly reviewed and updated. Besides the commercial options, and as an alternative for those with small budgets, there is a growing community developing DIY (do it yourself) initiatives, the participatory model, and open-source software platforms, such as those for camera traps (Nichols, 2020), drones (Anderson, 2020), bioacoustic sensors (Kim et al., 2017), and Internet of Things (Postcapes, 2019).

13.2.7 Data management

Big data in ecology refers to large volumes of data not readily handled by the usual tools and practices (Hampton et al., 2013). Advances in sensors have resulted in the generation of troves of wildlife locations, images, and sounds. Manually reviewing and analysing these data requires extensive resources. The advances in artificial intelligence, such as machine learning, may help discriminate useful data from those that are not [e.g. camera trap images, Chapter 11; (Tabak et al., 2019)], although accuracy may be different (e.g. lower) than that obtained with manual classification. These techniques can be applied: (1) in a computer where the data are stored; (2) in the cloud where the data can be uploaded; or (3) in an additional computing device physically close or embedded in the data collection device, which is known as edge computing (Hamilton, 2018). Edge computing is faster than processing the data on the cloud because there is no need to transmit all the information to be analysed (which avoids energetic expense) and because cloud saturation is avoided (O'Brien, 2019). In the age of the Internet of Things and with the progressive implementation of 5G, significant progress in edge computing is expected (Pan & McElhannon, 2018).

The development of specific algorithms for the detection and identification necessary to discriminate images (from camera traps or drones) or specific sounds (from bioacoustic sensors) requires high volumes of data for training to be effective (see details in Chapter 11). In this regard, the multidisciplinary collaboration between computer scientists, engineers, and conservationists; the development of data sharing platforms; and policies favouring these, could contribute to these advances taking place faster. For example, the Conservation AI initiative aims to harness machine learning for various conservation projects (https://conservationai.co.uk/). While possibly still far from accessible for the conservation technology community, the advances of quantum computing are likely to produce a dramatic breakthrough in artificial intelligence and increase the speed of data analysis (Neven, 2019).

13.3 New technological trends

This section describes some technological advances that are currently revolutionizing other sectors, but are yet to reach their full potential in conservation science—see also Allan et al. (2018) for a review.

13.3.1 Robots

Robots are machines controlled by computers that are used to perform jobs (Corke, 2011). They serve for different purposes, mainly aimed at saving efforts in routine industrial or domestic tasks or working in environments that are too dangerous or too difficult to access for humans. They are built from different materials (e.g. wood, metal, synthetics, or living tissues); require a power supply; and contain computer programming code that determines what to do, when, and how to do it. Robots can have sensing and actuation capabilities. Sensing is acquired once a unit is equipped with the appropriate instruments (e.g. vision, acoustic, or touch sensors) and facilitate their interaction with the environment, humans, and other robots. Actuation is performed in different ways, for example, by rolling, walking, running, jumping, and manipulating objects with different levels of dexterity. Some perform mechanical tasks such as vacuuming, gripping, pounding, or throwing objects. Despite

robot sensing capabilities being applied across sectors, there are still few examples of robot acting capabilities used in conservation (Grémillet et al., 2012).

Drones constitute an example of robots that have been successfully integrated as conservation technology tools (Chapter 3). They are currently mainly used as flying cameras that collect images, but their potential to perform other robotic tasks has recently started to be explored for applications such as gather water samples (Schwarzbach et al., 2014); trap insects (Albo Sanchez-Bedoya et al., 2013); obtain whale blow (Pirotta et al., 2017); collect data from biologged animals (Cliff et al., 2015; Dos Santos et al., 2015; Körner et al., 2010); record acoustic biotelemetry (Leonardo et al., 2013); intentionally direct animal movements (Penny et al., 2019); air pollution monitoring (Satterlee, 2016); spread water and non-hazardous chemicals to reduce air pollution (Deck, 2019) and deploy sensors in areas of interest (Shih et al., 2015). Drones are also being used in other sectors performing tasks that could be potentially transferred to conservation research or management. For example, they are used in agriculture to spread fertilizers and pesticides (Kale et al., 2015) and for mosquito control (Amenyo et al., 2014), which may have potential for invasive species management; in forestry for planting (US 9,852,644 B2; Salnikov et al., 2017), which could be applied to restoration tasks; for fishing and hunting (Hanna, 2015) which may contribute to wildlife captures in the conservation scenario; and for infrastructure manipulation (Acosta et al., 2020) which could be useful for wildlife management and gathering samples from locations that are difficult to access.

Unmanned ground vehicles (UGVs) can be used similarly to drones but in terrestrial environments, for example, to gather images or sounds; collect samples (Grémillet et al., 2012); gather data from marked animals; or for physically removing invasive species (Baskaran et al., 2017) and trash (Bogdon, 2018; Zapata-Impata et al., 2018). Unmanned aquatic robots in all their variants (linked to a boat by a tether, remotely operated, or autonomous), working at the surface or underwater, offer a new array of possibilities that has just recently started to be explored to study rivers, lakes, and oceans, obtaining images and gathering samples (Steimle &

Hall, 2006; Yuh et al., 2011). There are other robots with innovative locomotion abilities. For example, SlothBot (Notomista et al., 2019), which moves by switching between branching wires on a mesh, is designed for long-term environmental monitoring application, is solar powered, and energy efficient.

There are more futuristic applications of robots, still far from being economically or practically feasible in conservation biology, but that present potential. The advances in the use of drones for delivering goods or medical equipment could be transferrable to transport equipment, UGVs, or biological samples, which would support fieldwork in remote locations. Similarly, large UGVs can work as 'mule field assistants' for logistical assistance (Airsource Military, 2014). Robots and smart wearables may also contribute to safety, for example helping disabled people to conduct fieldwork (e.g. 4 × 4 wheelchairs) or for self-transportation (e.g. hoverboards/flyboard air).

13.3.2 Social networks, stream video platforms

Since 2000, data sharing has been primarily facilitated by social networks (e.g. Facebook, Instagram, Snapchat); websites dedicated to audio-visual contents (e.g. YouTube); an array of mobile phone apps (specific wildlife ones described in detail in Chapter 9); and messaging platforms (e.g. WhatsApp). The vast amount of information that users upload for various purposes, not necessarily related to science, may nevertheless have scientific interest for conservation. Posts, images, and videos can be analysed to extract data from animals, plants, or environmental features, to detect changes in the environment, and to study people's perception about conservation actions or topics. For example, Nekaris et al. (2013) analysed the comments and data posted on a viral YouTube video about slow lorises, a threatened and globally protected primates group. The webometric data allowed them to evaluate societal sentiments towards the species and served to identify a strong desire of people to have one of these animals as a pet, demonstrating the need for Web 2.0 sites to provide a mechanism that allows illegal animal material to be identified and policed.

The access and exchange of information can also open new possibilities for local communities exposed to human–wildlife conflicts to manage natural resources and protect biodiversity. For example, Lewis, Baird, and Sorice (2016) documented how people from Masai rural communities in northern Tanzania, on the border of Tarangire National Park, use mobile phones to alert herders about the presence of potentially dangerous animals, which helps to reduce the incidence and the severity of wildlife attacks.

Environmental authorities are increasingly investing resources in digital research to detect poaching networks, locate where the activities occur in the field (e.g. via mobiles phones), and collect evidence against offenders and consumers of wildlife-derived products that are often sold online. For example, information shared in social networks (Di Minin et al., 2019; Di Minin et al., 2018) and the dark web (Harrison et al., 2016; Roberts & Hernandez-Castro, 2017) proved to be useful to detect illegal trade of wildlife products.

13.3.3 Virtual and augmented reality

Like with social media, the recent advances in virtual reality (VR) and augmented reality (AR) contribute to disseminate conservation science. VR is characterized by the illusion of participation in a synthetic environment that relies on three-dimensional, stereoscopic, head-tracked displays, hand/body tracking, and binaural sound (Gigante, 1993). It is an immersive, multisensory experience commonly used for entertainment and educational purposes. Similarly, AR supplements the real world with virtual (computer-generated) objects that appear to coexist in the same space as the real world (Krevelen & Poelman, 2010). While there are similarities between VR and AR, VR creates an artificial environment to inhabit, and AR simulates artificial objects in the real environment (marxenlabs, 2020). None of these two technological developments provide new data, but they can substantially change how to visualize, contextualize, and interact with remote data.

VR has been used to recreate Australian forests, where experts are virtually immersed in assessing how suitable habitat was for koalas (*Phascolarctos cinereus*). In combination with data obtained by other means, this contributed to species distribution modelling and conservation (Leigh et al., 2019). Virtual tourism, in which the system user can feel embedded in different scenarios—such as a savanna or a coral reef—serves educational purposes, facilitating people to feel closer to nature and value it more. For example, the VR application apeAPP VR, designed to be viewed in full, immersive 360 degrees (when paired with a Google Cardboard-style headset) offers immersive tours of great ape habitats. It enables remote exploration of protected areas worldwide (e.g. Gashaka Gumti National Park in Nigeria, Tripa Peat Swamp in Indonesia) and learning about ape species through maps, images, audios, and other resources (Ape Alliance, 2020). Immersive VR field trips also constitute a useful tool for environmental education and important social issues such as climate change (Markowitz et al., 2018). VR can also potentially help scientists interact with remotely operated or remotely installed devices. For example, several VR software packages help in training drone pilots (DJI, 2020), and most drone systems can integrate VR headsets that serve to visualize the overflown area from the aircraft perspective in real-time and to pilot the system according to what the pilot sees as captured by the drone's frontal camera (first-person view flight mode). It is not difficult to imagine a scenario in which scientists operate mobile aerial, terrestrial, or aquatic robots deployed in remote locations using VR tools.

AR scenarios are particularly useful to help conservation managers and decision-makers visualize environmental characteristics in the field and to better interpret the results of predictive models. For example, Danado et al. (2003) describe an AR project developed in the Parque das Nações and the Tagus Estuary (Lisbon, Portugal) for environmental management that provides georeferenced information to the user and presents superimposing synthetic information over real images. The visualization includes water quality (using pollutant transport simulation models); superimposition of synthetic objects such as urban buildings and natural landscapes to visualize their characteristics and temporal evolution, and the projection of synthetic images that reveal the soil's composition at the user

location. AR also helps students learn about wildlife protection. For example, Conserv-AR (Phipps et al., 2016) is a virtual and AR mobile game that has been applied to enhance wildlife conservation at Murdoch University (Australia), specifically focusing on the Carnaby's Black Cockatoo (*Calyptorhynchus latirostris*), an endangered bird species of Western Australia. It takes the user on a field trip where the player moves within a dynamic map towards waypoints, receives queries to find targets, visualizes three-dimensional objects and text panels that contain environmental information, and interacts with game characters.

13.4 Integrating technologies

The integration of different technologies allows scientists to obtain information from different ecosystem components at matching spatiotemporal scales, which aids understanding of ecological processes and managing environmental threats, leading to more effective conservation strategies (see a review in Marvin et al., 2016). Conservation technology offers a range of tools that collect data from landscape to molecular scales: satellites, drones, and camera traps; acoustic sensors, animal-borne technologies, and physiological sensors; field labs and eDNA. This section reviews how the different available tools can be linked to increase efficacy of data collection efforts for research and management, first with human intervention and then without it.

13.4.1 Combining conservation technologies with human intervention

There are several examples where combining different technologies served to address conservation questions. For example, Rodríguez et al. (2012) used biologgers to obtain precise locations of hunting movements of reintroduced lesser kestrels (*Falco naumanni*), and drones to replicate the trajectories made by the marked individuals, collecting high-resolution aerial images that serve to characterize the habitat in quasi real-time. While animal-borne technologies had been used before for habitat selection studies, in this case spatiotemporal resolution of habitat data were dramatically better than traditional means by, for example, satellite imagery.

The combination of these two technologies allowed researchers to gather animal and habitat data at similar scales (temporal: hours; spatial: metres), which served to study short-term behavioural patterns in highly dynamic scenarios.

Technologies can also be physically integrated. For example, Wilson, Barr, and Zagorski (2017) attached acoustic sensors to drones to perform bird censuses. This allowed the researchers to work in areas difficult to access on foot and helped reduce the biases produced by the observer presence and habitat coverage. Different receivers have also been integrated into drones to collect data from tagged animals (Saunders 2016; Tremblay et al., 2017; Desrochers et al., 2018). This combination facilitated to increase the quantity and quality of data, that otherwise relies on the personnel deployed in difficult-to-access terrain. In a recent review, Buxton et al. (2018) provide numerous examples of how pairing camera traps with acoustic sensors serves to evaluate wildlife abundance, distribution, and behaviour across landscapes while monitoring human stressors at the same time. This technological combination turned out to be particularly useful to improve detection accuracy and helped to strengthen statistical inference at multiple survey scales.

The combination of different conservation technologies constitutes a valuable tool for the efficient management of protected areas and species (Marvin et al., 2016). As described in Chapter 10, SMART systems allow integrating, analysing, and visualizing data from different sources (such as rangers) or remotely deployed data collection devices (such as camera traps or bioacoustic sensors). For example, rhinoceros antipoaching can benefit from the combination of drones that provide aerial perspective; acoustic sensors that can detect gun-shots (González-Castano et al., 2009; Sarma & Baruah, 2015); motion sensors on fences that serve to detect unauthorized access (Cambron et al., 2015); camera traps to detect animals and humans; and animal-borne technologies to get real-time locations of animals (Firmat Banzi, 2014; Kalmár et al., 2019; Kamminga et al., 2018). The recent advances in artificial intelligence, such as machine learning, can facilitate the analyses of the vast amount of data collected from different sources, so that the end-users are provided with the relevant analytics (see Chapter 11).

The non-governmental organization Save the Elephants works towards facilitating elephant coexistence with local communities in Kenya by combining GPS collars with geofencing to detect when marked elephants enter conflictive areas such as agricultural fields. Their personnel automatically receive messages on smartphones that accelerate their learning when and where to chase off problematic individuals to avoid further damage (Save the Elephants, 2000).

13.4.2 Combining conservation technologies without human intervention: wireless sensor networks and Internet of Things (IoT)

IoT is one of the most important areas of future technology (Atzori et al., 2010). It is gaining widespread attention from diverse fields such as logistics, transportation, health, safety, traffic management, and city planning (Lee & Lee, 2015; Miorandi et al., 2012; Zhao et al., 2013). IoT concept is defined as 'a network of physical objects' (Patel & Patel, 2016). These physical objects are also referred to as 'things' and should: (1) have a physical embodiment (e.g. machines, animals, plants, or people); (2) be able to sense (i.e. sensors, e.g. temperature, light) and eventually also to perform actions (i.e. actuators; e.g. move); (3) be uniquely identifiable and addressable in a machine-readable way and; (4) be embedded with processors or micro-controllers that allow data processing and communication (Friedemann & Floerkemeir, 2010; Miorandi et al., 2012; Minerva et al., 2015; Patel & Patel, 2016; Atzori et al., 2017).

IoT is related and often overlaps with wireless sensor networks (WSN). WSN are spatially distributed networks of autonomous sensors built to collect physical or environmental data (e.g. temperature) in a certain area (Barrenetxea et al., 2008). IoT objectives are generally more complex than those of WSN and the 'things' often work as actuators in order to achieve goals without human intervention (Minerva et al., 2015). Besides, in IoT the 'things' have unique identifiers and are connected to the internet, which does not necessarily apply to WSN (Minerva et al., 2015).

The main advantage of WSNs is that they allow collecting data with minimum power consumption, so the networks can work autonomously for long periods (months or years). The data collected by the network nodes are transmitted among them using short-range connections in a smart way and then collected by a sink that forwards the data to the end-user (e.g. via internet or long-range radio). Many IoT applications often rely on short-range technologies such as Bluetooth, ZigBee, or WiFi, or eventually, on long-range ones such as cellular networks. Emerging LPWAN technologies such as LoRa or Sigfox offer better coverage with low energy consumption (Ayele et al., 2018). WSN can be customized in different ways so that data is sent to the end-user on a schedule, or when the results of local data processing indicate that an event of interest has taken place (Barrenetxea et al., 2008).

IoT—and generally WSN—contribute to wildlife conservation research in different ways (see a review about WSN in ecology in Porter et al., 2005; advanced sensors in ecology in Porter et al., 2009; and IoT in animal ecology in Guo et al., 2015). WSN are particularly useful to monitor large areas because numerous low-cost sensors can be deployed and connected with low power consumption, allowing researchers to detect changes in the environment in real-time (Porter et al., 2005, 2009; Othman & Shazali, 2012). Sensors gather data about physical-chemical parameters of interest, such as wind, humidity, and temperature, or they can be combined with wildlife data collection devices remotely deployed such as camera traps or acoustic sensors (Porter et al., 2005). WSNs have been used in research to support remote field-stations (Porter et al., 2009), conduct climate change studies (Fang et al., 2014), monitor wetlands (Xiaoying & Huanyan, 2011), detect changes in forests (Othman & Shazali, 2012), and monitor noise pollution (Santini et al., 2008).

WSN constitute a valuable tool for integrating animal-borne technologies. The integration of wildlife tags in WSNs enables data to be transmitted efficiently among the network nodes. In this way, WSNs conserve power, consequently increasing the autonomy and range of the tags, the time

they can work without human intervention. Another advantage of WSN is that once deployed (e.g. in a national park) WSN infrastructure can be used for diverse projects, favouring reliable and inexpensive data collection (Porter et al., 2005). One of the first studies of WSN applied for animal monitoring was the ZebraNet project (Juang et al., 2002; Zhang et al., 2004), where zebras in Masai Mara, Kenya were marked with transmitters in a WSN that allowed researchers to investigate their movements and interactions. WSN and IoT are often used for monitoring livestock and often combined with static sensors that record environmental variables (e.g. about climate) or with satellite images that enable studying cattle movements and their interactions with the landscape (Wark et al., 2007; Handcock et al., 2009; Stojkoska et al., 2018; McIntosh et al., 2020).

As mentioned, WSN constitute a useful solution to monitor large and even small animals that can carry tags with limited weight. They have been used to research the behaviour and spatial ecology of the Manx Shearwater (*Puffinus puffinus*) in Skomer Island, a UK National Nature Reserve (Naumowicz et al., 2008); for tracking and behaviour recognition of marked pigeons (Liu et al., 2017) and a wireless network of acoustic recorders aided to study yellow-bellied marmots (*Marmota flaviventris*) (Ali et al., 2009). They have also been used to monitor bats and birds with low-cost tags based on IoT system-on-chips using different communications modes (Toledo et al., 2018).

Several open-source options allow conservationists to incorporate IoT solutions at a lower cost than commercial ones. For example, IoT Mataki project offers a wirelessly enabled, low cost, and readily programmable tag developed by the Zoological Society London, University College London, and Microsoft Research (Institute of Zoology, ZSL, 2020). Arribada's Horizon platform provides the building blocks necessary to develop low-cost wireless sensors and biologging tags. It has a modular design that focuses on providing a central control board that breaks out popular input/output connectivity (SPI/I2C) that enables the researchers to use their own proprietary sensors or payloads (Shuttleworth Foundation, 2020).

WSNs are useful tools to collect, transmit and store vast volumes of environmental data that can be used to improve wildlife management (Jones et al., 2015). For example, organizing surveillance static and mobile sensors (e.g. drones) for detecting illegal activities that can affect endangered species in WSN optimizes power use and allows transmitting the information via local networks or the internet in real-time to the managers and rangers so that they can react on time (Kamminga et al., 2018; SMART, 2020). IoT can also contribute to managing other human–wildlife conflicts scenarios that require fast response. For example, the smart integration of sensors (cameras, motion sensors) installed around infrastructures and tags attached to animals has been suggested to divert animal intrusions in agricultural land (Bapat et al., 2017) and to avoid accidents along train lines and roads (Ramkumar & Sanjoy, 2015; Bhagyashree et al., 2019). When the sensors detect animals in dangerous proximity to the infrastructures, the network can provide early warning to drivers through smart phones or LED displays on the roads, or communicate directly to the vehicles making them stop to avoid accidents (Ramkumar & Sanjoy, 2015; Mohanasundaram & Dane, 2017; Kurain et al., 2018; Bhagyashree et al., 2019).

Most of the current IoT applications in conservation science exploit the idea that sensors are connected to each other sharing information, but IoT possibilities go far beyond this. The connected objects can also perform actions according to the information received from another sensor. This idea is exhibited in swarm theory, where drones or other autonomous vehicles can move in self-organized groups to acquire environmental or wildlife data and transfer the information to each other and the end-user in more efficient ways than if they operated without coordination (Allan et al., 2018). For example, long autonomy solar-powered drone swarms can act as data mules uploading raw or pre-processed (by edge computing) data from camera traps, acoustic recorders, or other remotely installed devices. In the Planet Project (Planet Consortium, 2013), an IoT platform combining static and mobile objects was deployed in Doñana National Park (Spain) for environmental pollution monitoring. The static sensors were integrated in environmental stations that collected data about water quality (among other variables) and the mobile sensors were fix-wing and rotary-wing drones and UGVs. A simulated abnormal value detected in the water by a static sensor triggered the flight of the drone, which in turn performed several actions: (1) gather images to visually investigate the occurrence of a pollution event; (2) drop specific sensors on the affected area to gather additional information on the water pollutants; (3) collect water samples that could be analysed in a lab; and (4) deploy a UGV equipped with video cameras for a closer inspection of the affected ground (Marrón et al., 2011; C.-Y. Shih et al., 2014; C. Shih et al., 2015).

In summary, IoT and WSNs formed by heterogeneous static and mobile sensors facilitate collecting and transmitting data in a power-efficient way, allowing scientists to monitor wildlife, large natural areas, detect environmental changes, and potentially react to them autonomously.

13.5 Problems and solutions in conservation technology

Conservation technologies substantially contribute to wildlife research and management, allowing species monitoring, habitat characterization, environmental change detection, and conservation threat exposure (Pimm et al., 2015; Lahoz-Monfort et al., 2019). However, the use of any technology is often a trade-off between risks, costs, data acquisition/processing, and issues related to logistics (Marvin et al., 2016). For example, some of the technological integration failures can result from short-term motivations to use technology in the first instance (e.g. to exploit the novelty or to publish the feasibility of an idea—Joppa, 2015); or to an unfounded excess of confidence in the tool to produce the desired results (the technology 'hype'; Arts et al., 2015; Lahoz-Monfort et al., 2019). These problems could be addressed by promoting more communication with other users, providers, and developers that may help to get a realistic idea about the actual possibilities, timeframe, training period, and total costs that should be invested for obtaining satisfactory results with a particular technology (Lahoz-Monfort et al., 2019, also for an example, see Chapter 11).

Conservationists must spend time familiarizing themselves with the new technologies (Marvin et al., 2016), whose learning curve varies among the different devices. As technologies are integrated into fieldwork, personnel need to have specialized skills, which has led to an increasing training offer by a range of companies and academic institutions via short specific courses (e.g. camera trapping from Wildlabs.net; Environmental DNA (eDNA) Methods from Natural Resources Training Group, Canada), or long-term programmes that cover several technologies (e.g. Wildlife Conservation Technology MSc from Liverpool John Moores University, UK).

Some authors have raised concern regarding the possible negative implications that conservation technology may have for nature and humans (Arts et al., 2015). Poachers or illegal fishermen/loggers can indeed buy cameras or drones for illicit purposes, and wildlife traffickers can access public apps or social networks to get information on the locations of individuals of endangered species (see Chapter 12, e.g. Sandbrook, 2015; Duffy et al., 2018). While this misuse of the technology is difficult—if not impossible—to control, there are some initiatives to address them, such as enforcing data confidentiality, discouraging the publication of locations of sensitive species, and adopting anti-hacking strategies that are aimed to mitigate these practices (Pimm et al., 2015). The effects of using surveillance technology (e.g. camera traps, drones, acoustic sensors) to gather information about people's activities (intentionally or unintentionally) continue to be discussed. Chapter 12 summarizes some of the major concerns on data privacy, civil liberties, and freedom, which are particularly concerning for local communities that may develop mistrust towards conservationists (Duffy, 2015)

Some of the main barriers to applying conservation technology are related to the difficulties that developers face during innovation. The sometimes short-term focus of research projects often leads to designs and prototypes that rarely scale up to become commercial and available for a broader conservation community (Joppa, 2015; Lahoz-Monfort et al., 2019). This is exacerbated by a range of problems, such as the lack of financial sustainability in the long term, or miscalculations about the potential market for the developed product (Iacona et al., 2019; Lahoz-Monfort et al., 2019). As an alternative to traditional business models, open-source software platforms, DIY initiatives, and participatory models are gaining popularity among users and developers (Pimm et al., 2015; Berger-Tal & Lahoz-Monfort, 2018). This networked science facilitates collaboration between researchers and engineers and may allow the broader population to access technology at lower costs (Pimm et al., 2015). This contributes to the democratization of progress, which is also important for conservation because low-income countries that host high biodiversity levels and short budgeted research teams have still limited access to some of these tools (Conway, 1986). Other solutions that have been suggested to promote open access collaboration include the 'hackathons' and 'codefests', where interdisciplinary experts meet to design novel technical solutions; prizes and competitions that aim to recognize and encourage the efforts of those actors to advance in specific conservation challenges (Pimm et al., 2015).

There is a general consensus among the conservation technology community that multidisciplinary collaboration and partnership between researchers, practitioners and technologists is good for the field (Arts et al., 2015; Joppa, 2015; Marvin et al., 2016). This should include not only those working in conservation, but also stakeholders from other fields (e.g. agriculture, health) that may have compatible interests (Allan et al., 2018). These synergies (1) help the development and implementation of solutions for specific needs; (2) boost the sharing of ideas and resources; (3) encourage market developments via business processes; and (4) the sustainability of resulting solutions (Joppa, 2015). For this purpose, several initiatives have been created (e.g. The Conservation Technology Working Group of the Society for Conservation Biology; Wildlabs.net; see Berger-Tal & Lahoz-Monfort, 2018 for a detailed list).

Conservation technology is poised to revolutionize the way biologists collect data, the speed and extend of complex analyses, and the means to feed critical results to stakeholders in time to address critical issues confronting wildlife. As often happens with innovation, the speed of the technology often exceeds the rules and ethical framework that guide its implementation, which can have knock-on effects for society (privacy violations, etc.). Nonetheless, the preceding chapters clarify the diverse, widespread, and flexible use of these developments and their role in the present and future efforts to conserve wildlife.

References

Acosta, J., De Cos, C. R., & Ollero, A. (2020). Accurate control of Aerial Manipulators outdoors. A reliable and self-coordinated nonlinear approach. *Aerospace Science and Technology*, 99, 105731.

Airsource Military (2014). *LS3 Robotic Pack Mule Field Testing by US Military*. Available at: https://www.youtube.com/watch?v=arIJm2lAfR8

Alberta Biodiversity Monitoring Institute (2020). *The Bioacoustic Unit: Discover Nature's Symphony*. http://bioacoustic.abmi.ca/

Albo Sanchez-Bedoya, C., Viguria Jimenez, L., Jimenez Bellido, A., et al. (2013). *Patent Number P201331941 Dispositivo De Captura De Muestras De Elementos Macroscópicos Presentes En El Entorno Con Múltiples Cámaras De Captura Y Almacenamiento*. Spain.

Ali, A. M., Yao, K., Collier, T. C., Taylor, C. E., Blumstein, D. T., & Girod, L. (2009). An empirical study of collaborative acoustic source localization. *Journal of Signal Processing Systems*, 57, 415–436.

Allan, B. M., Nimmo, D. G., Ierodiaconou, D., Vanderwal, J., Koh, L. P., & Ritchie, E. G. (2018). Futurecasting ecological research : the rise of technoecology. *Ecosphere*, 9, 1–11.

Alta Devices (2019). Solar power for unmanned systems. Available at: https://www.unmannedsystemstechnology.com/company/alta-devices/

Amenyo, J.-T., Phelps, D., Oladipo, O., et al. (2014). MedizDroids Project: ultra-low cost, low-altitude, affordable and sustainable UAV multicopter drones for mosquito vector control in malaria disease management. In IEEE Global Humanitarian Technology Conference (GHTC 2014) (pp. 590–596).

Anderson, C. (2020). DYI Drones, the leading community for personal UAVs. https://diydrones.com/

Anderson, K., & Gaston, K. J. (2013). Lightweight unmanned aerial vehicles will revolutionize spatial ecology. *Frontiers in Ecology and the Environment*, 11(3), 138–146.

Ape Alliance. (2020). VR apeAPP. https://4apes.com/25-home-page-slider/35-free-vr-apeapp

Arroyo, J., Valiente-Banuet, A., Aizen, M. A., et al. (2015). Beyond species loss : the extinction of ecological interactions in a changing world. *Functional Ecology*, 29, 299–307.

Arts, K., Van Der Wal, R., & Adams, W. M. (2015). Digital technology and the conservation of nature. *Ambio*, 44, 661–673.

Asay, N. (2019). *Ookla' S New 5G Map Tracks 5G Rollouts*. Available at: https://www.speedtest.net/insights/blog/ookla-global-5g-map/

Atzori, L., Iera, A., & Morabito, G. (2010). The Internet of Things: a survey. *Computer Networks*, 54(15), 2787–2805.

Atzori, L., Iera, A., & Morabito, G. (2017). Understanding the Internet of Things: definition, potentials, and societal role of a fast evolving paradigm. *Ad Hoc Networks*, 56, 122–140.

Augustin, A., Yi, J., Clausen, T., & Townsley, W. M. (2016). A study of Lora: long range & low power networks for the IoT. *Sensors (Switzerland)*, 16(9).

Ayele, E. D., Das, K., Meratnia, N., & Havinga, P. J. M. (2018). Leveraging BLE and LoRa in IoT network for wildlife monitoring system (WMS). In SIEEE World Forum on Internet of Things, WF-IoT 2018—Proceedings (Vol. 2018, pp. 342–348).

Ballard Power (2020). UAV fuel cell power, for a sustainable planet. http://search.ebscohost.com/login.aspx?direct=true&db=aph&AN=J0E377932159808&site=ehost-live

Bapat, V., Kale, P., Shinde, V., Deshpande, N., & Shaligram, A. D. (2017). WSN application for crop protection to divert animal intrusions in the agricultural land. *Computers and Electronics in Agriculture*, 133, 88–96.

Baran, M. J., Braten, M. N., Sahu, S., et al. (2019). Design rules for membranes from polymers of intrinsic microporosity for crossover-free aqueous electrochemical devices. *Joule*, 3(12), 2968–2985.

Barnosky, A. D., Matzke, N., Tomiya, S., et al. (2011). Has the Earth's sixth mass extinction already arrived? *Nature*, 471(7336), 51–57.

Barrenetxea, G., Ingelrest, F., Schaefer, G., & Vetterli, M. (2008). SensorScope : out-of-the-Box Environmental Monitoring. In 2008 International Conference on Information Processing in Sensor Networks, IPSN (p. 12).

Baskaran, B., Tadhakrishnan, K., Muthukrishnan, G., Prasath, N., Balaji, K. P., & Pandian, S. R. (2017). An autonomous UAV-UGV system for eradication of invasive weed Prosopius juliflora. In 2017 Conference on information and communication technology (pp. 5–10).

Belton, P. (2015). Game of drones: as prices plummet drones are taking off. Available at: https://www.bbc.com/news/business-30820399

Berger-Tal, O., & Lahoz-Monfort, J. J. (2018). Conservation technology: the next generation. *Conservation Letters*, 11(6), 1–6.

Bhagyashree, S. R., Sonal Singh, T., Kiran, J., & Padmini, L. S. (2019). Vehicle speed warning system and wildlife detection systems to avoid wildlife-vehicle collisions. *Lecture Notes in Electrical Engineering*, 545, 961–968.

Bitenc, J., Lindahl, N., Vizintin, A., Abdelhamid, M. E., Dominko, R., & Johansson, P. (2020). Concept and electrochemical mechanism of an Al metal anode—organic cathode battery. *Energy Storage Materials*, 24, 379–383.

Bodey, T. W., Cleasby, I. R., Bell, F., Parr, N., Schultz, A., Votier, S. C., & Bearhop, S. (2018). A phylogenetically controlled meta-analysis of biologging device effects on birds: deleterious effects and a call for more standardized reporting of study data. *Methods in Ecology and Evolution*, 9, 946–55.

Bogdon, C. (2018). *Warthog UGV Takes Out the Trash in Autotrans Project*. Available at: https://clear pathrobotics.com/blog/2018/11/warthog-ugv-mobile-manipulation-research-autotrans/

Bushaw, J., Ringelman, K., & Rohwer, F. (2019). Applications of unmanned aerial vehicles to survey mesocarnivores. *Drones*, 3(1), 28.

Buxton, R. T., Lendrum, P. E., Crooks, K. R., & Wittemyer, G. (2018). Pairing camera traps and acoustic recorders to monitor the ecological impact of human disturbance. *Global Ecology and Conservation*, 16, e00493.

Cambron, M. E., Brode, C., Butler, P., & Olszewkski, G. (2015). Poacher detection at fence crossing. Conference Proceedings—IEEE SOUTHEASTCON, 2015 (pp. 30–31).

Casella, E., Collin, A., Harris, D., et al. (2017). Mapping coral reefs using consumer-grade drones and structure from motion photogrammetry techniques. *Coral Reefs*, 36(1), 269–275.

Casella, E., Drechsel, J., Winter, C., Benninghoff, M., & Rovere, A. (2020). Accuracy of sand beach topography surveying by drones and photogrammetry. *Geo-Marine Letters*, 40(2), 255–268.

Chen, B. X. (2019). The reason your photos are about to get a lot better. Available at: https://www.nytimes.com/2019/10/15/technology/personaltech/google-pixel-photography.html

chinadaily.com.cn (2018). New satellite constellation promises free internet access. Available at: https://www.telegraph.co.uk/china-watch/technology/free-internet-access/

Cliff, O., Fitch, R., Sukkarieh, S., Saunders, D. L., & Heinsohn, R. (2015). Online localization of radio-tagged wildlife with an autonomous aerial robot system In Robotics: Science and Systems, Rome, Italy.

Colorado State University (2020). Sound and light ecology team. Available at: https://www.soundandlight ecologyteam.colostate.edu/

Conway, W. (1986). Can technology aid species preservation? In Wilson, E. O. (ed.). *Biodiversity*. National Academic Press, Washington DC (pp. 263–268).

Corke, P. (2011). Springer tracts in advanced robotics. In Corke, P. (ed.). *Robotics, Vision and Control*. Springer, Berlin, Heidelberg (p. 495).

Cracknell, A. P. (2017). UAVs: regulations and law enforcement. *International Journal of Remote Sensing*, 38(8–10), 3054–3067.

Crane, L. (2019). SpaceX is launching 60 satellites to start its global internet scheme. Available at: https://www.newscientist.com/article/2204394-spacex-is-launching-60-satellites-to-start-its-global-internet-scheme/

Danado, J., Dias, E., Romão, T., et al. (2003). Mobile augmented reality for environmental management (MARE). In Association, T. E. (ed.), Eurographics 2003 Conference, Granada, Spain.

Deans, D. H. (2020). How 5G technology will continue to evolve mobile voice services. Available at: https://www.telecomstechnews.com/news/2020/jan/23/how-5g-technology-will-evolve-mobile-voice-services/

Deck, J. (2019). These drones are spraying pollution out of the sky in Thailand. Available at: https://www.globalcitizen.org/en/content/drones-air-pollution-bangkok-thailand/

Desrochers, A., Tremblay, J., Aubry, Y., Chabot, D., Pace, P., & Bird, D. (2018). Estimating wildlife tag location errors from a VHF receiver mounted on a drone. *Drones*, 2(4), 44.

Di Minin, E., Fink, C., Hiippala, T., & Tenkanen, H. (2019). A framework for investigating illegal wildlife trade on social media with machine learning. *Conservation Biology*, 33(1), 210–213.

Di Minin, E., Fink, C., Tenkanen, H., & Hiippala, T. (2018). Machine learning for tracking illegal wildlife trade on social media. *Nature Ecology and Evolution*, 2(3), 406–407.

Dirzo, R., & Raven, P. H. (2003). Global state of biodiversity and loss. *Annual Review of Environment and Resources*, 28(1), 137–167.

Ditmer, M. A., Vincent, J. B., Werden, L. K., et al. (2015). Bears show a physiological but limited behavioral response to unmanned aerial vehicles. *Current Biology*, 25(17), 2278–2283.

DJI (2020). DJI simulator. Available at: https://www.dji.com/es/simulator

Dore, K. M., Hansen, M. F., Klegarth, A. R., et al. (2020). Review of GPS collar deployments and performance on nonhuman primates. *Primates*, 61(3), 373–387.

Dorning, J., & Harris, S. (2019). The challenges of recognising individuals with few distinguishing features: identifying red foxes *Vulpes vulpes* from camera-trap photos. *PLoS One*, 14(5), 1–23.

Dos Santos, G. A. M., Barnes, Z., Lo, E., et al. (2015). Small unmanned aerial vehicle system for wildlife radio collar tracking. In Proceedings of the 11th IEEE International Conference on Mobile Ad Hoc and Sensor Systems MASS 2014, (1), 761–766.

Duffy, J. P., Cunliffe, A. M., DeBell, L., et al. (2018). Location, location, location: considerations when using lightweight drones in challenging environments. *Remote Sensing in Ecology and Conservation*, 4(1), 7–19.

Duffy, R. (2015). Responsibility to protect? Ecocide, interventionism and saving biodiversity. In the Political Studies Association Conference (p. 16). Sheffield.

Elias, A. R., Golubovic, N., Krintz, C., & Wolski, R. (2017). Where's the bear? Automating wildlife image processing using IoT and edge cloud systems. Proceedings of the 2017 IEEE/ACM 2nd International Conference on Internet-of-Things Design and Implementation, IoTDI 2017 (Part of CPS Week) (pp. 247–258).

Fang, S., Xu, L. Da, Member, S., et al. (2014). An integrated system for regional environmental monitoring and management based on Internet of Things. *IEEE Transactions on Industrial Informatics*, 10(2), 1596–1605.

Firmat Banzi, J. (2014). A sensor based anti-poaching system in Tanzania. *International Journal of Scientific and Research Publications*, 4(4), 1–7.

Forlani, G., Dall'Asta, E., Diotri, F., Di Cella, U. M., Roncella, R., & Santise, M. (2018). Quality assessment of DSMs produced from UAV flights georeferenced with on-board RTK positioning. *Remote Sensing*, 10(2), 311.

Friedemann, M., & Floerkemeir, C. (2010). From the internet to the internet of things. In Heidelberg, S. B. (ed.). *From Active Data Management to Event-Based Systems and More*. Springer, Berlin, Germany (pp. 242–259).

Fuchs, C. (2008). The implications of new information and communication technologies for sustainability. *Environment, Development and Sustainability*, 10(3), 291–309.

Fudickar, A. M., Wikelski, M., & Partecke, J. (2012). Tracking migratory songbirds: accuracy of light-level loggers (geolocators) in forest habitats. *Methods in Ecology and Evolution*, 3(1), 47–52.

Garmin (2020). Garmin technical support. https://support.garmin.com/en-US/?faq=aZc8RezeAb9LjCDpJplTY7

Gigante, M. A. (1993). 1—Virtual reality: definitions, history and applications. In Earnshaw, R. A, Gigante, M. A., & Jones, H. B. T.-V. R. S. (eds.). *Virtual Reality Systems*. Academic Press, Boston, MA (pp. 3–14).

González-Castano, F. J., Vales Alonso, J., Costa-Montenegro, E., et al. (2009). Acoustic sensor planning for gunshot location in national parks: a pareto front approach. *Sensors*, 9(12), 9493–9512.

Greenville, A. C., & Emery, N. J. (2016). Gathering lots of data on a small budget. *Science*, 353(6306), 1360–1361.

Grémillet, D., Puech, W., Garçon, V., Boulinier, T., & Le Maho, Y. (2012). Robots in ecology: welcome to the machine. *Open Journal of Ecology*, 2(2), 49–57.

Guo, S., Qiang, M., Luan, X., et al. (2015). Application of the Internet of Things (IoT) to animal ecology. *Integrative Zoology*, 10(6), 572–578.

Hamilton, E. (2018). What is edge computing: the network edge explained. Available at: https://www.cloudwards.net/what-is-edge-computing/

Hampton, S. E., Strasser, C. A., Tewksbury, J. J., et al. (2013). Big data and the future of ecology. *Frontiers in Ecology and the Environment*, 11(3), 156–162.

Handcock, R. N., Swain, D. L., Bishop-Hurley, G. J., et al. (2009). Monitoring animal behaviour and environmental interactions using wireless sensor networks, GPS collars and satellite remote sensing. *Sensors*, 9(5), 3586–3603.

Hanna, E. (2015). Should unmanned aerial systems (drones) be used for hunting? *Journal of Unmanned Vehicle Systems*, 72, 69–72.

Harrison, J. R., Roberts, D. L., & Hernandez-Castro, J. (2016). Assessing the extent and nature of wildlife trade on the dark web. *Conservation Biology*, 30(4), 900–904.

Harrison, L. (2017). The future of file sharing. Available at: https://www.thruinc.com/the-future-of-file-sharing/

Hebblewhite, M., & Haydon, D. T. (2010). Distinguishing technology from biology: a critical review of the use of GPS telemetry data in ecology. *Philosophical Transactions of the Royal Society of London. Series B, Biological Sciences*, 365(1550), 2303–2312.

Heiniger, J., & Gillespie, G. (2018). High variation in camera trap-model sensitivity for surveying mammal species in northern Australia. *Wildlife Research*, 45(7), 578–585.

Hodgson, J. C., & Koh, L. P. (2016). Best practice for minimising unmanned aerial vehicle disturbance to wildlife in biological field research. *Current Biology*, 26(10), R404–R405.

Hodgson, J. C., Mott, R., Baylis, S. M., et al. (2017). Drones count wildlife more accurately and precisely than humans. *Methods in Ecology and Evolution*, 2018, 1–19.

Iacona, G., Ramachandra, A., McGowan, J., et al. (2019). Identifying technology solutions to bring conservation into the innovation era. *Frontiers in Ecology and the Environment*, 1–8.

Institute of Zoology, ZSL (2020). Mataki, tracking behaviour in the wild. http://mataki.org/about/

IRNAS. (2018). Animal conservation with LoRaWAN—turtles, fish and more. Available at: https://www.irnas.eu/animal-conservation-with-lorawan-turtles-fish-and-more/

Jiménez López, J., & Mulero-Pázmány, M. (2019). Drones for conservation in protected areas: present and future. *Drones*, 3(1), 10.

Jones, C., Warburton, B., Carver, J., & Carver, D. (2015). Potential applications of wireless sensor networks for wildlife trapping and monitoring programs. *Wildlife Society Bulletin*, 39(2), 341–348.

Joppa, L. N. (2015). Technology for nature conservation: an industry perspective. *Ambio*, 44(S4), 522–526.

Juang, P., Oki, H., Wang, Y., Martonosi, M., Peh, L. S., & Rubenstein, D. (2002). Energy-efficient computing for wildlife tracking: design tradeoffs and early experiences with zebranet. *ACM SIGPLAN Notices*, 37, 96–107.

Kale, S. D., Khandagale, S. V, Gaikwad, S. S., Narve, S. S., & Gangal, P. V. (2015). Agriculture drone for spraying fertilizer and pesticides. *International Journal of Advanced Research in Computer Science and Software Engineering*, 5(12), 804–807.

Kalmár, G., Wittemyer, G., Völgyesi, P., Rasmussen, H. B., Maróti, M., & Lédeczi, Á. (2019). Animal-Borne Anti-Poaching System. In Proceedings of the 17th Annual International Conference on Mobile Systems, Applications, and Services (pp. 91–102).

Kamminga, J., Ayele, E., Meratnia, N., & Havinga, P. (2018). Poaching detection technologies—a survey. *Sensors*, 18, 1–27.

Kay, W. P., Naumann, D. S., Bowen, H. J., et al. (2019). Minimizing the impact of biologging devices: using computational fluid dynamics for optimizing tag design and positioning. *Methods in Ecology and Evolution*, 10(8), 1222–1233.

Kenward, R. (2001). *A Manual for Wildlife Radiotagging*. Associated Press, London, UK.

Kim, M., Sanzeni, F., & Conaty, D. B. (2017). DIY bioacoustics. Available at: https://www.hackster.io/65001/diy-bioacoustics-ae28a3

Koh, L., & Wich, S. (2012). Dawn of drone ecology: low-cost autonomous aerial vehicles for conservation. *Tropical Conservation Science*, 5(2), 121–132.

Körner, F., Speck, R., Göktoğan, A. H., & Sukkarieh, S. (2010). Autonomous airborne wildlife tracking using radio signal strength. In The 2010 IEEE/RSJ International Conference on Intelligent Robots and Systems, Taipei (pp. 107–12).

Krevelen, D. W. F. van, & Poelman, R. (2010). Augmented reality: technologies, applications, and limitations. *The International Journal of Virtual Reality*, 9(2), 1–20.

Kurain, N. S., Poojasree, S., & Priyadharrshini, S. (2018). Wildlife vehicle collision avoidance system. *SSRG Intrnational Journal of Electronics and Communication Engineering (SSRG-IJECE)*, 5(3), 14–17.

Lahoz-Monfort, J. J., Chadès, I., Davies, A., et al. (2019). A call for international leadership and coordination to realize the potential of conservation technology. *BioScience*, XX(X), 1–10.

Langridge, M., & Edwards, L. (2020). Future batteries, coming soon: charge in seconds, last months and power over the air. Available at: http://www.pocket-lint.com/news/130380-future-batteries-coming-soon-charge-in-seconds-last-months-and-power-over-the-air

Lee, I., & Lee, K. (2015). The Internet of Things (IoT): applications, investments, and challenges for enterprises. *Business Horizons*, 58(4), 431–440.

Leigh, C., Heron, G., Wilson, E., et al. (2019). Using virtual reality and thermal imagery to improve statistical modelling of vulnerable and protected species. *PLoS One*, 14(12), 1–17.

Leonardo, M., Jensen, A. M., Coopmans, C., McKee, M., & Chen, Y. (2013). A miniature wildlife tracking UAV payload system using acoustic biotelemetry. In International Design Engineering Technical Conferences and Computers and Information in Engineering Conference. ASME (p. V004T08A056).

Lewis, A. L., Baird, T. D., & Sorice, M. G. (2016). Mobile phone use and human–wildlife conflict in northern Tanzania. *Environmental Management*, 58(1), 117–129.

Lisovski, S., Bauer, S., Briedis, M., et al. (2020). Light-level geolocator analyses: a user's guide. *Journal of Animal Ecology*, 89(1), 221–236.

Liu, Xiaohan, Yang, T., & Yan, B. (2017). Internet of Things for wildlife monitoring. In 2015 IEEE/CIC International Conference on Communications in China—Workshops, CIC/ICCC 2015 (pp. 62–66). China.

Liu, X., Gao, H., Ward, J. E., et al. (2020). Power generation from ambient humidity using protein nanowires. *Nature*, 578, 550–554.

Long, R., MacKay, P., Zielinski, W., & Ray, J. (2008). *Noninvasive Survey Methods for Carnivores*. Island Press, Washington DC.

LoRa Alliance. (2020). What is the LoRaWAN® specification? https://lora-alliance.org/about-lorawan

Lyons, M., Brandis, K., Callaghan, C., McCann, J., Mills, C., Ryall, S., & Kingsford, R. (2018). Bird interactions with drones from individuals to large colonies. *Australian Field Ornithology*, 35, 51–56.

Markowitz, D. M., Laha, R., Perone, B. P., Pea, R. D., & Bailenson, J. N. (2018). Immersive virtual reality field trips facilitate learning about climate change. *Frontiers in Psychology*, 9, 2364.

Marrón, P. J., Shih, C. Y., Figura, R., Fu, S., & Soleymani, R. (2011). Challenges in the planning, deployment, maintenance and operation of large-scale networked heterogeneous cooperating objects. In UBICOMM 2011—5th International Conference on Mobile Ubiquitous Computing, Systems, Services and Technologies (pp. 338–343).

Marvin, D. C., Koh, L. P., Lynam, A. J., et al. (2016). Integrating technologies for scalable ecology and conservation. *Global Ecology and Conservation*, 7, 262–275.

marxenlabs. (2020). What is virtual reality. Available at: https://www.marxentlabs.com/what-is-virtual-reality/

McIntosh, M., Cibils, A. F., Nyamuryekunge, S., et al. (2020). A test of LoRa WAN real-time GPS tracking on beef cattle in desert pastures. In Proceedings of the 73rd Annual Society for Range Management Meeting. Denver, CO (pp. 1–2).

Meek, P. D., Ballard, G. A., Fleming, P. J. S., Schaefer, M., Williams, W., & Falzon, G. (2014). Camera traps can be heard and seen by animals. *PLoS One*, 9(10), e110832.

Mesquita, G. P. M., Rodríguez, J. D., Wich, S. A. W., & Mulero-Pázmány, M. (2020). Measuring disturbance at swift breeding colonies due to the visual aspects of a drone : a quasi-experiment study. *Current Zoology*, zoaa038. https://doi.org/10.1093/cz/zoaa038

Millspaugh, J. J., Kesler, D. C., Kays, R. W., et al. (2012). Wildlife radiotelemetry and remote monitoring. In Silvy, N. J. (Ed.), *The Wildlife Techniques Manual*. Johns Hopkins University Press, Baltimore, MD (p. 1136).

Minerva, R., Biru, A., & Rotondi, D. (2015). *Towards a Definition of the Internet of Things (IoT)*. IEEE Internet Initiative, Torino, Italy.

Miorandi, D., Sicari, S., De Pellegrini, F., & Chlamtac, I. (2012). Internet of things: vision, applications and research challenges. *Ad Hoc Networks*, 10(7), 1497–1516.

Mohanasundaram, S. R., & Dane, D. (2017). Prevention and monitoring of wildlife by using wireless networks. *IJARIIE-ISSN(O)*, 3(2), 2506–2511.

Mulero-Pázmány, M., Barasona, J. Á., Acevedo, P., Vicente, J., & Negro, J. J. (2015). Unmanned Aircraft Systems complement biologging in spatial ecology studies. *Ecology and Evolution*, 5(21), 4808–4818.

Mulero-Pázmány, M., Jenni-Eiermann, S., Strebel, N., Sattler, T., Negro, J. J., & Tablado, Z. (2017). Unmanned aircraft systems as a new source of disturbance for wildlife: a systematic review. *PLoS One*, 12(6), 1–17.

Mulero-Pázmány, M., Stolper, R., Van Essen, L. D., Negro, J. J., & Sassen, T. (2014). Remotely piloted aircraft systems as a rhinoceros anti-poaching tool in Africa. *PLoS One*, 9(1), 1–10.

Mustafa, O., Barbosa, A., Krause, D. J., Peter, H.-U., Vieira, G., & Rümmler, M.-C. (2018). State of knowledge: Antarctic wildlife response to unmanned aerial systems. *Polar Biology*, 21, 2387–2398.

National Coordination Office for Space-Based Positioning Navigation and Timing (2020). Official U.S. government information about the Global Positioning System (GPS) and related topics. Available at: https://www.gps.gov/systems/gps/performance/accuracy/

Naumowicz, T., Freeman, R., Heil, A., et al. (2008). Autonomous monitoring of vulnerable habitats using a wireless sensor network. REALWSN 2008—Proceedings of the 2008 Workshop on Real-World Wireless Sensor Networks (pp. 51–55).

Nekaris, B. K. A. I., Campbell, N., Coggins, T. G., Rode, E. J., & Nijman, V. (2013). Tickled to death: analysing public perceptions of 'cute' videos of threatened species (slow lorises—*Nycticebus* spp.) on Web 2.0 Sites. *PLoS One*, 8(7), 1–10.

Neven, H. (2019). Computing takes a quantum leap forward. Available at: https://www.blog.google/technology/ai/computing-takes-a-quantum-leap-forward/

Nichols, W. (2020). How to make a DSLR camera trap housing. Available at: https://www.naturettl.com/how-to-make-a-dslr-camera-trap-housing/

Niedballa, J., Sollmann, R., Courtiol, A., & Wilting, A. (2016). camtrapR: an R package for efficient camera trap data management. *Methods in Ecology and Evolution*, 7(12), 1457–1462.

Notomista, G., Emam, Y., & Egerstedt, M. (2019). The SlothBot: a novel design for a wire-traversing robot. *IEEE Robotics and Automation Letters*, 4(2), 1993–1998.

O'Brien, C. (2019). Why edge computing is key to a high-speed future. https://innovator.news/why-edge-computing-is-key-to-a-high-speed-future-77b61cd64a17

Omeyer, L. C. M., Fuller, W. J., Godley, B. J., Snape, R. T. E., & Broderick, A. C. (2019). The effect of biologging systems on reproduction, growth and survival of adult sea turtles. *Movement Ecology*, 7, 2.

Oneweb (2019). OneWeb network update. Available at: https://www.oneweb.world/technology

Othman, M. F., & Shazali, K. (2012). Wireless sensor network applications: a study in environment monitoring system. *Procedia Engineering*, 41, 1204–1210.

Pan, J., & McElhannon, J. (2018). Future edge cloud and edge computing for internet of things applications. *IEEE Internet of Things Journal*, 5(1), 439–449.

Patel, K. K., & Patel, S. M. (2016). Internet of Things—IoT: definition, characteristics, architecture, enabling technologies, application & future challenges. *International Journal of Engineering Science and Computing*, 6(5), 1–10.

Pearce, J. M. (2012). Building research equipment with free, open-source hardware. *Science*, 337, 1303–1304.

Peniche, G., Vaughan-Higgins, R., Carter, I., Pocknell, A., Simpson, D., & Sainsbury, A. (2011). Long-term health effects of harness-mounted radio transmitters in red kites (*Milvus milvus*) in England. *Veterinary Record*, 169, 311.

Penny, S. G., White, R. L., Scott, D. M., MacTavish, L., & Pernetta, A. P. (2019). Using drones and sirens to elicit avoidance behaviour in white rhinoceros as an anti-poaching tactic. *Proceedings of the Royal Society B: Biological Sciences*, 286(1907), 20191135.

Phipps, L., Alvarez, V., De Freitas, S., Wong, K., Baker, M., & Pettit, J. (2016). Conserv-AR: a virtual and augmented reality mobile game to enhance students' awareness of wildlife conservation in Western Australia. In Dyson, L. E., Ng, W., & Fergusson, J. (eds.). *Mobile Learning Futures—Sustaining Quality Research and Practice in Mobile Learning*. The University of Technology, Sydney, Australia (pp. 214–217).

Pimm, S. L., Alibhai, S., Bergl, R., et al. (2015). Emerging technologies to conserve biodiversity. *Trends in Ecology & Evolution*, 30(11), 685–696.

Pirotta, V., Smith, A., Ostrowski, M., et al. (2017). An economical custom-built drone for assessing whale health. *Frontiers in Marine Science*, 4, 1–12.

Planet consortium (2013). 'Planet project'. Available at: http://www.planet-ict.eu/

Porter, J., Arzberger, P., Braun, H.-W., et al. (2005). Wireless sensor networks for ecology. *BioScience*, 55(7), 561–572.

Porter, J., Nagy, E., Kratz, T. K., Hanson, P., Collins, S. L., & Arzberger, P. (2009). New eyes on the world: advanced sensors for ecology. *BioScience*, 59(5), 385–397.

Postcapes (2019). *IoT DIY Projects Handbook*. Available at: https://www.postscapes.com/diy-iot-projects/

Priestley, E. (2020). How much have mobile phone prices increased over the last decade? Available at: https://www.gadget-cover.com/blog/how-much-have-mobile-phone-prices-increased-over-the-last-decade/

Puehringer-Sturmayr, V., Loretto, M. C. A., Hemetsberger, J., et al. (2020). Effects of bio-loggers on behaviour and corticosterone metabolites of Northern Bald Ibises (*Geronticus eremita*) in the field and in captivity. *Animal Biotelemetry*, 8(1), 1–13.

Rach, M. M., Gomis, H. M., Granado, O. L., Malumbres, M. P., Campoy, A. M., & Martín, J. J. S. (2013). On the design of a bioacoustic sensor for the early detection of the red palm weevil. *Sensors (Switzerland)*, 13(2), 1706–1729.

Ramkumar, R., & Sanjoy, D. (2015). A multiple sensor automated warning system for roadkill prevention. *Environmental Science—An Indian Journal*, 11(6), 187–193.

Ramos, E. A., Maloney, B. M., Magnasco, M. O., & Reiss, D. (2018). Bottlenose dolphins and antillean manatees respond to small multi-rotor unmanned aerial systems. *Frontiers in Marine Science*, 5, 316.

RBD-CSIC. (2020). CSIC-ICTS. Available at: http://icts.ebd.csic.es/en/web/icts-ebd/home

Rebolo-Ifrán, N., Graña Grilli, M., & Lambertucci, S. A. (2019). Drones as a threat to wildlife: YouTube complements science in providing evidence about their effect. *Environmental Conservation*, 46, 205–210.

Rebolo-Ifrán, N., Grilli, M. G., & Lambertucci, S. A. (2019). Drones as a threat to wildlife: YouTube complements science in providing evidence about their effect. *Environmental Conservation*, 46(3), 205–210.

Roberts, D. L., & Hernandez-Castro, J. (2017). Bycatch and illegal wildlife trade on the dark web. *Oryx*, 51(3), 393–394.

Rodríguez, A., Negro, J. J., Mulero, M., Rodríguez, C., Hernández-Pliego, J., & Bustamante, J. (2012). The eye in the sky: combined use of unmanned aerial systems and GPS data loggers for ecological research and conservation of small birds. *PLoS One*, 7(12), e50336.

Salnikov, V., Filin, A., & Burema, H. (2017). Hybrid airship-drone farm robot system for crop dusting, planting, fertilizing and other field jobs. US 9852644 B2. USA. Available at: https://patentimages.storage.googleapis.com/c3/01/23/7a528443ab8654/US9852644.pdf

Sandbrook, C. (2015). The social implications of using drones for biodiversity conservation. *Ambio*, 44(S4), 636–647.

Santini, S., Ostermaier, B., & Vitaletti, A. (2008). First experiences using wireless sensor networks for noise pollution monitoring. In Proceedings of the Workshop on Real-World Wireless Sensor Networks—REALWSN'08 (pp. 61–65).

Sarma, T., & Baruah, V. (2015). Real time poaching detection: a design approach. 2015 International Conference on Industrial Instrumentation and Control, ICIC 2015 (ICIC) (pp. 922–924).

Satterlee, L. (2016). Climate drones: a new tool for oil and gas air emission monitoring. *Environmental Law Reporter*, 46(12), 42–58.

Saunders, D. (2016). Wildlife Drones poster.

Save the elephants. (2000). Geofencing project. Available at: https://www.savetheelephants.org/project/geo-fencing/

Schwarzbach, M., Laiacker, M., Mulero-Pázmány, M., & Kondak, K. (2014). Remote water sampling using flying robots. In 2014 International Conference on Unmanned Aircraft Systems (ICUAS) (pp. 72–76).

Sheetz, M. (2019). Here's why Amazon is trying to reach every inch of the world with satellites providing internet. Available at: https://www.cnbc.com/2019/04/05/jeff-bezos-amazon-internet-satellites-4-billion-new-customers.html

Shih, C.-Y., Capitán, J., Marrón, P. J., et al. (2014). On the cooperation between mobile robots and wireless sensor networks. In Khamis, A., & Khelil, A. (eds.). *Cooperative Robots and Sensor Networks. Studies in Computational Intelligence*. Springer, Berlin Heidelberg, Germany (Vol. 554, pp. 31–52).

Shih, C., Jos, P., Schwarzbach, M., et al. (2015). MASS 2015 PLANET: an integrated framework for automating cooperating objects deployment and management using autonomous unmanned vehicles. MASS 2015.

Shuttleworth Foundation. (2020). Low-cost open access biologging. Available at: https://arribada.org/about/

Silva, R., Afán, I., Gil, J. A., & Bustamante, J. (2017). Seasonal and circadian biases in bird tracking with solar GPS-tags. *PLoS One*, 12(10), 1–19.

Silvy, N. J., Lopez, R. R., & Peterson, M. J. (2012). Wildlife Marking Techniques. In Silvy, N. J. (ed.). *The Wildlife Techniques Manual*. Johns Hopkins University Press, Baltimore, MD, (pp. 230–257).

SMART (2020). SMART-Spatial Monitoring and Reporting Tool. Available at: https://smartconservationtools.org/

Smart Parks (2020). First LoRaWAN® elephant collar successfully deployed in Liwonde. Available at: https://www.smartparks.org/news/first-lorawan-elephant-collar-successfully-deployed-in-liwonde/

Steimle, E. T., & Hall, M. L. (2006). Unmanned surface vehicles as environmental monitoring and assessment tools. In OCEANS 2006, 18–26 September 2006, IEEE. Available at: https://ieeexplore.ieee.org/document/4099104

Stojkoska, B. R., Bogatinoska, D. C., Scheepers, G., & Malekian, R. (2018). Real-time Internet of Things architecture for wireless livestock tracking. *Telfor Journal*, 10(2), 74–79.

Sutherland, W. J., Adams, W. M., Aronson, R. B., et al. (2009). One hundred questions of importance to the conservation of global biological diversity. *Conservation Biology*, 23(3), 557–567.

Tabak, M. A., Norouzzadeh, M. S., Wolfson, D. W., et al. (2019). Machine learning to classify animal species in camera trap images: applications in ecology. *Methods in Ecology and Evolution*, 10(4), 585–590.

Tafintsev, N., Gerasimenko, M., Moltchanov, D., Akdeniz, M., Yeh, S. P., Himayat, N., … Valkama, M. (2019). Improved Network Coverage with Adaptive Navigation of mmWave-Based Drone-Cells. In 2018 IEEE Globecom Workshops, GC Workshops 2018—Proceedings, (December), (pp. 1–7). Available at: https://doi.org/10.1109/GLOCOMW.2018.8644097

The Things Network (2020). The things network. Available at: https://www.thethingsnetwork.org/u/TheThingsNetwork

Toledo, S., Orchan, Y., Shohami, D., Charter, M., & Nathan, R. (2018). Physical-layer protocols for lightweight wildlife tags with internet-of-things transceivers. In 19th IEEE International Symposium on a World of Wireless, Mobile and Multimedia Networks, WoWMoM 2018 (pp. 1–4). IEEE. Available at: https://doi.org/10.1109/WoWMoM.2018.8449778

Tremblay, J. A., Desrochers, A., Aubry, Y., Pace, P., & Bird, D. M. (2017). A low-cost technique for radio-tracking wildlife using a small standard unmanned aerial vehicle. *Journal of Unmanned Vehicle Systems*, 7, 2016–2021.

Tung, L. (2018). Future SD cards: expect monster 128TB storage, plus zippier data transfers. Available at: https://www.zdnet.com/article/future-sd-cards-expect-monster-128tb-storage-plus-zippier-data-transfers/

US National Park Services. (2020). Cellular & internet access in national parks. Available at: https://www.nps.gov/aboutus/broadband.htm

Vas, E., Lescröel, A., Duriez, O., Boguszewski, G., & Grémillet, D. (2015). Approaching birds with drones : first experiments and ethical guidelines. *Biology Letters*, 11(2), 20140754.

Vermeulen, C., Lejeune, P., Lisein, J., Sawadogo, P., & Bouché, P. (2013). Unmanned aerial survey of elephants. *PLoS One*, 8(2), e54700.

Voltaic (2016). Solar powered camera trap. Available at: https://blog.voltaicsystems.com/solar-powered-camera-trap/

Wark, T., Corke, P., Sikka, P., et al. (2007). Transforming agriculture through pervasive wireless sensor networks. *IEEE Pervasive Computing*, 6(2), 50–57.

Welbourne, D. J., Claridge, A. W., Paull, D. J., & Ford, F. (2020). Camera-traps are a cost-effective method for surveying terrestrial squamates: a comparison with artificial refuges and pitfall traps. *PLoS One*, 15(1), e0226913.

Welbourne, D. J., MacGregor, C., Paull, D., & Lindenmayer, D. B. (2015). The effectiveness and cost of camera traps for surveying small reptiles and critical weight range mammals: a comparison with labour-intensive complementary methods. *Wildlife Research*, 42(5), 414–425.

Wilson, A. M., Barr, J., & Zagorski, M. E. (2017). The feasibility of counting songbirds using unmanned aerial vehicles. *The Auk*, 134, 350–362.

X Development LLC. (2019). Loon. Expanding internet connectivity with stratospheric balloons. Available at: https://x.company/projects/loon/

Xiaoying, S., & Huanyan, Q. (2011). Design of wetland monitoring system based on the Internet of Things. *Procedia Environmental Sciences*, 10 (B), 1046–1051.

Yuh, J., Marani, G., & Blidberg, D. R. (2011). Applications of marine robotic vehicles. *Intelligent Service Robotics*, 4(4), 221–231.

Zapata-Impata, B. S., Shah, V., Platt, R., & Singh, H. (2018). Autonomous trash picking mobile robot.

Zhang, P., Sadler, C. M., Lyon, S. A., & Martonosi, M. (2004). Hardware design experiences in ZebraNet. In Proceedings of the 2nd International Conference on Embedded Networked Sensor Systems, SenSys. Baltimore, MD, USA. Available at: https://doi.org/10.1145/1031495.1031522

Zhao, J., Zheng, X., Dong, R., & Shao, G. (2013). The planning, construction, and management toward sustainable cities in China needs the Environmental Internet of Things. *International Journal of Sustainable Development & World Ecology*, 20(3), 195–198.

Index

Notes Tables, figures, and boxes are indicated by an italic t, f, and b following the page number

Abbreviations used
EO satellites - Earth Observing satellites
PAM - passive acoustic monitoring

A

accelerometers 108t, 111, 112f, 113
  video cameras and 114
accessibility, data collection software 183, 184–6t, 187
accuracy, computer vision 234–5
Acinonyx jubatus (cheetah), accelerometers 112f, 113
acoustic mark-capture-recapture methods 58
acoustic sensors 5, 53–77, 256t
  camera traps and 69
  composition of 60, 61t
  data analysis software 62–3t
  future directions 261t
  limitations & constraints 261t
  see also passive acoustic monitoring (PAM)
acoustic transmitters 111
active infrared sensors (AIRs) 81
active learning 234
active radio frequency identification 110–11
active remote sensing 14
activity patterns, acoustic sensors 54–5
Adélie penguin (*Pygoscelis adeliae*), video cameras 114
Advanced Very High Resolution Radiometer (AVHRR) 16
adverse weather, acoustic sensors 53

aerial images 255
aerial line transects 2–3
aerial survey space, computer vision 232
AES (Areas off Ecological Significance) 115
African elephant (*Loxodonta africana*), livestock depredation by camera trapping 88
agouti (*Dasyprocta* species), camera trapping 92
agriculture, robots 267
AI *see* artificial intelligence (AI)
Ailuropoda melanoleuca (giant panda), wildlife corridors by camera trapping 91
AIRs (active infrared sensors) 81
Air Shepherd 42
AIV (Avian Influenza Virus) 168
Alaotran gentle lemur (*Hapalemur alaotrensis*), drone case study 42–4, 43f, 44f
algorithms, computer vision 230, 235
All for Earth (Microsoft) 7
all-round detection zone, camera trapping 96
Amazon Web Services, SMART Connect 208
Anabat Insight 62t
Anacapa toolkit 164

ancient ecosystems, environmental DNA 165
Android 182–3
Animal Behaviour Pro 186t, 187, 189
animal-bourne technologies 105–28, 108t, 257t
  accelerometers 111, 112f, 113
  animal welfare 259
  audio recorders 113–14
  case studies 115–16
  data analyses 115
  definition 105
  future directions 120–2, 261t
  historical use 105
  limitations & constraints 116–19, 261t
  non-tracking technologies 111, 113–15
  radio-frequency identification 110–11
  satellite tracking 107–10
  social impact & privacy 119–20
  very-high frequency tracking 106–11, 107f
  video cameras 114
  wireless sensor networks 271
animal counts, drones 38, 39–41
animal density 35
  drones 40